Advanced Mobile Robotics

Advanced Mobile Robotics

Volume 3

Special Issue Editor

DaeEun Kim

MDPI • Basel • Beijing • Wuhan • Barcelona • Belgrade

Special Issue Editor
DaeEun Kim
Yonsei University
Korea

Editorial Office
MDPI
St. Alban-Anlage 66
4052 Basel, Switzerland

This is a reprint of articles from the Special Issue published online in the open access journal *Applied Sciences* (ISSN 2076-3417) from 2018 to 2019 (available at: https://www.mdpi.com/journal/applsci/special_issues/Advanced_Mobile_Robotics).

For citation purposes, cite each article independently as indicated on the article page online and as indicated below:

LastName, A.A.; LastName, B.B.; LastName, C.C. Article Title. *Journal Name* **Year**, *Article Number*, Page Range.

Volume 3
ISBN 978-3-03921-946-9 (Pbk)
ISBN 978-3-03921-947-6 (PDF)

Volume 1-3
ISBN 978-3-03921-942-1 (Pbk)
ISBN 978-3-03921-943-8 (PDF)

Cover image illustrated by Danho Kim.

Contents

About the Special Issue Editor

DaeEun Kim received his BE and MS from the Department of Computer Science and Engineering of Seoul National University, South Korea, and the University of Michigan at Ann Arbor, USA, respectively. He received his Ph.D. from the University of Edinburgh, UK, in 2002. From 2002 to 2006 he was a research scientist at the Max Planck Institute for Human Cognitive and Brain Sciences in Munich, Germany. Currently he is a professor in Yonsei University in Seoul, Korea. His research interests are in the areas of biorobotics, autonomous robots, artificial intelligence, artificial life, neural networks and neuroethology.

 *applied sciences*

Article

Study on Path Planning Method for Imitating the Lane-Changing Operation of Excellent Drivers

Guoqing Geng [1,*], Zhen Wu [1], Haobin Jiang [1,2], Liqin Sun [1] and Chen Duan [3]

[1] School of Automobile and Traffic Engineering, Jiangsu University, Zhenjiang 212013, China; wz_laughing@163.com (Z.W.); jianghb@ujs.edu.cn (H.J.); slq@ujs.edu.cn (L.S.)
[2] Automotive Engineering Research Institute, Jiangsu University, Zhenjiang 212013, China
[3] Electrical and Computer Engineering, Wayne State University, Detroit, MI 48202, USA; duanchen0206@hotmail.com
*   Correspondence: ggq@ujs.edu.cn; Tel.: +86-0511-8878-2845

Received: 22 April 2018; Accepted: 15 May 2018; Published: 18 May 2018

Abstract: Lane-changing is an important operation of an autonomous vehicle driving on the road. Safety and comfort are fully considered by excellent drivers in lane-changing operation. However, only the kinematic and dynamic constraints are taken into account in the traditional path planning methods, and the path generated by the traditional methods is very different from the actual trajectory of the vehicle driven by the excellent driver. In this paper, a path planning method for imitating the lane-changing operation of excellent drivers is proposed. Five experienced drivers are invited to do the lane-changing test, and the lane-changing trajectories data under different conditions are recorded. The excellent driver lane-changing model is established based on the genetic algorithm (GA) and back propagation (BP) neural network trained by the data of the lane-changing tests. The proposed approach can plan out an optimized lane change path according to the vehicle condition by learning the excellent drivers' driving routes. The results of simulations verify that the path generated by the proposed algorithm is basically same as the track selected by the excellent drivers under same conditions, which can reflect the characteristics of the operations of the excellent driver. While applying safe lane-changing to autonomous vehicle, it can improve the ride comfort of the vehicle and therefore reduce the probability of motion sickness of the passengers caused by improper operation during lane change.

Keywords: path planning; lane change; excellent driver model; neural networks; autonomous vehicle

---

1. Introduction

In recent years, with the continuous increase in vehicle ownership, the problem of traffic safety has been deteriorating. Smart vehicles and unmanned driving that can effectively improve traffic safety are being vigorously developed and applied. Path planning that generates a driving route from the initial point to the destination is one of the keys features of unmanned driving technology. For autonomous vehicles, the constraints of vehicle kinematic and dynamics are important to be studied in path planning algorithm in addition to the obstacles avoidance.

Path planning methods for autonomous vehicles have been widely studied. There are some advanced path planning methods for autonomous road vehicles, such as artificial potential field methods [1,2] and optimal control [3,4]. The obstacles, road structures, and vehicle dynamics are considered in these proposed methods. However, there is a sudden change in the curvature of the trajectory generated by these methods, and it is often necessary to smooth the generated path and increase the workload. Many studies about curvature–continuous path planning methods have been conducted [5–7]. In these studies, some continuous curves such as Bezier curve are used for trajectory planning. Choi et al [7] proposed a practical path planning algorithm based on Bezier curves for autonomous vehicles operating under waypoints and corridor constraints. The path planning algorithm combines a set of low-degree of Bezier curve segments smoothly to generate the reference trajectory. However, since the planned path has difficulty meeting all comfort requirements of different passengers, it may still increase the probability of the motion sickness. Michael and Brandon [8] found that the proportion of people with motion sickness riding smart vehicles is much higher than traditional vehicles. Other studies have proven this conclusion [9,10]. As a result, conventional path panning algorithm is not sufficient to meet motion comfort requirements in these scenarios. However, under any condition, an experienced driver is always able to find an optimal path to keep the vehicle moving steadily. Thus, it is necessary to study the driving paths of experienced drivers.

Lane-changing is an important part of autonomous driving behavior in arterial road traffic [11], which involves changes in both longitudinal and lateral velocity as well as movement in the presence of other moving vehicles [12]. Therefore, many studies have been carried out on lane change of autonomous vehicle [13,14]. Dubins path, the shortest path for a wheel-drive robot consisting of a set of two circular arcs and line segments [15], is one of the most well-known and widely studied methods to generate a smooth path [16]. Chop et al [17] presented a path planning method using circle curve as the lane-changing path for autonomous vehicle. However, the path has a fatal drawback that the curvature at the joint nodes connecting the lines and arcs is discontinuous.Ren et al [18] presented a lane-changing trajectory generating method based on the vehicle lateral acceleration during lane changing meeting the constraints of positive and negative trapezoid. Wang et al. [19] used the seven-order polynomial as the path expression of the lane-changing path. The location of lane changing ends is determined according to the average lane changing time assuming that the longitudinal speed is constant. However, the characteristics of different drivers in the actual lane changing operation were not considered in these studies.

The artificial neural network is a powerful, nonlinear, and adaptive mathematical model [20]. It has been used extensively and successfully in various fields, including image processing [21], pattern recognition [22] and voice recognition [23]. It is difficult to describe the characteristics of the actual driver's lane-changing operation by accurate mathematical modeling. Thus, the neural network is utilized to establish the lane-changing path model. In previous studies, by purposely propagating output-layer errors into hidden-layers, and deriving the optimal weights with gradient descent optimization [24], Back Propagation (BP) neural network is widely used to minimize errors. However, the further development and application of BP neural network is limited by the drawback that it easily falls into local optimal solutions. Many researchers have attempted to use different types of evolutionary algorithms, such as Genetic Algorithm [25], Particle Swarm Optimization (PSO) [26], and Simulated Annealing (SA) [27], to optimize the weight and threshold of the BP neural network in training process. Yu and Xu [28] presented a short-term load forecasting model of natural gas based on Genetic Algorithm and Back Propagation (GA-BP) neural network. Wang et al. [29] proposed a wind speed forecasting model based on GA-BP neural network. They found that the accuracy and the learning speed of the BP neural network can be improved significantly with optimization through the genetic algorithm.

To solve the problem that the traditional path planning methods for lane-changing do not consider the actual driving characteristic of lane-changing, five excellent drivers were invited to do lane-changing tests. The trajectories under different modes were recorded. The paths were fitted by polynomial curve by comparing different curves. The routes of lane-changing considering the feature of the intermediate state and the final position were obtained. The excellent driver lane-changing model was established based on GA-BP neural networks trained by the optimal trajectory database obtained by the experiment. The path planning method for lane-changing based on the excellent driver lane-changing model was proposed. It can meet the requirements of different types of passengers for riding comfort and reduce the probability of motion sickness.

The structure of the paper is as follows. Section 2 presents the lane-changing test, and gives an approach to transform the data of Global Position System (GPS) into geographic coordinate system. The fitting of driver's lane-changing path based on six-order polynomial is shown in Section 3.

In Section 4, the excellent driver model based on GA-BP neural networks is highlighted. Simulation results are discussed in Section 5. Finally, Section 6 presents some concluding remarks.

2. Acquisition of Ideal Path

2.1. Lane-Changing Test

Lane changing is one of the important operations in driving. The driver will plan out an ideal path to make the vehicle move smoothly before starting to change lane. To study the characteristics of the path that the excellent driver planned at the time of lane changing, driving tests with excellent drivers were carried out. Figure 1 shows the test vehicle used in this experiment. It is equipped with GPS device to record the travel path and S-Motion biaxial optical speed sensor to obtain the yaw rate and lateral acceleration of the vehicle. Due to the difference of driving experience and driving habits, different drivers usually have different driving characteristics. An evaluation questionnaire about the types of driver was designed. After each experiment, the passenger appraised the ride experience. Thus, the drivers were divided into three types based on the driving characteristics: aggressive, intermediate and conservative. To track the difference in lane-changing paths between different drivers, five experienced drivers were invited to participate in the experiment. Information of the five drivers is shown in Table 1. During driving, the drivers were required to do the lane-changing operation to meet the demand of obstacle avoidance or overtaking as well as others operation. In this experiment, the test was divided into two working conditions: obstacle avoidance lane change and free lane change. The single lane change test was conducted as the obstacle avoidance condition by changing the pile position to simulate different obstacle distance. Figure 2 shows the single lane change test environment.

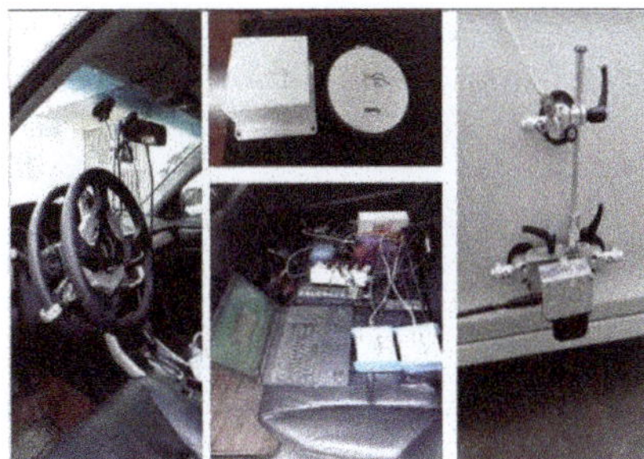

Figure 1. Test vehicle.

Figure 2. Single lane change test.

Table 1. Information of drivers.

Driver Number	Gender	Age	Driving Age (Years)
Driver 1	Female	55	33
Driver 2	Male	28	10
Driver 3	Male	53	31
Driver 4	Male	46	22
Driver 5	Male	53	21

2.2. Data Processing

Owing to the data recoded by GPS being longitude, latitude and elevation, it is difficult to directly reflect the vehicle's actual running path in the geodetic coordinate system. In geographic coordinate system, the origin of the coordinates lies in the centroid of the carrier, and its X_g axis, Y_g axis and Z_g axis are the east, north and sky directions, respectively, of the carrier's location. To accurately describe the travel path, it is necessary to transform the coordinates of the geodetic coordinate system into the geographic coordinate system. However, a direct conversion of coordinates between the geodetic coordinate system and the geographic coordinate system is hard to process. Therefore, the Earth Cartesian coordinate system is introduced into the transformation. The relationships between the geodetic coordinate system, the geographic coordinate system and the Earth Cartesian coordinate system are shown in Figure 3.

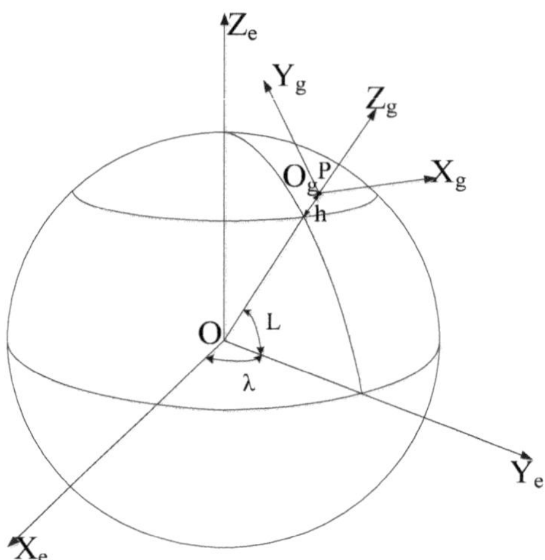

Figure 3. Relationships between the three different coordinate systems.

The coordinates of the point P is (L, λ, h) in geodetic coordinate system. The coordinates of the point P (x_e, y_e, z_e) in the Cartesian coordinate system can be obtained from Equation (1).

$$\begin{cases} x_e = (R_N + h) \cos L \cos \lambda \\ y_e = (R_N + h) \cos L \sin \lambda \\ z_e = [R_N (1 - e)^2 + h] \sin L \end{cases} \quad (1)$$

where R_N is the radius of curvature of the ellipsoid, $R_N = R_e(1 + e\sin2L)$. e is eccentricity of ellipsoid, $e = (R_e - R_p)/R_e$. R_e is the long radius of the ellipse and R_p is the short radius of the ellipsoid.

The travel path of the vehicle is a spatial curve connected by many spatial points in geodetic coordinate system. The spatial curve needs to be projected to the $x_g - y_g$ plane in the geographic coordinate system. To facilitate the analysis of the driving path, the coordinate origin of the geographical coordinate system is set at the initial record point of the driving track. Thus, the coordinates of each sampling point in the Earth Cartesian coordinate system needs to be transformed by Equation (2).

$$\begin{cases} \Delta x(yz)_{e1} = x(yz)_{e1} - x(yz)_{e1} \\ \Delta x(yz)_{e2} = x(yz)_{e2} - x(yz)_{e1} \\ \Delta x(yz)_{e3} = x(yz)_{e3} - x(yz)_{e1} \\ \quad \cdots\cdots \\ \Delta x(yz)_{en} = x(yz)_{en} - x(yz)_{e1} \end{cases} \tag{2}$$

Then, the coordinates of the Earth Cartesian coordinate system are transformed into the coordinates of the geographic coordinate system using Equation (3).

$$\begin{bmatrix} x_{gi} & y_{gi} & z_{gi} \end{bmatrix} = \begin{bmatrix} \Delta x_{ei} & \Delta y_{ei} & \Delta z_{ei} \end{bmatrix} \begin{bmatrix} -\sin\lambda & \cos\lambda & 0 \\ -\sin L\cos\lambda & -\sin L\sin\lambda & \cos L \\ \cos L\cos\lambda & \cos L\sin\lambda & \sin L \end{bmatrix} \tag{3}$$

With the above transformations, the driving path can be shown accurately in two-dimensional plane of the geographic coordinate system using the coordinate (x_g, y_g).

3. Fitting of the Test Path

Function fitting refers to search a curve expression by adjusting some undetermined coefficients in this function to minimize the difference between the function and the known set of coordinate points (x_i, y_i). According to above process, the coordinates of the ideal path in the geographical coordinate system have been obtained. For the convenience of calculation, only the two-dimensional coordinate (x_g, y_g) is needed to be fitted without considering the vertical motion. Figure 4 is a schematic diagram of the lane-changing path.

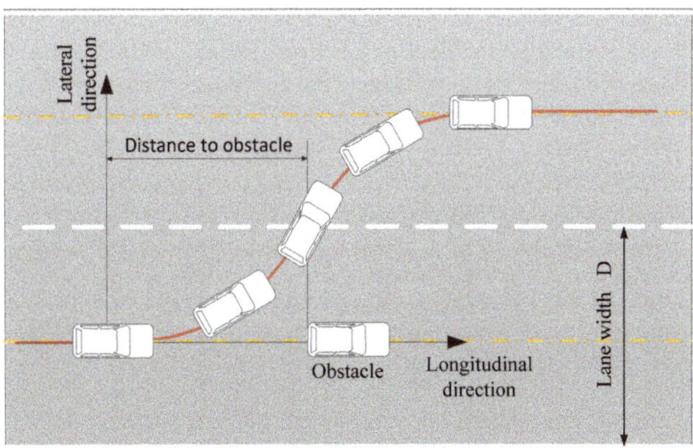

Figure 4. The schematic diagram of the lane-changing path.

It is necessary to satisfy the requirement of curvature continuity of the path curve to ensure the vehicle moves smoothly. Once the shape of the curve is known, the curvature–continuous fitting function should be determined. An easement curve is a curve of continuous curvature, usually set between a straight line and a circular curve or between two circular curves with different radius. There will be no mutation of the curvature of the path curve. Therefore, it can improve the comfort and stability of the vehicle considerably. There are usually two types of transition curves: clothoid and cubic parabola. In engineering practice, the swirl curve is obtained by the method of point selection and lofting. However, it is difficult to accurately fit the swirl curve, as its mathematical equation is too complicated, which will lead to high computation cost. The expression of polynomial curve is succinct, and all derivatives are continuous. It can accurately fit the driving path under different operation conditions by changing the coefficients of each item. In this paper, the polynomial curve is adopted as the lane-changing path.

Polynomial equations are set as follows:

$$y = a_n x^n + a_{n-1} x^{n-1} + \cdots + a_1 x + a_0 \tag{4}$$

$$k = \frac{y''}{(1 + y'^2)^{\frac{3}{2}}} \tag{5}$$

where y' is the first derivative and y'' is the second derivative of the polynomial curve. k is the curvature of a point on the curve.

The center of the mass of vehicle at the beginning of lane-changing is set as the origin of the coordinate system for simplifying the system with proposer assumptions. Thus, the polynomial constant term is zero. According to the analysis of lane changing operation, it is known that, at the initial and final state of the lane-changing operation, the vehicle's heading direction should be parallel to the lane line and the steering angle shall be zero. To make the unmanned vehicle able to travel straight along the lane, it is required that the steering wheel angular speed be zero when the lane-changing operation ends. Therefore, the first derivative and the second derivative of the polynomial curve at the initial point and the final point are zero. The final point of lane-changing is set as (x_f, y_f) and y_f is the lane width of the current driving road. Thus, the position of the vehicle is at the middle of the lane when lane-changing is completed. While the initial and final positions of lane-changing are determined, the trajectory selected by different types of drivers will also be different, which has impact on the riding comfort of the lane-changing process. Therefore, it is necessary to consider the intermediate state of the vehicle during lane-changing path planning. It is noticed that the entire lane-changing process can be divided into three phases: collision avoidance, rotation and adjustment [30]. It can be seen that the steering wheel angular speed during the obstacle avoidance and the rotation phase are faster than that during the adjustment phase by analyzing the steering wheel angular speed during the lane-changing process. Therefore, the vehicle position state (x_m, y_m) at the end of the rotation phase is adopted as the intermediate state constraint of the lane-changing operation. By substituting the state constraint of the initial point, final point and intermediate point into Equation (4), Equation (6) can be obtained as follows:

$$\begin{cases} y_0 = a_n x_0^n + \cdots + a_5 x_0^5 + a_4 x_0^4 + a_3 x_0^3 + a_2 x_0^2 + a_1 x_0 \\ y_f = a_n x_f^n + \cdots + a_5 x_f^5 + a_4 x_f^4 + a_3 x_f^3 + a_2 x_f^2 + a_1 x_f \\ \dot{y}(x_0) = n a_n x_0^{n-1} + \cdots + 5 a_5 x_0^4 + 4 a_4 x_0^3 + 3 a_3 x_0^2 + 2 a_2 x_0 + a_1 = 0 \\ \dot{y}\left(x_f\right) = n a_n x_f^{n-1} + \cdots + 5 a_5 x_f^4 + 4 a_4 x_f^3 + 3 a_3 x_f^2 + 2 a_2 x_f + a_1 = 0 \\ \ddot{y}(x_0) = n \cdot (n-1) a_n x_0^{n-2} + \cdots + 20 a_5 x_0^3 + 12 a_4 x_0^2 + 6 a_3 x_0 + 2 a_2 = 0 \\ \ddot{y}\left(x_f\right) = n \cdot (n-1) a_n x_f^{n-2} + \cdots + 20 a_5 x_f^3 + 12 a_4 x_f^2 + 6 a_3 x_f + 2 a_2 = 0 \\ y_m = a_n x_m^n + \cdots + a_5 x_m^5 + a_4 x_m^4 + a_3 x_m^3 + a_2 x_m^2 + a_1 x_m \end{cases} \tag{6}$$

Substituting $x_0 = y_0 = 0$ and $y_f = D$ into Equation (6), Equation (7) is as follows:

$$\begin{cases} D = a_n x_f^n + \cdots + a_5 x_f^5 + a_4 x_f^4 + a_3 x_f^3 \\ 0 = n a_n x_f^{n-1} + \cdots + 5 a_5 x_f^4 + 4 a_4 x_f^3 + 3 a_3 x_f^2 \\ 0 = n \cdot (n-1) a_n x_f^{n-2} + \cdots + 20 a_5 x_f^3 + 12 a_4 x_f^2 + 6 a_3 x_f \\ y_m = a_n x_m^n + \cdots + a_5 x_m^5 + a_4 x_m^4 + a_3 x_m^3 \end{cases} \tag{7}$$

From Equation (7), it is known that the number of polynomial coefficients need to be determined is $n - 2$, and the number of constraint equation is only4. To determine the unique expression of the lane-changing path, the equation (8) needs to be satisfied.

$$n - 2 = 4 \tag{8}$$

Thus, $n = 6$. The expression of the lane-changing trajectory is as follows:

$$y = a_6 x^6 + a_5 x^5 + a_4 x^4 + a_3 x^3 \tag{9}$$

The expression of the optimal lane-changing path under different conditions can be obtained with the middle point coordinates and the final distance of lane-changing determined.

4. Path Planning Method Based on Excellent Driver Lane-Changing Model

Lane-changing is an important and complex operation in vehicle driving. Trajectory has great influence on the comfort of the driverless vehicle. Optimal lane-changing trajectories of several excellent drivers under different conditions were obtained from previous research. The lane-changing model of excellent driver was established based on the GA-BP neural network trained and tested by the testing path data. The path planning method for imitating the lane-changing operation of excellent drivers is proposed. It can generate an optimal lane-changing trajectory according to the driving conditions of vehicles and the types selected by the passengers therefore improve the comfort of the autonomous vehicles. Figure 5 shows the framework of the path planning method for imitating the lane-changing operation of excellent drivers.

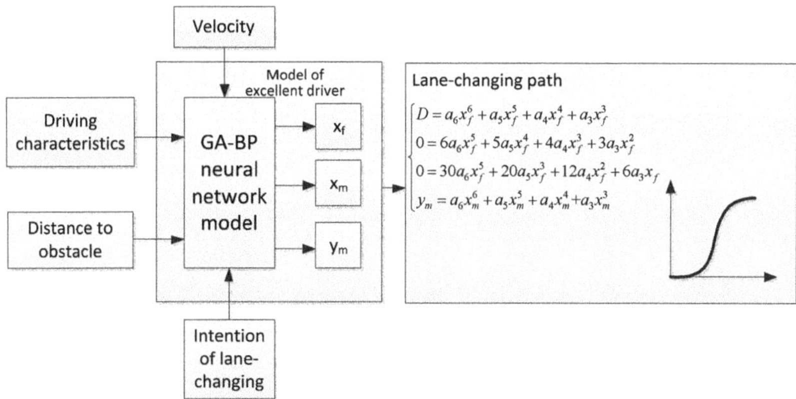

Figure 5. The framework of the path planning method for imitating the lane-changing operation of excellent drivers.

4.1. GA-BP Neural Networks

BP neural network is a multi-unit feed forward neural network, which can be trained by error back propagation. The error of the actual output value and the desired output value of the network is minimized by adjusting the weights and thresholds based on the gradient descent method. There are three units in the BP neural network: input layers, output layers and hidden layers. In the training process, there are two stages of forward and back propagation. In the stage of forward propagation, the input information is transmitted from the input layer through the hidden layer to the output layer. The state of each layer only affects the next layer state of neurons. The back propagation is introduced in the case the error is larger than the threshold. The weights of each layer of neurons are modified to minimize the error.

Genetic algorithm is a global optimization random search algorithm. It can get the individual with high adaptation degree by simulating the phenomena of selection, crossover and mutation in the genetic process. To solve the problem that BP neural network algorithm is easy to fall into local optimal solutions, the GA is used to optimize the initial weights and thresholds of BP neural network. The population of GA is generated based on the weights and thresholds of BP neural network. With going through the selection, crossover and mutation process, the optimal individuals are selected as the initial weights and thresholds of BP neural network. With GA, the convergence speed of the BP neural network can be improved significantly and the possibility of falling into local optimal solutions can be reduced. Figure 6 shows the GA-BP neural network algorithm flow chart. The specific algorithm is as follows:

(1)   Initialization of the population

The initial population of scale P, $X = (X_1, X_2, \ldots , X_p)^T$, is randomly generated. The individual code, $X_i = (x_1, x_2, \ldots , x_s)$, utilizes the real number coding method. The length coding is as follows:

$$s = n \times m + m \times l + m + l \tag{10}$$

where m is hidden layer nodes, n is input layer nodes and l is output layer nodes.

(2)   Determination of the fitness function

In the GA-BP model, the use of a fitness function F is based on the error of the output layer. The function F is defined as:

$$F_i = k \cdot \sum_{j=1}^{l} (o_j - y_j)^2 \quad (i = 1, 2, 3, \cdots , p) \tag{11}$$

where y_j is the expected output. o_j is the actual output based on the weights and thresholds generated in Step 1. k is compensation factor.

(3)   Selection operation

This paper uses the roulette method to select the operator. The probability of each individual is calculated as follows

$$f_i = 1 / F_i \tag{12}$$

$$p_i = \frac{f_i}{\sum_{j=1}^{P} f_j} \tag{13}$$

(4)   Crossover operation

The crossover operation between the chromosome k and the chromosome l in the gene j is as follows

$$\begin{cases} x_{kj} = x_{kj} \cdot (1 - b) + x_{lj} \cdot b \\ x_{lj} = x_{lj} \cdot (1 - b) + x_{kj} \cdot b \end{cases} \tag{14}$$

where b is a random number in [0, 1].

(5)   Mutation operation

The mutation operation of the chromosome i in the gene j is as follows

$$x_{ij} = \begin{cases} x_{ij} + (x_{ij} - x_{max}) \cdot f(g) & r > 0.5 \\ x_{ij} + (x_{min} - x_{ij}) \cdot f(g) & r \leq 0.5 \end{cases} \tag{15}$$

$$f(g) = r_2 \times \left(1 - \frac{g}{G_{\max}}\right) \tag{16}$$

where $x_{\min}$ and $x_{\max}$ are the minimum and maximum values of the x_{ij}, respectively. r is a random number in [0, 1]. r_2 is a random number. g represents the current number of iterations and $G_{\max}$ is the maximum number of evolutions.

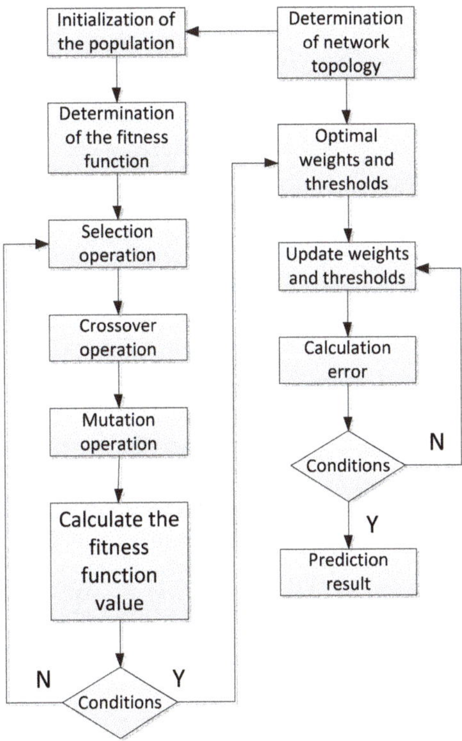

Figure 6. Flow chart of GA-BP neural network algorithm.

4.2. Excellent Driver Lane-Changing Model

The lane-changing trajectories chosen by different drivers under the same working conditions will be different due to the various driving habits and personalities of each driver. Figure 7 shows the lane-changing paths of different drivers at the same speed. It can be observed that the driving characteristic is one of the most important factors that influence the travel route. Through the analysis of the operating habits of different drivers, the driver types are divided into aggressive, intermediate and conservative. In this research, 1 represents aggressive type, 0 represents conservative type, and 0.5 means intermediate type. Figure 8 shows the paths of the free lane change and the obstacle avoidance lane change. In the figure, it is shown that different steering intentions also create a considerable impact on the vehicle driving path. In this paper, 1 represents the intention of obstacle avoidance lane change and 0 is for free lane change. In addition, the speed and the distance from obstacle also have an impact on the choice of the lane-changing path. Figure 9 shown the lane-changing paths of one driver at different speeds. Therefore, the excellent driver lane-changing model in this paper has four inputs, namely driver type, steering intention, vehicle speed and distance from obstacle. According to the above analysis, the optimal lane-changing path under different conditions can be obtained when the characteristic point coordinates (x_m, y_m) and the final distance x_f of lane-changing are determined. Thus, the model proposed in the paper has three outputs: x_m, y_m, and x_f.

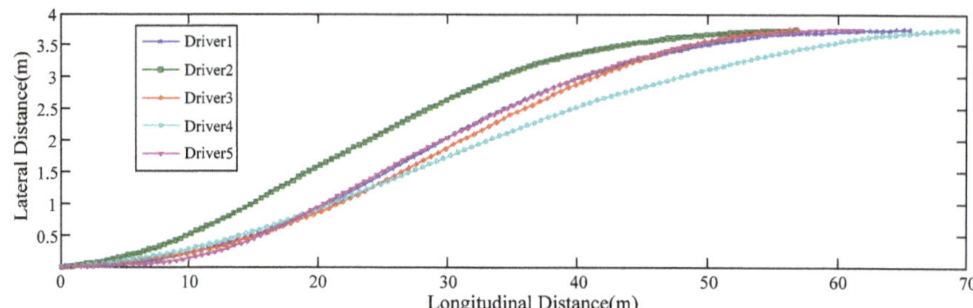

Figure 7. The lane-changing paths of different drivers at the same speed.

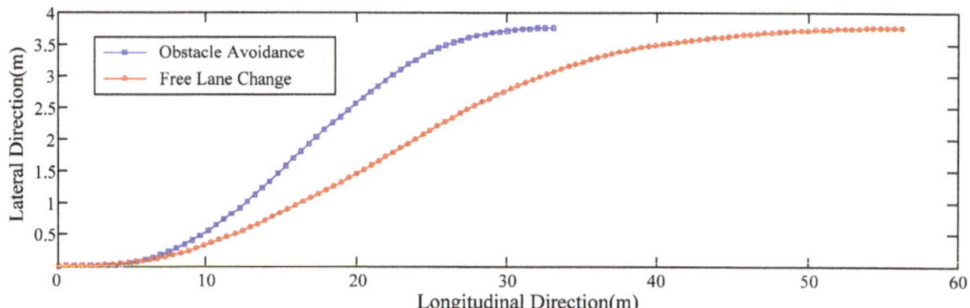

Figure 8. The lane-changing paths with different intention.

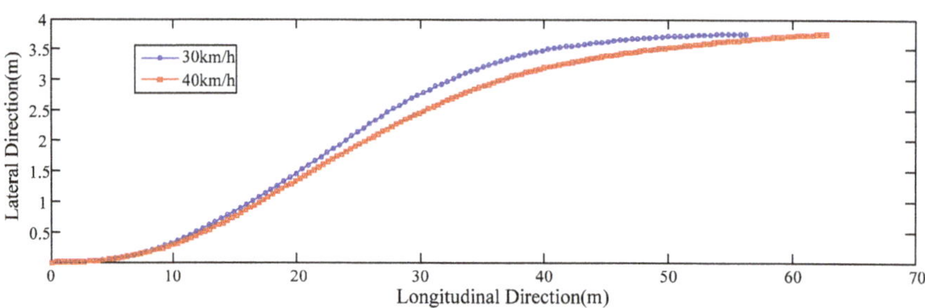

Figure 9. The lane-changing paths of one driver at different speeds.

The experimental data are shown in Table 2. The steering intention is determined based on distance from obstacle. The distance from obstacle under free lane change condition is set to 100 m. Therefore, there are 300 experimental data. Ninety percent of the data were selected at random as the training data, while the remaining are the testing data.

Table 2. Information of experiment data.

Classification	Information
Number of drivers	5
Velocity (km/h)	30, 35, 40, 45, 50
Distance from obstacle (m)	30, 35, 40, 45, 50, 55, 60, 65, 70, 75, 80, 100 (no obstacle)

The effectiveness of the neutral network was estimated using the GA-BP error. The error is defined as the difference between the simulation output and experimental output as Equation (17). For the testing in this research, the error is shown in Figure 10. In the figure, it can be observed that the GA-BP neural network is with high accuracy.

$$error = output\_simu - output\_testing \tag{17}$$

where output_simu is the output value of the GA-BP neural network and output_testing is the value of the testing date.

Figure 11 shows the mean square error variation curve of the GA-BP neural networks model. The mean square error is 0.009 after training, and it can meet the precision requirements.

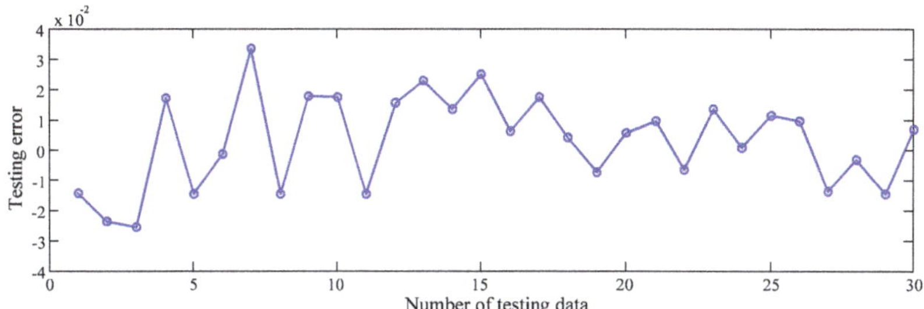

Figure 10. The testing error of the GA-BP neural network.

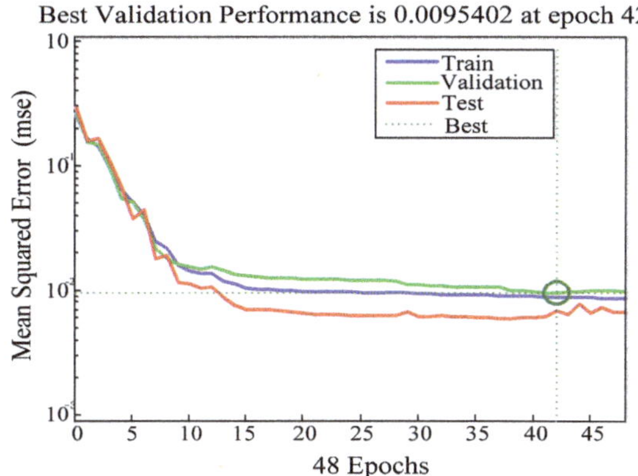

Figure 11. The mean square error variation curve of the GA-BP neural networks model.

5. Simulation and Analysis

To study the performance of the path planning algorithm based on the excellent driver model proposed in the paper, simulation experiments under different working conditions were carried out in MATLAB environments. The scenarios of obstacle avoidance steering and free lane-changing steering at the speed of 30 km/h and 40 km/h were simulated. In each simulation test, the aggressive and conservative types were selected as driver type, respectively. Then, the lane-changing trajectory generated by the proposed method was compared with the actual driving trajectory under the same working condition. The effectiveness of the proposed algorithm can be quantitatively evaluated by calculating the deviation between the actual value and the simulation value. The change of the error is also presented.

5.1. Obstacle Avoidance Simulation

Obstacle avoidance simulation test conditions: The speed is set to 30 km/h and 40 km/h, respectively; lane width is 3.75 m; and the distance to obstacle at the beginning of steering is 35 m.

Figures 12 and 13 are the simulation results at the speed of 30 km/h under the obstacle avoidance conditions. Figure 12a shows the simulation trajectory and actual trajectory of the conservative driver under the obstacle avoidance conditions. Figure 13a shows the trajectory of the aggressive driver. The red curve represents the lane-changing track obtained by the lane-changing path generation algorithm proposed in this paper. The blue curve represents the actual running track recorded in the real vehicle lane-changing test under the same condition. It can be seen in Figures 12a and 13a that the trajectory obtained from simulation is basically consistent with the trajectory in real vehicle testing. It also meets the requirements of vehicle safety obstacle avoidance. Figures 12b and 13b are the lateral error between the simulated and real trajectories of conservative drivers and aggressive drivers. The maximum lateral deviations are 0.056 m and 0.17 m, respectively. It means that the algorithm developed in this paper is with high accuracy and imitate the lane-changing operation of excellent drivers.

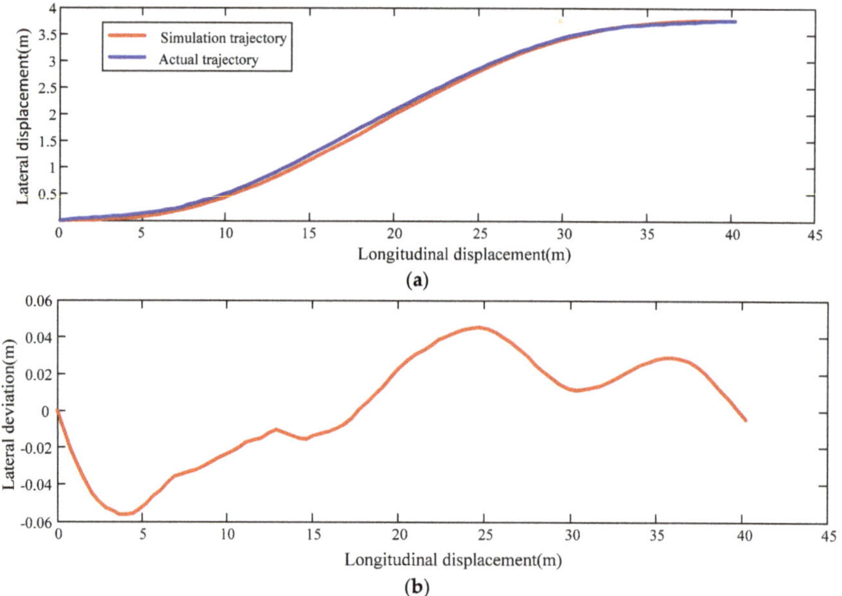

Figure 12. The simulation results at the speed of 30 km/h under the obstacle avoidance conditions of the conservative driver: (**a**) simulation trajectory and actual trajectory; and (**b**) lateral deviations between the simulated and real trajectories.

Figures 14 and 15 are the simulation results at the speed of 40 km/h under the obstacle avoidance conditions. Figure 14a shows the simulation trajectory and actual trajectory of the conservative driver under the obstacle avoidance conditions. Figure 15a shows the two paths of the aggressive driver. Figures 14a and 15a show that the simulation path and the actual path are very close to each other. It means the simulation trajectory can reflect the characteristics of excellent drivers' lane-changing trajectories under the same working conditions. Figures 14b and 15b are the lateral deviation between the simulated and real trajectories of conservative drivers and aggressive drivers at the speed of 40 km/h. The maximum lateral deviations are 0.087 m and 0.075 m, respectively, which means the method can keep high accuracy with the increased speed.

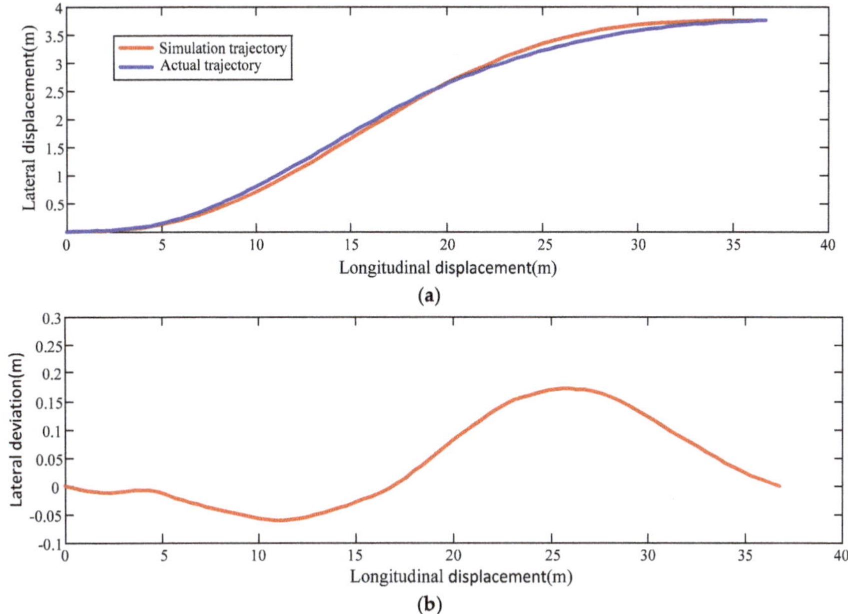

(a)

(b)

Figure 13. The simulation results at the speed of 30 km/h under the obstacle avoidance conditions of the aggressive driver: (**a**) simulation trajectory and actual trajectory; and (**b**) lateral deviations between the simulated and real trajectories.

As can be observed in Figures 12a, 13a, 14a and 15a, the longitudinal distance of the lane-changing trajectory is increased with the increased speed, and the longitudinal distance of the conservative driver at the same speed is shorter than that of the aggressive driver at same speed. This is because, in the process of obstacle avoidance, the conservative drivers usually keep a larger safety distance, and a larger steering wheel angle will be input at the early stage of the lane-changing. Therefore, the vehicle will travel to the target lane as soon as possible. Aggressive drivers often choose a smaller safety distance, so the lane-changing operation is more stable, making the terminal distance of lane changing longer.

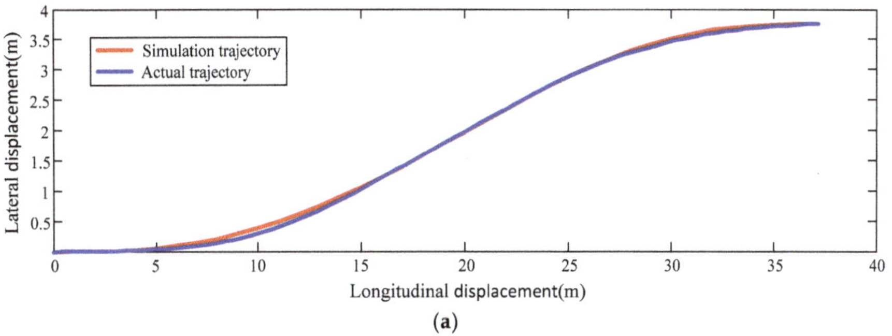

(a)

Figure 14. *Cont.*

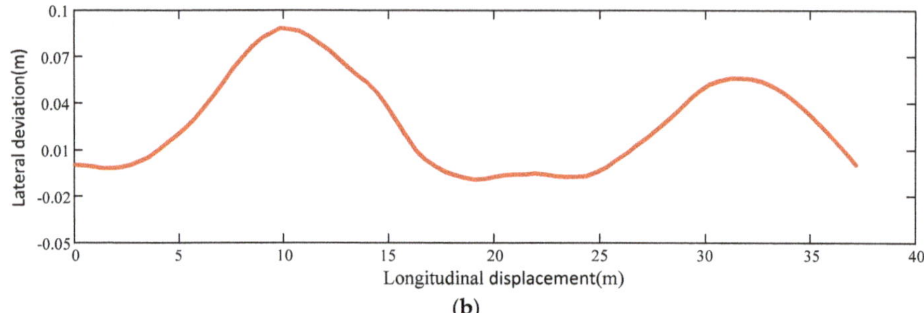

Figure 14. The simulation results at the speed of 40 km/h under the obstacle avoidance conditions of the conservative driver: (**a**) simulation trajectory and actual trajectory; and (**b**) lateral deviations between the simulated and real trajectories.

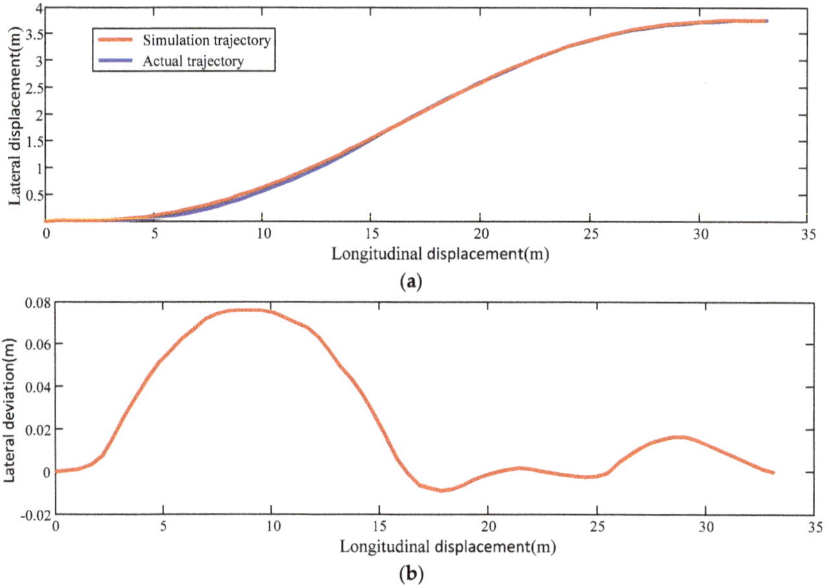

Figure 15. The simulation results at the speed of 30 km/h under the obstacle avoidance conditions of the aggressive driver: (**a**) simulation trajectory and actual trajectory; and (**b**) lateral deviations between the simulated and real trajectories.

5.2. Free Lane-Changing Simulation

Free lane-changing simulation test conditions: The speed is 30 km/h and 40 km/h, respectively; lane width is 3.75 m; and there is no obstacle.

Figures 16 and 17 are the simulation results at the speed of 30 km/h under the free lane-changing conditions. Figure 16a shows the simulation trajectory and actual trajectory of the aggressive driver under the free lane-changing conditions. Figure 17a shows the two trajectories of the conservative driver. The red curve represents the lane change track obtained by the lane-changing path generation algorithm proposed in this paper. The blue curve represents the test trajectory. In Figures 16a and 17a, it can be seen that the coincidence of the trajectory obtained by the simulation and the actual trajectory of the test is high. Figures 16b and 17b are the lateral error between the simulation and testing of aggressive drivers and conservative drivers, respectively. The maximum lateral deviations are 0.15 m and 0.074 m. Under the free lane-changing condition, the algorithm in this paper can also imitate the route of experienced drivers when do lane-changing operation.

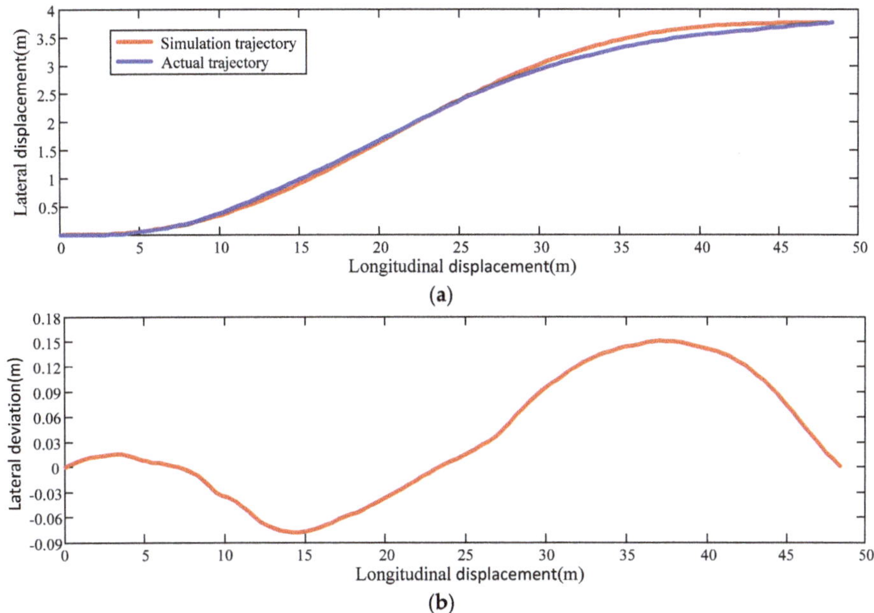

Figure 16. The simulation results at the speed of 30 km/h under the free lane-changing conditions of the aggressive driver: (**a**) simulation trajectory and actual trajectory; and (**b**) lateral deviations between the simulated and real trajectories.

Figures 18 and 19 are the simulation results at the speed of 40 km/h under the free lane-changing conditions. Figure 18a shows the simulation trajectory and actual trajectory of the aggressive driver under the obstacle avoidance conditions. The two trajectories of the conservative driver at 40 km/h are presented in Figure 19a. The two diagrams show that the features of the route chosen by experienced driver can be represented by the simulation path. The lateral error between the simulation and test are shown in Figures 18b and 19b. The maximum lateral deviations are 0.088 m and 0.16 m, respectively, which means the algorithm is with high accuracy under the free lane-changing condition.

In Figures 16a, 17a, 18a and 19a, it can be seen the longitudinal distance chosen by the same type of drivers increases with the speed under the free lane-changing condition. However, the longitudinal distance chosen by the conservative driver under the free lane-changing condition is significantly longer than that of aggressive driver. Because a driver conducts the lane-changing operation according to his own habit when there is no obstacle under free condition, for the conservative driver, to reduce the lateral acceleration when changing lanes, the input steering wheel angle at the beginning of lane change will be reduced, which increases the longitudinal distance of the entire lane change, whereas, for aggressive drivers, they will choose to complete the lane change within shortest possible time, which requires a large steering wheel angle input, resulting in a smaller longitudinal distance of the lane-changing path.

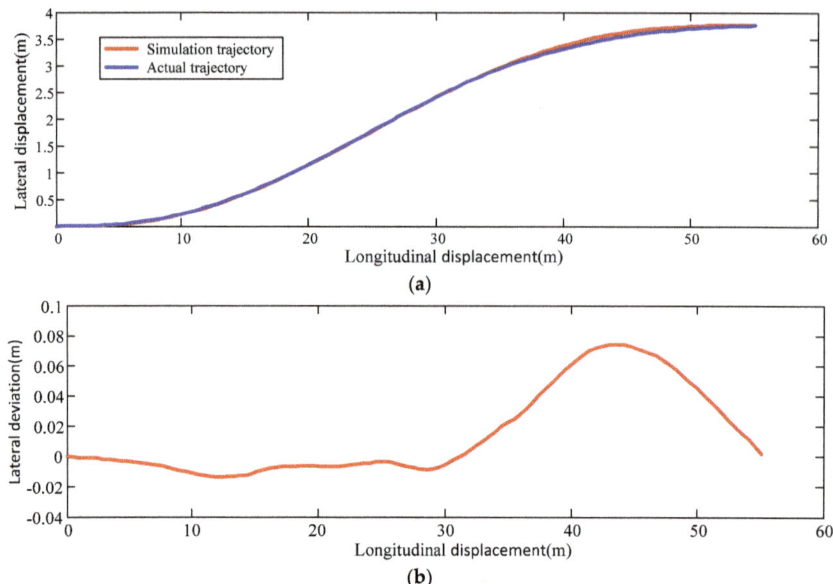

Figure 17. The simulation results at the speed of 30 km/h under the free lane-changing conditions of the conservative driver: (**a**) simulation trajectory and actual trajectory; and (**b**) lateral deviations between the simulated and real trajectories.

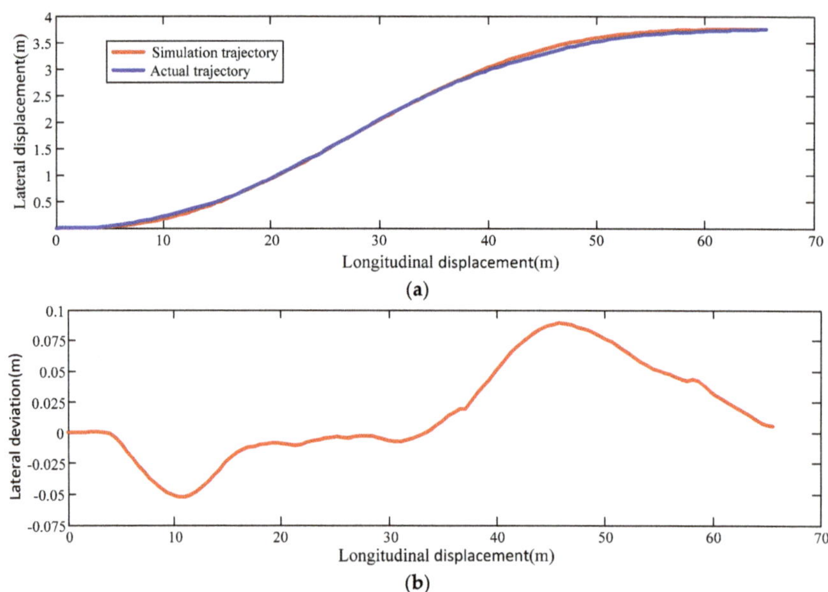

Figure 18. The simulation results at the speed of 40 km/h under the free lane-changing conditions of the aggressive driver: (**a**) simulation trajectory and actual trajectory; and (**b**) lateral deviations between the simulated and real trajectories.

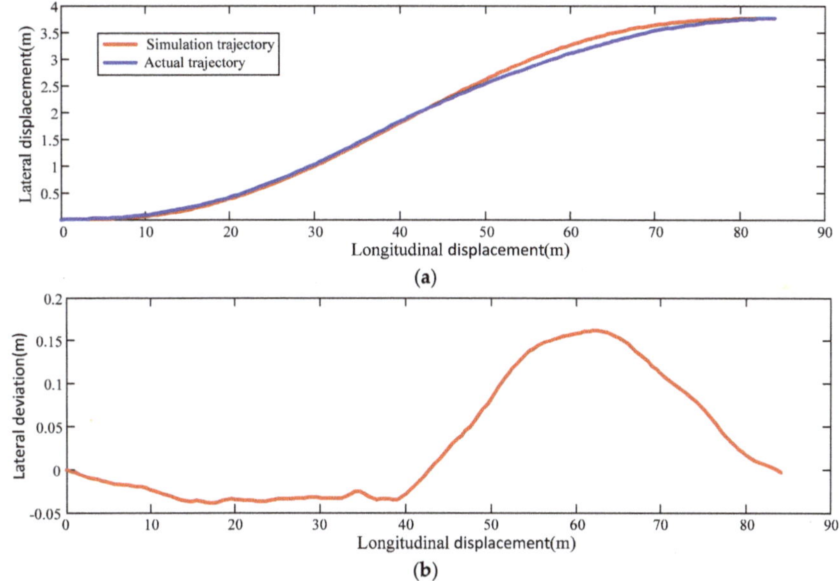

Figure 19. The simulation results at the speed of 40 km/h under the free lane-changing conditions of the conservative driver: (**a**) simulation trajectory and actual trajectory; and (**b**) lateral deviations between the simulated and real trajectories.

6. Conclusions

A new lane-changing path planning method for unmanned vehicles is proposed in this paper. It is developed to generate a path that can make the vehicle finish the lane-changing operation smoothly and safely based on the information of vehicle speed, steering intention, obstacle and driving style. First, to obtain the lane-changing paths of the excellent drivers under different conditions, many real vehicle tests were carried out. Then, the final point and intermediate feature point of each lane change trajectory were extracted, with the expression of each path obtained. Next, the excellent driver model based on the GA-BP neural network was established and trained with experimental data. Finally, the path planning algorithm for imitating the lane-changing operation of excellent drivers was developed and presented.

The path developed in this paper can imitate real excellent driver lane-changing path to meet the kinematic constraints of the vehicle and avoid the obstacles at the same time. Simulation results verifies that the path planning method proposed in this research can generate an optimal lane-changing trajectory according to the vehicle driving condition and the driving styles selected by the passengers, which is basically the same with the trajectory selected by the excellent driver under the same conditions and the lateral error between the two trajectories is small. Therefore, the lane-changing path planning algorithm proposed in this paper can imitate the excellent driver to change the lane, improve the stability of the unmanned vehicle, improve the riding comfort of passengers, and reduce the probability of passengers suffering from motion sickness.

Author Contributions: G.G. and Z.W. conceived of and designed the method. G.G. and L.S. performed the experiments and analyzed the experimental data. Finally, Z.W. wrote the paper with the help of G.G., H.J. and C.D.

Acknowledgments: This work was financially supported by The National Natural Science Fund (No. U1564201 and No. 51675235), Six Talent Peaks Project Fund of Jiangsu Province (No. 2015-XNYQC-002), and Senior Talent Fund of Jiangsu University (No. 15JDG075).

Conflicts of Interest: The authors declare no conflict of interest.

References

1. Rasekhipour, Y.; Khajepour, A.; Chen, S.K. A Potential Field-Based Model Predictive Path-Planning Controller for Autonomous Road Vehicles. *IEEE Trans. Intell. Transp. Syst.* **2017**, *18*, 1255–1267. [CrossRef]
2. Korayem, M.H.; Nekoo, S.R. The SDRE control of mobile base cooperative manipulators: Collision free path planning and moving obstacle avoidance. *Robot. Auton. Syst.* **2016**, *86*, 86–105. [CrossRef]
3. Gao, Y.Q.; Gray, A.; Tseng, H.E. A tube-based robust nonlinear predictive control approach to semiautonomous ground vehicles. *Veh. Syst. Dyn.* **2014**, *52*, 802–823. [CrossRef]
4. Gao, Y.Q.; Lin, T.; Borrelli, F. Predictive Control of Autonomous Ground Vehicles With Obstacle Avoidance on Slippery Roads. In Proceedings of the ASME 2010 Dynamic Systems and Control Conference, Cambridge, MA, USA, 12–15 September 2010; pp. 265–272.
5. Donatelli, M.; Giannelli, C.; Mugnaini, D. Curvature continuous path planning and path finding based on PH splines with tension. *Comput. Aided Des.* **2017**, *88*, 14–30. [CrossRef]
6. Giannelli, C.; Mugnaini, D.; Sestini, A. Path planning with obstacle avoidance by G1 PH quintic splines. *Comput. Aided Des.* **2016**, *76*, 47–60. [CrossRef]
7. Choi, J.W.; Curry, R.E.; Elkaim, G.H. Curvature-continuous trajectory generation with corridor constraint for autonomous ground vehicles. In Proceedings of the 49th IEEE Conference on Decision and Control, Hilton Atlanta Hotel, Atlanta, GA, USA, 15–17 December 2010; pp. 7166–7171.
8. Meyer, G.; Beiker, S. *Road Vehicle Automation 2*; Springer International Publishing: Cham, Switzerland, 2015; ISBN 978-3-319-19077-8.
9. Elbanhawi, M.; Simic, M.; Jazar, R. Improved maneuvering of autonomous passenger vehicles: Simulations and field results. *J. Vib. Control* **2017**, *23*, 1954–1983. [CrossRef]
10. Diels, C.; Bos, J.E. Self-driving carsickness. *Appl. Ergon.* **2016**, *53*, 374–382. [CrossRef] [PubMed]
11. Cao, P.; Hu, Y.; Miwa, T. An optimal mandatory lane change decision model for autonomous vehicles in urban arterials. *J. Intell. Transp. Syst.* **2017**, *21*, 271–284. [CrossRef]
12. Butakov, V.A.; Ioannou, P. Personalized driver/vehicle lane change models for ADAS. *IEEE Trans. Veh. Technol.* **2015**, *64*, 4422–4431. [CrossRef]
13. Nilsson, J.; Brännström, M.; Coelingh, E. Lane change maneuvers for automated vehicles. *IEEE Trans. Intell. Transp. Syst.* **2017**, *18*, 1087–1096. [CrossRef]
14. Nilsson, J.; Brannstrom, M.; Coelingh, E. Longitudinal and lateral control for automated lane change maneuvers. In Proceedings of the American Control Conference, Palmer House Hilton, Chicago, IL, USA, 1–3 July 2015; pp. 1399–1404.
15. Dubins, L.E. On Curves of Minimal Length with a Constraint on Average Curvature, and with Prescribed Initial and Terminal Positions and Tangents. *Am. J. Math.* **1957**, *79*, 497–516. [CrossRef]
16. Lamiraux, F.; Lammond, J.P. Smooth motion planning for car-like vehicles. *IEEE Trans. Robot. Autom.* **2001**, *17*, 498–501. [CrossRef]
17. Choi, Y.G.; Lim, K.I.; Kim, J.H. Lane change and path planning of autonomous vehicles using GIS. In Proceedings of the 12th International Conference on Ubiquitous Robots and Ambient Intelligence, KINTEX, Goyang, Korea, 28–30 October 2015; pp. 163–166.
18. Ren, D.B.; Zhang, J.Y.; Zhang, J.M. Trajectory planning and yaw rate tracking control for lane changing of intelligent vehicle on curved road. *Sci. China Technol. Sc.* **2011**, *54*, 630–642. [CrossRef]
19. Wang, C.; Zheng, C.Q. Lane change trajectory planning and simulation for intelligent vehicle. *Adv. Mater. Res.* **2013**, *671*, 2843–2846. [CrossRef]
20. Yang, T.; Asanjan, A.A.; Faridzad, M. An enhanced artificial neural network with a shuffled complex evolutionary global optimization with principal component analysis. *Inf. Sci.* **2017**, *418*, 302–316. [CrossRef]
21. Zhang, X.Y.; Yin, F.; Zhang, Y.M. Drawing and recognizing Chinese characters with recurrent neural network. *IEEE Trans. Pattern Anal. Mach. Intell.* **2018**, *99*, 849–862. [CrossRef] [PubMed]
22. Jain, A.K.; Duin, R.P.W.; Mao, J. Statistical pattern recognition: A review. *IEEE Trans. Pattern Anal. Mach. Intell.* **2000**, *22*, 4–37. [CrossRef]

23. Saon, G.; Soltau, H.; Nahamoo, D. Speaker adaptation of neural network acoustic models using i-vectors. In Proceedings of the IEEE Workshop on Automatic Speech Recognition and Understanding (ASRU), Olomouc, Czech Republic, 8–13 December 2013; pp. 55–59.
24. Li, J.C.; Zhao, D.L.; Ge, B.F. A link prediction method for heterogeneous networks based on BP neural network. *Physics A* **2017**, *495*, 1–17. [CrossRef]
25. Sharifian, A.; Ghadi, M.J.; Ghavidel, S. A new method based on Type-2 fuzzy neural network for accurate wind power forecasting under uncertain data. *Renew. Energy* **2018**, *120*, 220–230. [CrossRef]
26. Xue, H.; Bai, Y.; Hu, H. Influenza activity surveillance based on multiple regression model and artificial neural network. *IEEE Access.* **2018**, *6*, 563–575. [CrossRef]
27. Al-Qutami, T.A.; Ibrahim, R.; Ismail, I. Virtual multiphase flow metering using diverse neural network ensemble and adaptive simulated annealing. *Expert Syst. Appl.* **2018**, *93*, 72–85. [CrossRef]
28. Yu, F.; Xu, X.Z. A short-term load forecasting model of natural gas based on optimized genetic algorithm and improved BP neural network. *Appl. Energy* **2014**, *134*, 102–113. [CrossRef]
29. Wang, S.X.; Zhang, N.; Wu, L. Wind speed forecasting based on the hybrid ensemble empirical mode decomposition and GA-BP neural network method. *Renew. Energy* **2016**, *94*, 629–636. [CrossRef]
30. Bin, W.; Zhu, X.; Shen, J. Analysis of driver emergency steering lane changing behavior based on naturalistic driving data. *J. Tongji Univ.* **2017**, *45*, 554–561.

 *applied*
sciences

Article

MPC and PSO Based Control Methodology for Path Tracking of 4WS4WD Vehicles

Qifan Tan [1,*], Penglei Dai [2] , Zhihao Zhang [3] and Jay Katupitiya [3]

[1]   School of Mechanical, Electronic and Control Engineering, Beijing Jiaotong University, Beijing 100044, China
[2]   School of Mechanical and Mechatronic Engineering, University of Technology Sydney,
     Sydney, NSW 2007, Australia; penglei.dai@uts.edu.au
[3]   School of Mechanical and Manufacturing Engineering, The University of New South Wales,
     Sydney, NSW 2052, Australia; zhihao.zhang1@unsw.edu.au (Z.Z.); j.katupitiya@unsw.edu.au (J.K.)
*   Correspondence: 14116355@bjtu.edu.cn; Tel.: +86-186-0124-6239

Received: 31 May 2018; Accepted: 13 June 2018; Published: 19 June 2018

Abstract: Four wheel steering and four wheel drive (4WS4WD) vehicles are over-actuated systems with superior performance. Considering the control problem caused by the system nonlinearity and over-actuated characteristics of the 4WS4WD vehicle, this paper presents two methods to enable a 4WS4WD vehicle to accurately follow a predefined path as well as its reference trajectories including velocity and acceleration profiles. The methodologies are based on model predictive control (MPC) and particle swarm optimization (PSO), respectively. The MPC method generates the virtual inputs in the upper controller and then allocates the actual inputs in the lower controller using sequential quadratic programming (SQP), whereas the PSO method is proposed as a fully optimization based method for comparison. Both methods achieve optimization of the steering angles and wheel forces for each of four independent wheels simultaneously in real time. Simulation results achieved by two different controllers in following the reference path with varying disturbances are presented. Discussion about two methodologies is provided based on their theoretical analysis and simulation results.

Keywords: 4WS4WD vehicle; force control; MPC; PSO; path tracking

---

1. Introduction

With the development of Autonomous Ground Vehicles (AGVs) in the last few decades, the demand for accuracy, maneuverability and controllability in vehicle's navigation is ever increasing. For example, an AGV may be required to follow a path accurately under unstructured and uneven terrain conditions, where a significant amount of wheel slip and unpredictable disturbance forces occur at the vehicle's wheels. The 4WS4WD vehicle, with four wheels that can be steered and driven independently, is a revolutionary platform that has great potential to perform high maneuverability and flexibility in harsh environments.

The main challenge in the control of 4WS4WD control is the number of control inputs (four steering angles and four drive torques), which results in an over-actuated system, where only three outputs including its degree of freedom (DOF) in the longitudinal, lateral and angular directions of the vehicle are concerned. How to allocate all eight control inputs to achieve high path following performance has not yet been effectively solved. The control allocation is proposed to handle the control problems of over-actuated systems [1]. Generally, the control allocation can be treated as an optimization problem.

For controller designing, a model of the vehicle under control is generally required to facilitate the selection of future control inputs. A dynamic model describes the states of the vehicle based on the forces applied. However, the development of a detailed vehicle dynamic model is always a challenging task due to the uncertainty of parameters and the complex disturbances from the external forces. The majority of existing control methodologies for 4WS4WD vehicles are proposed based on the linearized dynamic model, which lead to the loss of input degree of freedom [2,3]. Meanwhile, some other methods use nonlinear dynamic models in the control of 4WS4WD vehicles [4,5], where control inputs are subjected to some relationship constraints to simplify the controller design. However, in practice, the four independent wheels may interact with different terrain conditions, where different slip, wheel forces and terrain disturbances are generated on the corresponding contact patches. Hence, it is desirable to make four wheels individually controlled, thereby limiting and/or overcoming different slip and disturbance on different contact patch.

There is an abundance of literature that presents kinematic modelling of ground vehicles [6–8], in which the vehicles are assumed to operate at low speeds to reduce the dynamic effects. Most vehicle kinematic models are developed based on the non-integrable kinematic constraints, known as non-holonomic constraints. As a result, the wheel slip has to be ignored with the assumption of zero relative velocity between wheels and terrain [9,10]. To utilize the kinematic model's advantage of keeping the steering control relatively independent of velocity control [11], it is desirable to incorporate wheel slip in the vehicle kinematic model so as to facilitate the accurate vehicle control in complex terrain conditions where the no-slip assumption is not applicable.

To realize force control, a novel type of 4WS4WD vehicle with force sensors at each wheel has been designed as shown in Figure 1. This vehicle possesses the characteristics of independent steering and drive control at each wheel and force measurements. In this work, the dynamic model is partitioned into a hierarchy of three levels to facilitate the incorporation of force sensors and overcome uncertainties in the model. The force sensors at the drive unit allow the measurement of actual force data, while models of the drive unit and tire are used for simulation.

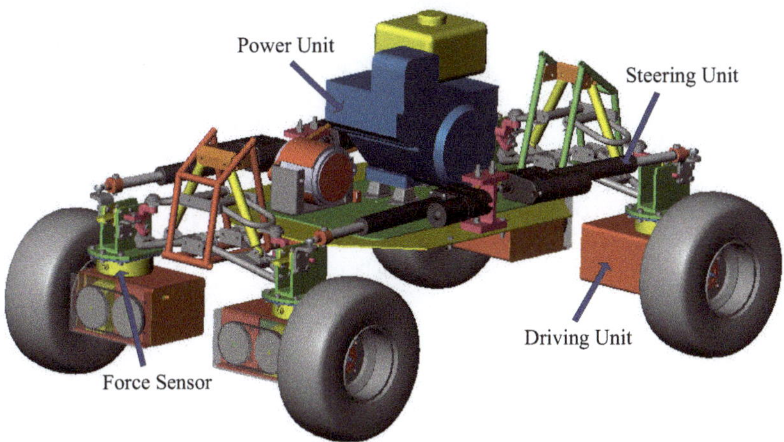

Figure 1. The 4WS4WD (four wheel steering and four wheel drive) vehicle.

Model predictive control (MPC) is selected for its ability in handling linear constraints and time-varying systems as well as its good performance in tracking problems. Particle swarm optimization (PSO) is also selected for its fast searching speed in global optimization. MPC and PSO have been successfully applied in the controller design of real-time control systems [12–17]. In this work, MPC and PSO based control methods are proposed to realize the controlling aim of achieving good path following performance as well as high motion quality via vehicle steering control and independent force control at four wheels.

As novel contributions, the MPC methodology is applied to achieve precise path tracking of 4WS4WD vehicle. Based on the MPC theory, an offline control law is proposed to guarantee the stability of the upper controller. An sequential quadratic programming (SQP) based control allocation is developed to control the 4WS4WD vehicle in the lower controller. The inclusion of full independent force control and steering control on all four wheels enable the maximization of performance. In this work, comparison of MPC and PSO on the same vehicle model is provided, in which the proposed PSO control methodology is a further refinement of the PSO methodology presented previously in [18]. The PSO control methodology in this paper simplifies the derivation and gives an algorithm in a more general form, which facilitates the comparison with other control methods. In addition, both methodologies are compared with the kinematic model based method proposed in [19].

The paper is organized as follows: Section 2 describes the 4WS4WD vehicle modelling. The MPC and PSO control methodologies are presented in Section 3. In Section 4, simulation setup and the reference path are presented. The result of the two controllers are compared. In Section 5, the discussion about two methodologies are provided based on their theoretical analysis and simulation results. Finally, conclusions are provided in Section 6.

2. Vehicle System Modelling

In order to develop suitable control methodology, the 4WS4WD vehicle must be modelled for controller design and simulation. The model used for this work is a fully dynamic model that consists of three components, vehicle body dynamics to estimate vehicle body movement, drive unit dynamics to model the tire force acting on the drive unit in vehicle coordinate frame and a tire dynamic model to model the force generated by the tire in a wheel coordinate frame. The separation of dynamic model allows force sensors to be incorporated easily and reduces uncertainties by obtaining actual force measurements. In the end, an offset error model is proposed for evaluating the tracking performance and is used in the controller design.

2.1. Vehicle Body Dynamic Model

As shown in Figure 2, the vehicle body dynamic model describes the motion of the vehicle by representing each drive unit as a pair of forces, with F_{SI} in the longitudinal and F_{SL} in the lateral direction of the corresponding wheel. The reason for separating the dynamics of the vehicle at the drive unit boundary is to allow force sensors to be incorporated to measure the forces acting on each wheel, instead of relying on the tire model to estimate them. To facilitate the notations, the four wheels are numbered by 1, 2, 3 and 4 in circles. In particular, the force acting at each force sensor is decomposed into two components (i.e., F_{Sli} and F_{SLi}, $i = 1, \ldots, 4$) normal to each other.

The equation of the vehicle body dynamics can be expressed as

$$acc = \begin{bmatrix} a_r \\ a_l \\ \gamma \end{bmatrix} = M^{-1} C \begin{bmatrix} R_1 F_{AS1} \\ R_2 F_{AS2} \\ R_3 F_{AS3} \\ R_4 F_{AS4} \end{bmatrix}, \tag{1}$$

where the acceleration vector denoted by acc consists of longitudinal acceleration a_l, radial acceleration a_r and angular acceleration γ.

The mass matrix is represented by M:

$$M = \begin{bmatrix} M - 4m_d & 0 & 0 \\ 0 & M - 4m_d & 0 \\ 0 & 0 & J_z \end{bmatrix}, \tag{2}$$

where M is the mass of the vehicle, J_z is the vehicle body inertia, and m_d is the mass of the drive unit.

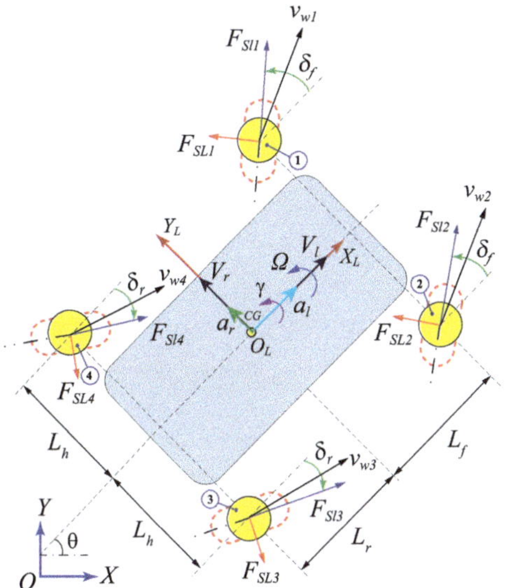

Figure 2. Vehicle body dynamic model.

The matrix C is decided by vehicle dimension, written as

$$C = \begin{bmatrix} 1 & 0 & 1 & 0 & 1 & 0 & 1 & 0 \\ 0 & 1 & 0 & 1 & 0 & 1 & 0 & 1 \\ -L_h & L_f & L_h & L_f & L_h & -L_r & -L_h & -L_r \end{bmatrix}. \tag{3}$$

The force vectors F_{ASi} are defined as

$$F_{ASi} = \begin{bmatrix} F_{Sli} & F_{SLi} \end{bmatrix}^T, \quad i = 1, \ldots, 4. \tag{4}$$

The transfer matrices R_i from wheel frame to vehicle local coordinate frame $X_L - O_L - Y_L$ are presented in

$$R_i = \begin{bmatrix} \cos \delta_i & -\sin \delta_i \\ \sin \delta_i & \cos \delta_i \end{bmatrix}, \quad i = 1, \ldots, 4. \tag{5}$$

The lateral components F_{SLi} are determined by the lateral forces acting on tires, while the longitudinal components F_{Sli} are viewed as intermediate control inputs of vehicle system, which can be achieved by controlling the torques (i.e., $T_i, i = 1, \ldots, 4$) applied on wheels. To simplify the coupled issue of left and right wheel steerings, the steering angles are constrained by

$$\begin{aligned} \delta_1 &= \delta_2 = \delta_f, \\ \delta_3 &= \delta_4 = \delta_r. \end{aligned} \tag{6}$$

Therefore, the vehicle system in this work considers six control inputs (i.e., δ_f, δ_r and $T_i, i = 1, \ldots, 4$) in total.

2.2. Driving Unit Dynamic Model

While the force sensors provide measurement of the current forces acting on each wheel, controller design requires prediction of future values based on control inputs. The drive unit dynamic model as shown in Figure 3 is described in this section.

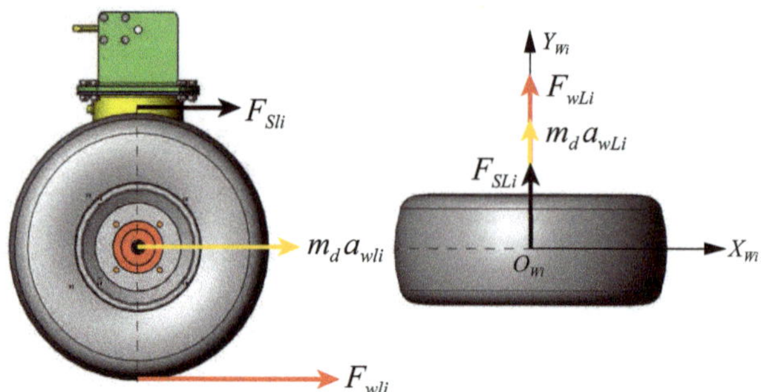

Figure 3. Driving unit dynamic model.

In each driving unit, the force sensor mounted on the wheel hub rotate with the wheel. Force measurement is in the direction of the steering angle. According to Figure 3, the dynamics of driving unit can be expressed by

$$F_{ASi} = \begin{bmatrix} F_{Sli} & F_{SLi} \end{bmatrix}^T = F_{wi} - \begin{bmatrix} m_d a_{wli} & m_d a_{wLi} \end{bmatrix}^T,$$ (7)

where a_{wli} and a_{wLi} denote the longitudinal and lateral accelerations of each driving unit i in the wheel coordinate system ($X_{Wi} - O_{Wi} - Y_{Wi}$), respectively. Using the transfer matrices R_i and the accelerations a_l, a_r and γ given in vehicle body dynamic model, a_{wli} and a_{wLi} can be deduced as

$$
\begin{aligned}
\begin{bmatrix} a_{wl1} & a_{wL1} \end{bmatrix} &= \begin{bmatrix} a_l - L_h\gamma & a_r + L_f\gamma \end{bmatrix} R_1, \\
\begin{bmatrix} a_{wl2} & a_{wL2} \end{bmatrix} &= \begin{bmatrix} a_l + L_h\gamma & a_r + L_f\gamma \end{bmatrix} R_2, \\
\begin{bmatrix} a_{wl3} & a_{wL3} \end{bmatrix} &= \begin{bmatrix} a_l + L_h\gamma & a_r - L_r\gamma \end{bmatrix} R_3, \\
\begin{bmatrix} a_{wl4} & a_{wL4} \end{bmatrix} &= \begin{bmatrix} a_l - L_h\gamma & a_r - L_r\gamma \end{bmatrix} R_4.
\end{aligned}
$$ (8)

In Equation (7), F_{wi} represents the force vector acting on each wheel, which is written as

$$F_{wi} = \begin{bmatrix} F_{wli} & F_{wLi} \end{bmatrix}^T.$$ (9)

2.3. Tire Model

The actual generation of wheel forces are from the contact between tires and the ground. The wheel forces depend on the surface friction, load, and the intrinsic properties of the tire. To analyze the wheel forces F_{wi} in Equation (9), the tire model is built as shown in Figure 4.

Considering the generation of wheel forces and external disturbances, F_{wi} can be expressed as

$$F_{wi} = \begin{bmatrix} F_{wli} & F_{wLi} \end{bmatrix}^T = \begin{bmatrix} F_{li} - F_{gli} & F_{Li} - F_{gLi} \end{bmatrix}^T,$$ (10)

where F_{li} and F_{Li} ($i = 1, \ldots, 4$) are longitudinal and lateral forces caused by wheel slip, respectively. F_{gli} and F_{gLi} denote the terrain disturbances corresponding to F_{li} and F_{Li}.

According to [20], the longitudinal force can be considered proportional to the slip ratio in small range, which is expressed by

$$F_{li} = \begin{cases} k_{li} F_{Ni} \frac{\varsigma_i}{0.1}, & |\varsigma_i| \le 0.1, \\ k_{li} F_{Ni}, & |\varsigma_i| > 0.1, \end{cases}$$ (11)

where F_{Ni} is the weight force on wheel i. In the simulation, the load transfer due to the longitudinal and lateral accelerations and roll angle are taken into consideration. F_{Ni} can be calculated by considering longitudinal and lateral load transfer according to [21]. The longitudinal slip stiffness k_{li} is determined by the tire type and terrain condition. ς_i is the longitudinal slip ratio of wheel i presented in [22]:

$$\varsigma_i = \begin{cases} \dfrac{R_w\omega_i - v_{wi}}{R_w\omega_i}, & R_w\omega_i > v_{wi} > 0, \\[2mm] \dfrac{R_w\omega_i - v_{wi}}{v_{wi}}, & R_w v_{wi} > \omega_i > 0, \\[2mm] 0, & R_w v_{wi} = \omega_i \geq 0, \end{cases} \tag{12}$$

where R_w and ω_i denote the tire radius and angular velocity of wheel i as shown in Figure 4b. v_{wi} represents the actual velocity of the wheel i, which can be calculated by the vehicle geometry as well as velocities V_l, V_r and Ω shown in Figure 2. Note that the model is proposed only considering the vehicle is moving forward. Thus, v_{wi} and ω_i are set to be nonnegative. As is discussed in Equation (12), the slip ratio is zero when the vehicle is resting.

Finally, as per Figure 4b, the wheel dynamic equation indicating the relationship between T_i and F_{li} can be written as

$$T_i = (F_{li} + F_{ri})R_w + \dot{\omega}_i J_w, \tag{13}$$

where J_w represents the wheel inertia, and F_{ri} is the rolling resistance calculated by

$$F_{ri} = F_{Ni}(K_{r0} + K_{r1}V_l^2), \tag{14}$$

where $K_{r0} = 0.015$ and $K_{r1} = 7 \times 10^{-6}$ s^2/m^2 are the parameters for common car tires [23].

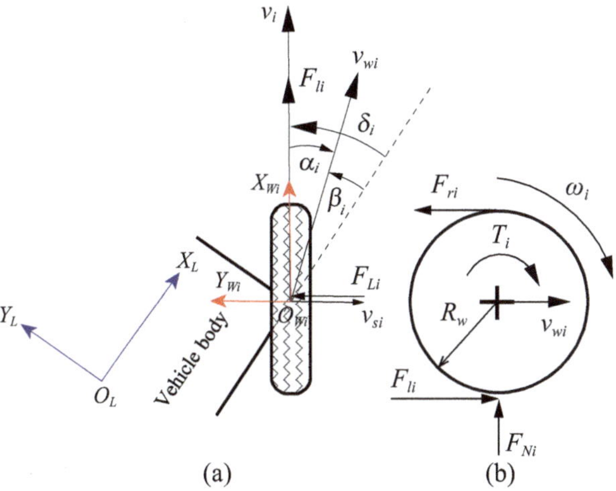

Figure 4. Tire model. (a) Top view of the tire; (b) lateral view of the tire.

The steering motion of wheels result in the lateral slip velocities v_{si}, which causes the actual direction of wheel velocity v_{wi} to differ from the wheel center plane by the slip angle α_i. According to the tire lateral characteristic curve [24], a linear model for F_{Li} is given in

$$F_{Li} = \begin{cases} -k_{Li}F_{Ni}\frac{\alpha_i}{5}, & |\alpha_i| \leq 5°, \\ -k_{Li}F_{Ni}, & \alpha_i > 5°, \\ k_{Li}F_{Ni}, & \alpha_i < -5°, \end{cases} \tag{15}$$

where k_{Li} is the lateral slip stiffness of the wheel i. Note that, due to the sign conventions in $X_{Wi} - O_{Wi} - Y_{Wi}$ and $X_L - O_L - Y_L$, a negative slip angle causes a positive lateral force and vice versa.

As shown in Figure 4a, the slip angle α_i can be calculated by steering angle δ_i and side slip angle β_i,

$$\alpha_i = \beta_i - \delta_i, \tag{16}$$

where the expression of β_i is available in our previous work presented in [25].

2.4. Offset Model

An offset error model is proposed for the evaluation of the performance of the controllers. The reference position (RP) of the vehicle on the reference path is defined as the normal projection of the heading of the vehicle at the centre of gravity on to the reference path, as shown in Figure 5. The tracking error in vehicle coordinate frame consists of lateral offset error l_{os} and heading error θ_{os}, expressed as

$$\widetilde{pos} = \begin{bmatrix} 0 & l_{os} & \theta_{os} \end{bmatrix}^T \tag{17}$$

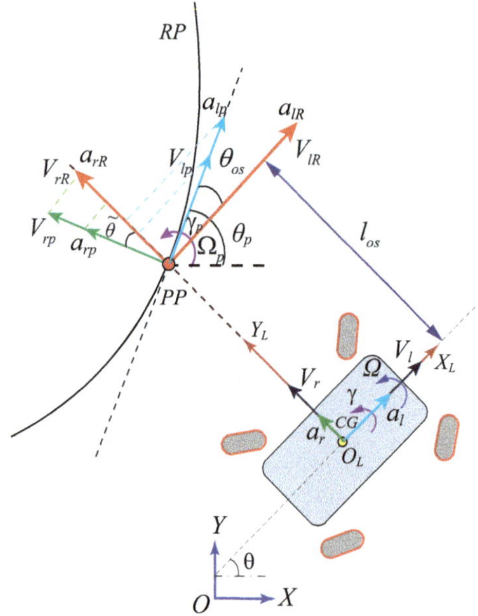

Figure 5. Offset model.

The position errors are always in the lateral direction of the vehicle. The longitudinal position error is set to 0. The heading error can be found by

$$\theta_{os} = \theta - \theta_{ref}, \tag{18}$$

where θ_{ref} is the reference heading direction of the RP, tangential to the reference path.

The velocity error states $\widetilde{vel}$ is the difference between vehicle velocity vel and reference velocity states vel_{ref} in the vehicle coordinate frame. $\widetilde{vel}$ can be expressed as

$$\widetilde{vel} = \begin{bmatrix} V_{lerr} & V_{rerr} & \Omega_{err} \end{bmatrix}^T = vel - vel_{ref}, \tag{19}$$

where V_{lerr}, V_{rerr} and Ω_{err} are the errors in longitudinal, lateral, and angular velocity, respectively. Reference velocity states are defined as

$$vel_{ref} = \begin{bmatrix} V_{lref} & V_{rref} & \Omega_{ref} \end{bmatrix}^T, \tag{20}$$

where reference angular velocity Ω_{ref} is the rate of change of the heading of RP on the reference path.

The reference longitudinal velocity V_{lref} and lateral velocity V_{rref} in vehicle coordinate frames can be derived from the reference path coordinate frame by the equations:

$$\begin{aligned} V_{lref} &= V_{lp} \cos(\theta_{os}) - V_{rp} \sin(\theta_{os}), \\ V_{rref} &= V_{lp} \sin(\theta_{os}) + V_{rp} \cos(\theta_{os}), \end{aligned} \tag{21}$$

where V_{lp} is the reference longitudinal velocity, taken as tangential to the reference path. V_{rp} is the reference radial velocity normal to the reference path at RP.

Finally, the acceleration error states $\widetilde{acc}$ are written as:

$$\widetilde{acc} = \begin{bmatrix} a_{lerr} & a_{rerr} & \gamma_{err} \end{bmatrix}^T = acc - acc_{ref}, \tag{22}$$

where acc is the difference between vehicle acceleration acc and reference acceleration acc_{ref}. The reference acceleration vector acc_{ref} is defined as

$$acc_{ref} = \begin{bmatrix} a_{lref} & a_{rref} & \gamma_{ref} \end{bmatrix}^T, \tag{23}$$

where a_{lref} and a_{rref} are, respectively, the reference longitudinal and lateral acceleration at point RP, measured in the direction and normal to the vehicle heading, and γ_{ref} is the reference angular acceleration of the point RP.

The reference acceleration in vehicle coordinate frame can be derived from the path coordinate frame by

$$\begin{aligned} a_{lref} &= a_{lp} \cos(\theta_{os}) - a_{rp} \sin(\theta_{os}), \\ a_{rref} &= a_{lp} \sin(\theta_{os}) + a_{rp} \cos(\theta_{os}), \end{aligned} \tag{24}$$

where a_{lp} and a_{rp} are the reference acceleration tangential and normal to the reference path.

3. Control Methodology

Two methods of finding the optimal control inputs for steering angles (i.e., δ_f and δ_r) and drive forces F_{Sli}, $(i = 1, ..., 4)$ are developed. The first method used model predictive control (MPC) with a sequential quadratic programming (SQP) solver. The second method is particle swarm optimization (PSO). The objective of the controller is to accurately follow a predetermined path through multiple terrain types.

3.1. MPC-SQP Method

3.1.1. Offset Model Linearization

In order to be applied in the MPC controller design, the offset model is linearized by a feedback linearization method. By defining the state vector $x = [x_1, x_2, x_3, x_4, x_5]$, as $x_1 = V_{lerr}$, $x_2 = l_{os}$, $x_3 = \dot{l}_{os}$, $x_4 = \theta_{os}$ and $x_5 = \dot{\theta}_{os}$, the offset model can be written in the following form:

$$
\begin{aligned}
\dot{x}_1 &= a_r - (a_{lp} \cos x_4 - a_{rp} \sin x_4), \\
\dot{x}_2 &= x_3, \\
\dot{x}_3 &= a_r - (a_{lp} \sin x_4 + a_{rp} \cos x_4), \\
\dot{x}_4 &= x_5, \\
\dot{x}_5 &= \gamma - \gamma_{ref}.
\end{aligned}
\tag{25}
$$

In this model, a_r, a_l and γ are considered as the inputs while other constants including reference accelerations and velocities can be obtained from the vehicle states. The nonlinearity of the model comes from trigonometric terms of x_4. For accurate path tracking problems, the yaw error is assumed to vary smoothly within $[-10° \ 10°]$. Then, the model can be linearized by feedback linearization. Define the input vector $u = [a_r \ a_l \ \gamma]^T$; then, the new input vector at time t_k is obtained as

$$
v(x, t)\big|_{t=t_k} = u - \begin{bmatrix} a_{lp} \cos x_4 - a_{rp} \sin x_4 \\ a_{lp} \sin x_4 + a_{rp} \cos x_4 \\ \gamma_{ref} \end{bmatrix}_{x_4 = x_4(t_k)},
\tag{26}
$$

and the offset model is expressed as

$$
\dot{x} = A \cdot x + v(x, t)\big|_{t=t_k},
\tag{27}
$$

where

$$
A = \begin{bmatrix}
0 & 0 & 0 & 0 & 0 \\
0 & 0 & 1 & 0 & 0 \\
0 & 0 & 0 & 0 & 0 \\
0 & 0 & 0 & 0 & 1 \\
0 & 0 & 0 & 0 & 0
\end{bmatrix}.
$$

It can be seen that the model is time-varying but can be treated as a linear model at each sampling step. The output vector is denoted by y_c and the equation is written as

$$
y_c = C_c x,
\tag{28}
$$

where

$$
C_c = \begin{bmatrix}
1 & 0 & 0 \\
0 & 1 & 0 \\
0 & 0 & 0 \\
0 & 0 & 1 \\
0 & 0 & 0
\end{bmatrix}.
$$

3.1.2. Model Predictive Control

To develop the MPC controller, the offset model as given in Equations (27) and (28) needs to be discretized and its discrete-time state vector form can be written as

$$
\begin{aligned}
x_d(k+1) &= A_d x_d(k) + B_d u_d(k), \\
y_d(k) &= C_d x_d(k),
\end{aligned}
\tag{29}
$$

where $x_d(k)$, $y_d(k)$ and $u_d(k)$ are the state vector, output vector and input vector, respectively. The coefficient matrices A_d, B_d and C_d are updated after discretization.

To eliminate undesirable oscillations, embedded integrator vectors $\Delta x_d(k) = x_d(k+1) - x_d(k)$, $\Delta u(k) = u_d(k+1) - u_d(k)$ are defined, thereby an augmented state-space model can be expressed as

$$
\begin{aligned}
\begin{bmatrix} \Delta x_d(k+1) \\ y_d(k+1) \end{bmatrix} &= \begin{bmatrix} A_d & O^T \\ C_d A_d & I \end{bmatrix} \begin{bmatrix} \Delta x_d(k) \\ y_d(k) \end{bmatrix} + \begin{bmatrix} B_d \\ C_d B_d \end{bmatrix} \Delta u(k), \\
y_d(k) &= \begin{bmatrix} O & I \end{bmatrix} \begin{bmatrix} \Delta x_d(k) \\ y_d(k) \end{bmatrix},
\end{aligned}
\tag{30}
$$

where O is a zero matrix, and I represents the identity matrix.

Defining the new state vector $x(k) = [\Delta x_d(k)^T \ y_d(k)^T]^T$, the augmented model can be written in the following matrix form:

$$
\begin{aligned}
x(k+1) &= A x(k) + B \Delta u(k), \\
y(k) &= C x(k),
\end{aligned}
\tag{31}
$$

where

$$
A = \begin{bmatrix} A_d & O^T \\ C_d A_d & I \end{bmatrix}, \quad B = \begin{bmatrix} B_d \\ C_d B_d \end{bmatrix}, \quad C = \begin{bmatrix} O^T & I \end{bmatrix}.
$$

Theorem 1. *Given a discrete time system following the form of Equation (31), the asymptotic stabilization of the closed-loop system can be realized by substituting the first item of ΔU^* as the control input $\Delta u(k)$, when ΔU^* is the optimal solution of the following optimization problem:*

$$
\begin{aligned}
\arg\min_{\Delta U} J &= (R_s - Y)^T (R_s - Y) + \Delta U^T \bar{R} \Delta U, \\
\text{s.t.} \quad y_{N_p}(\Delta U) &= 0,
\end{aligned}
\tag{32}
$$

where Y denotes the predicted output sequence, ΔU denotes the future input sequence, and N_p is the prediction horizon. R_s is the sequence of the control target vector. $y_{N_p}(\Delta U)$ represents the error between the final predicted output and target.

Proof of Theorem 1. To prove the stability, the Lyapunov function $V(x_k)$ is defined equal to the value of the objective function J_k subjected to its optimal solution, which can be expressed as

$$
\begin{aligned}
V(x_k) &= \min J_k \\
&= \sum_{i=1}^{N_p} y_{k+i}^T y_{k+i} + \sum_{i=0}^{N_p-1} \Delta u_{k+i}^T r_w \Delta u_{k+i},
\end{aligned}
\tag{33}
$$

where, $\Delta u_k, ..., \Delta u_{k+N_p-1}$ are obtained by the optimal solution of future inputs, and $y_k, ..., y_{k+N_p}$ represent the corresponding error sequence between future outputs and target. r_w is the nonnegative gain matrix.

According to the definition in Equation (33), the Lyapunov function at the next sample $k+1$ is written as

$$V(x_{k+1}) = \sum_{i=1}^{N_p} y_{k+1+i}^T y_{k+1+i} + \sum_{i=0}^{N_p-1} \Delta u_{k+1+i}^T r_w \Delta u_{k+1+i}. \tag{34}$$

To facilitate the comparison between two neighboring Lyapunov function values, a intermediate function $\bar{V}$ is defined, which is formed by evaluating $V(x_{k+1})$ with a defined inputs sequence, which is obtained by shifting the optimal inputs sequence of $V(x_{ki})$ one step forward, and setting its last input Δu_{k+N_p} as zero. It is obvious that the objective function value of non-optimal inputs sequence has to be no less than $V(x_{ki+1})$, which can be expressed as

$$V(x_{k+1}) \leq \bar{V}, \tag{35}$$

thereby,

$$V(x_{k+1}) - V(x_k) \leq \bar{V} - V(x_k). \tag{36}$$

Since $\bar{V}$ shares the same future inputs sequence and the predictive outputs sequence with $V(x_k)$ for the sample time $k+1, ..., k+N_p-1$, it can be easily derived that the difference between these two functions is

$$\bar{V} - V(x_k) = y_{k+N_p}^T y_{k+N_p} - y_{k+1}^T y_{k+1} - \Delta u_k^T r_w \Delta u_k. \tag{37}$$

As is given in Equation (32), the optimization problem is subjected to the constraint $y_{k+N_p} = 0$. Then, it can be obtained that

$$\bar{V} - V(x_k) \leq -y_{k+1}^T y_{k+1} - \Delta u_k^T r_w \Delta u_k. \tag{38}$$

Then, the monotonicity of the Lyapunov function can be obtained by

$$V(x_{k+1}) - V(x_k) \leq -y_{k+1}^T y_{k+1} - \Delta u_k^T r_w \Delta u_k < 0, \tag{39}$$

which can prove the asymptotic stability of the system. $\square$

The model predictive control algorithm is realized by receding optimization. In order to apply the MPC efficiently, we assume that the predicted outputs sequence is in a finite prediction horizon N_p and the inputs sequence is in a control horizon N_c, which is less than N_p. The sequences mentioned above can be expressed in the matrix form:

$$\begin{aligned} Y &= \begin{bmatrix} y(k+1|k) & y(k+2|k) & ... & y(k+N_p|k) \end{bmatrix}^T, \\ \Delta U &= \begin{bmatrix} \Delta u(k) & \Delta u(k+1) & ... & \Delta u(k+N_c-1) \end{bmatrix}^T, \end{aligned} \tag{40}$$

where $y(k+n|k)$ denotes the predicted outputs at time $k+n$ based on the states at time k. Based on Theorem 1 and the assumption above, the corollary about finite-time unconstrained MPC can be obtained as follows.

Corollary 1. *Given the system without input and output constraints, the prediction and control horizon are N_p and N_c, respectively. Then, the following feedback control law $\Delta u(k) = Kx(k)$ can asymptotically stabilize the closed-loop system, where*

$$K = [1\ 0\ \dots\ 0]^T \overbrace{}^{N_c} (-\Xi[\Gamma - (\Psi\Xi)^{-1}(CA^{N_p} + \Psi\Xi\Gamma)]), \tag{41}$$

and,

$$F = \begin{bmatrix} CA \\ CA^2 \\ \vdots \\ CA^{N_p} \end{bmatrix}, \quad \Psi = \begin{bmatrix} CA^{N_p-1}B \\ CA^{N_p-2}B \\ \vdots \\ CA^{N_p-N_c}B \end{bmatrix}^T,$$

$$\Phi = \begin{bmatrix} CB & 0 & \cdots & 0 \\ CAB & CB & \cdots & 0 \\ CA^2B & CAB & \cdots & 0 \\ \vdots & & & \\ CA^{N_p-1}B & CA^{N_p-2}B & \cdots & CA^{N_p-N_c}B \end{bmatrix}, \tag{42}$$

$$\Xi = (\Phi^T\Phi + \bar{R})^{-1},$$

$$\Gamma = \Phi^T F.$$

Proof of Corollary 1. According to Equations (31) and (40), the predicted output sequence Y can be expressed by

$$Y = Fx(k) + \Phi\Delta U. \tag{43}$$

Meanwhile, the final item of Y can be expressed as

$$y(k + N_p|k) = CA^{N_p}x(k) + \Psi\Delta U. \tag{44}$$

By substituting Equations (43) and (44) into Theorem 1, the optimization problem turns into the following form:

$$\arg\min_{\Delta U} J = (R_s - Fx(k_i) - \Phi\Delta U)^T (R_s - Fx(k_i) - \Phi\Delta U)$$
$$+ \Delta U^T \bar{R}\Delta U, \tag{45}$$
$$s.t. \qquad \Psi\Delta U + CA^{N_p}x(k_i) = 0,$$

where $\bar{R}$ is a weighting matrix.

Note that in the application of offset model based path tracking, the target vector R_s should be zero all the time. It can be seen that J meets an equality constrained quadratic programming. Then, the objective function is expanded by Lagrange expression and simplified by omitting the constant term,

$$J = 2\Delta U^T \Phi^T Fx(k) + \Delta U^T(\Phi^T\Phi + \bar{R})\Delta U$$
$$+ \xi^T(\Psi\Delta U + CA^{N_p}x(k)), \tag{46}$$

where ξ is the Lagrange multiplier.

According to the Lagrange multiplier method, the optimal control input vector ΔU^* can be found by solving the following equation system. The solution is obtained by taking the first partial derivatives of J with respect to the vectors ΔU and λ, and then equating these derivatives to zero:

$$\frac{\partial J}{\partial \Delta U} = \Delta U^T(\Phi^T\Phi + \bar{R}) + \Phi^T Fx(k) + \Psi^T\xi = 0,$$
$$\frac{\partial J}{\partial \xi} = \Psi\Delta U + CA^{N_p}x(k) = 0. \tag{47}$$

Then, its optimal solution can be obtained:

$$\Delta U^* = -\Xi[\Gamma - (\Psi\Xi)^{-1}(CA^{N_p} + \Psi\Xi\Gamma)]x(k), \tag{48}$$

where

$$\Xi = (\Phi^T\Phi + \bar{R})^{-1},$$
$$\Gamma = \Phi^T F.$$

According to the receding horizon control principle, the first increment of ΔU^* is applied as the control inputs. Then, the control law in Equation (41) can be obtained. $\square$

It can be seen that Corollary 1 gives an offline solution of the MPC algorithm, which significantly improves its computing efficiency. Based on Corollary 1, the desired accelerations $\tilde{A} = [\tilde{a}_r, \tilde{a}_l, \tilde{\gamma}]^T$ can be obtained by integrating the optimal solution $\Delta u(k)^*$.

3.1.3. Sequential Quadratic Programming Based Control Allocation

Considering the vehicle body with the mass M and inertia J in this work, the command force vector τ is defined as

$$\tau = \begin{bmatrix} \tau_l \\ \tau_r \\ \tau_\gamma \end{bmatrix} = \begin{bmatrix} M & 0 & 0 \\ 0 & M & 0 \\ 0 & 0 & J \end{bmatrix} \tilde{A}, \tag{49}$$

where τ_l, τ_r and τ_γ are the longitudinal and lateral forces and the moment about a vertical axis of the vehicle, respectively.

According to the vehicle body dynamic model given in Equation (1), the actual actuating forces come from the longitudinal forces F_{Sl} and lateral forces F_{SL} on each wheel. Let the command forces τ produced jointly by the wheel forces and steering angles be expressed as

$$\tau(F_{Sl}, F_{SL}, \delta) = B_u(\delta)F_{Sl} + B_w(\delta)F_{SL}, \tag{50}$$

where the i-th column of $B_u(\delta)$ and $B_w(\delta)$ can be written as

$$B_u^i(u_{\delta i}) = \begin{bmatrix} \cos\delta_i \\ \sin\delta_i \\ -L_{yi}\cos\delta_i + L_{xi}\sin\delta_i \end{bmatrix}, \quad B_w^i(u_{\delta i}) = \begin{bmatrix} -\sin\delta_i \\ \cos\delta_i \\ L_{yi}\sin\delta_i + L_{xi}\cos\delta_i \end{bmatrix}.$$

(L_{xi}, L_{yi}) represents the location of each wheel in a coordinate system with its origin at the centre of gravity and positive x-axis forward.

On a 4WS4WD vehicle, the drive forces F_{Sl} and the steering angles δ are the direct inputs while the lateral forces F_{SL} obtained from sensors are considered as a measured disturbance. In this work, the steering angles δ are composed of δ_f and δ_r, which represent the front and rear steering inputs, respectively. Thus, the control problem is reduced to obtaining the feasible solution of Equation (50). In order to facilitate the computation, a slack vector s is defined by

$$s = \tau - B_u(\delta)F_{Sl} + B_w(\delta)\hat{F}_{SL}, \tag{51}$$

which denotes the error between the commanded and actual generalized forces. The slack variable s guarantees that there always exists a feasible solution in the following optimization [26].

In order to solve this control allocation problem, the objective function is defined with respect to F_{Sl}, δ and s,

$$J(F_{Sl}, \delta, s) = F_{Sl}^T Q_f F_{Sl} + (\delta - \delta^*)^T Q_\delta (\delta - \delta^*) + s^T Q_s s,$$

$$\text{s.t.} \quad s = \tau - B_u(\delta) F_{Sl} + B_w(\delta) \hat{F}_{SL},$$

$$s \in \mathcal{B}_s,$$

$$F_{Sl} \in \mathcal{B}_{F_{Sl}},$$

$$\delta \in \mathcal{B}_\delta,$$

$$(52)$$

where $\mathcal{B}_s$, $\mathcal{B}_{F_{Sl}}$ and $\mathcal{B}_\delta$ are the search bounds of each variable.

In this function, the first term minimizes the magnitudes of the drive forces; the second term is used to ensure the steering angle to search around its previous value. By penalizing the slack variable s in the third term, the actual generalized force vector coincides as much as possible with the commanded forces τ. The matrix $Q_f \in I^{4\times4}$, $Q_\delta \in I^{2\times2}$ and $Q_s \in I^{3\times3}$ are used to tune the objective. The search bounds of all variables (i.e. F_{Sl}, δ and s) are specified by the constraints.

Based on the objective function presented above, the control allocation is converted to a nonlinear constrained optimization problem. Using the sequential quadratic programming (SQP), the optimal solution can be computed efficiently and reliably by standard numerical software.

3.2. PSO-Based Method

For the PSO-based method, an objective function including vehicle error states is proposed using the sliding surfaces. In Sliding Mode Control (SMC), the time-varying sliding surface is normally defined by the scalar equation $s(x;t) = 0$, in which $s(x;t)$ is expressed by [27],

$$s(x;t) = (\frac{d}{dt} + \lambda)^{n-1} \tilde{x},$$

$$(53)$$

where λ is a positive constant and $\tilde{x}$ is the error state vector.

According to the idea of SMC, the problem of maintaining $\tilde{x} = 0$ is transformed into keeping $s = 0$. In this work, the scalar quantities composed of vehicle error states are introduced into the definition of objective function. Instead of designing the switching control law in SMC, the optimization is used to maintain the scalar quantities on the sliding surface $s = 0$. The vehicle error state vectors including $\widetilde{pos}$, $\widetilde{vel}$ and $\widetilde{acc}$ are following the definitions in the offset model. Therefore, to track the trajectories of the RP, the vector $s = [s_l, s_r, s_a]$ needs to be defined correspondingly. Note that vector s follows a different definition than that in SQP. It is defined to facilitate the notations in the local boundary part.

For $\widetilde{pos}$ presented in Equation (17), its first component, which represents the position error in the longitudinal direction, is always zero. Therefore, according to Equation (53), the longitudinal scalar quantity s_l can be obtained as

$$s_l = a_{lerr} + \lambda_l V_{lerr},$$

$$(54)$$

where a_{lerr} and V_{lerr} are the longitudinal components of $\widetilde{acc}$ and $\widetilde{vel}$, which are given in Equations (22) and (19), respectively.

To maintain the vehicle on the RP geometrically, another first order scalar quantity s_r is designed as

$$s_r = V_{rerr} + \lambda_r l_{os},$$

$$(55)$$

which aims to minimize the offset errors l_{os} and V_{rerr}.

The angular acceleration error $\tilde{\gamma}$ is included to consider the effects of forces and yaw movements of vehicle, and thus a second order scalar quantity is chosen as

$$s_a = \gamma_{err} + 2\lambda_a \Omega_{err} + \lambda_a^2 \theta_{os},$$

$$(56)$$

where θ_{os}, Ω_{err} and γ_{err} are given in Equations (17), (19) and (22), respectively.

Then, the problem of following the trajectories of RP is transformed into maintaining s at 0. Using the linear scalarization, an objective function is defined as

$$J_{\min}(F_{Sl}, \delta) = C_l \, |s_l| + C_r \, |s_r| + C_a \, |s_a| \,, \tag{57}$$

where C_l, C_r and C_a are the weighting coefficients strictly positive and constrained by

$$C_l + C_r + C_a = 1. \tag{58}$$

The variables of objective function in Equation (57) consist of the forces F_{Sli}, steering angles δ_f and δ_r, which can be written as a variable vector,

$$v_{obj} = \begin{bmatrix} \delta_f & \delta_r & F_{Sl1} & F_{Sl2} & F_{Sl3} & F_{Sl4} \cdot \end{bmatrix}^T \tag{59}$$

First invented by Kennedy and Eberhart (1995), PSO has been successfully applied to solve problems featuring nonlinearity, non-differentiable, and multiple optima. PSO is found to be capable of generating high quality solutions with more stable and faster convergence characteristics, and shorter calculation times than other stochastic methods [17]. For standard PSO at time t, the updating velocity $v_i(t)$ and position $x_i(t)$ of the i-th particle are presented in the following equations:

$$\begin{aligned} v_i(t+1) &= \zeta v_i(t) + \phi_1 \eta_1 (p_b - x_i(t)) + \phi_2 \eta_2 (g_b - x_i(t)), \\ x_i(t+1) &= x_i(t) + v_i(t+1), \end{aligned} \tag{60}$$

where $v_i(t)$ and $x_i(t)$ are vectors in multi-dimensional space. p_b and g_b denote the local optimal position and the global optimal position, respectively. The particle inertia weight is represented by ζ. The particle cognitive acceleration and social acceleration are denoted by ϕ_1 and ϕ_2, which are defined as positive constants. η_1 and η_2 are two stochastic parameters within [0 1].

The search space of PSO in this work is defined as a six-dimensional space corresponding to the dimension of v_{obj}. Therefore, the particle position vector $x_i(i)$ in Equation (60) represents a possible solution of the objective function in Equation (57).

3.3. Boundary Definition

Given that both methods are reduced to the optimization problems, the definition of the search space determines the quality of the solution.

3.3.1. Global Boundaries

In this work, the global boundaries can be assigned based on the properties of each actuator, which is defined as

$$\mathcal{B}_g = \begin{cases} \delta_{\min} \leq \delta_f \leq \delta_{\max}, \\ \delta_{\min} \leq \delta_r \leq \delta_{\max}, \\ F_{d\min} \leq F_{S}li \leq F_{d\max}, \end{cases} \tag{61}$$

where $\delta_{\min}$ and $\delta_{\max}$ are the minimum and maximum steering angles of each wheel. $F_{d\min}$ and $F_{d\max}$ are the minimum and maximum drive forces provided by the driving unit, which can be calculated by

$$F_{d\min} = -\frac{T_{\max}}{R_w}, \quad F_{d\max} = \frac{T_{\max}}{R_w}, \tag{62}$$

where $T_{\max}$ is the maximum torque that can be achieved by the driving unit of the vehicle.

3.3.2. Local Boundaries

To realize real-time optimization, the computing time obtaining optimal values for variables $[F_{Sl}, \delta]$ needs to be constrained within the sample time of controller T_s. According to the properties of each actuator, the maximum variations of $[F_{Sl}, \delta]$ within T_s can be obtained. In this work, to improve the computing efficiency, the local boundaries are also determined by the states of s.

According to a_r in Equation (1), when $s_l < 0$, the forces F_{Sli} need to be increased, thereby dragging s_l in Equation (54) towards the surface $s_l = 0$. Similarly, when $s_l > 0$, the forces F_{Sli} need to be decreased. Therefore, the local boundaries of F_{Sli} can be written as

$$\mathfrak{B}^P_{F_{Sli}} = \begin{cases} F_{Sli}(t) \leq F_{Sli}(t + T_s) \leq F_{Sli}(t) + \Delta F_{d\,max}, & s_l < 0, \\ F_{Sli}(t) - \Delta F_{d\,max} \leq F_{Sli}(t + T_s) \leq F_{Sli}(t), & s_l \geq 0, \end{cases} \tag{63}$$

where $\Delta F_{d\,max}$ is the maximum change of F_{Sli} that can be achieved within T_s. At the time $t + T_s$, $F_{Sli}(t + T_s)$ are the possible optimal solutions.

For the steering angles δ_f and δ_r in v_{obj}, they affect the vehicle lateral and angular motions in a coupled way. To solve this problem, an allocating method is specified that the vehicle lateral error determines searching direction of δ_f, while δ_r is related to vehicle angular error. When $s_r < 0$, δ_f needs to be increased to drive s_r back to the surface $s_r = 0$, and vice versa. Thus, the local boundary of δ_f is summarized as,

$$\mathfrak{B}^P_{\delta_f} = \begin{cases} \delta_f(t) \leq \delta_f(t + T_s) \leq \delta_f(t) + \Delta\delta_{max}, & s_r < 0, \\ \delta_f(t) - \Delta\delta_{max} \leq \delta_f(t + T_s) \leq \delta_f(t), & s_r \geq 0, \end{cases} \tag{64}$$

where $\Delta\delta_{max}$ is the maximum change of steering angle that can be achieved within T_s. At the time $t + T_s$, the possible optimal solution of front steering angle is $\delta_f(t + T_s)$.

Similarly, based on s_r, the local boundary of δ_r is defined as

$$\mathfrak{B}^P_{\delta_r} = \begin{cases} \delta_r(t) - \Delta\delta_{max} \leq \delta_r(t + T_s) \leq \delta_r(t), & s_a < 0, \\ \delta_r(t) \leq \delta_r(t + T_s) \leq \delta_r(t) + \Delta\delta_{max}, & s_a \geq 0. \end{cases} \tag{65}$$

For the PSO in particular, its particle velocity boundaries define the range of speed that particles can achieve to search for the optimal solution. To improve search performance in PSO, the absolute maximum particle velocity is normally set as a certain percentage of particle position range [28]. According to Equations (63)–(65), the particle velocity boundaries can be obtained as

$$\begin{aligned} \mathfrak{B}^V_{F_{Sli}} &= \sigma_F \begin{bmatrix} -\Delta F_{d\,max} & \Delta F_{d\,max,} \end{bmatrix} \\ \mathfrak{B}^V_{\delta_f} &= \sigma_{\delta_f} \begin{bmatrix} -\Delta\delta_{max} & \Delta\delta_{max,} \end{bmatrix} \\ \mathfrak{B}^V_{\delta_r} &= \sigma_{\delta_r} \begin{bmatrix} -\Delta\delta_{max} & \Delta\delta_{max,} \end{bmatrix} \end{aligned} \tag{66}$$

where σ_F, σ_{δ_f} and σ_{δ_r} are the particle velocity coefficients for F_{Sli}, δ_f and δ_r, respectively.

4. Simulation

4.1. Simulation Setup

The working process of the vehicle control system is illustrated in Figure 6. The dashed box at the top shows the operation of the vehicle dynamic model. Based on the actual inputs (i.e., T_i and δ_i) as well as the vehicle velocity states vel, the tire model can generate the wheel forces, which is then used in driving unit model. The driving unit model considers the wheel inertia and gives the actual forces acting on the vehicle body. The dynamic states of the vehicle such as positions and velocities are simulated by running the proposed vehicle body dynamic model in Matlab (R2017a, MathWorks, Natick, MA, United States) using a Runge–Kutta method based solver with a time step of 8.33 μs.

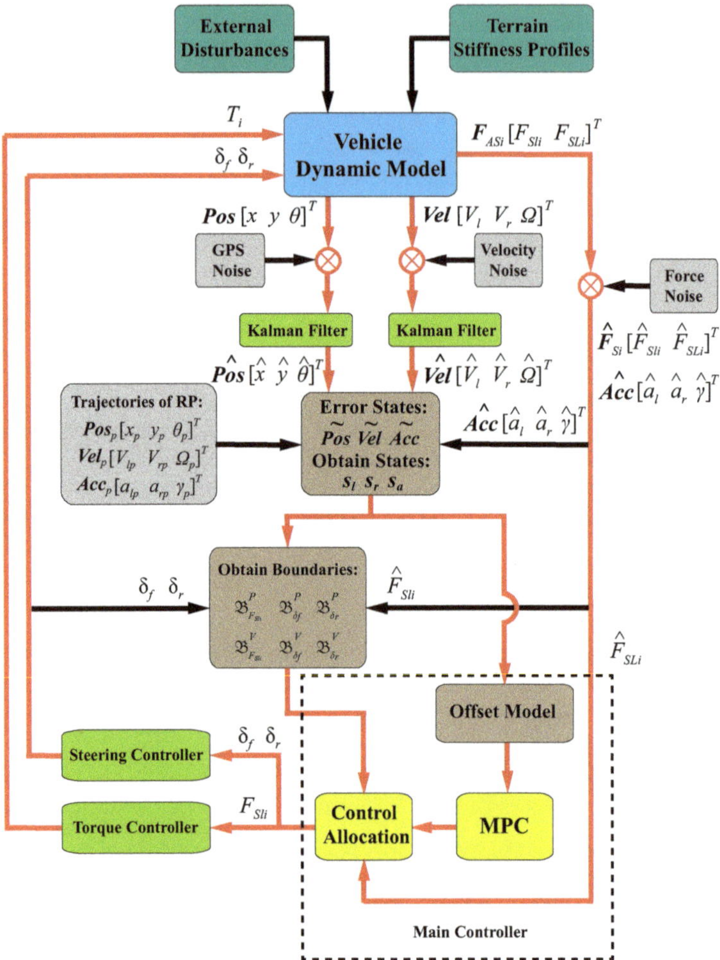

Figure 6. Flowchart of the vehicle control system.

In the simulation, both external disturbances and state measurement noises are involved in validating the robustness of the proposed control methodology. The measured states are compared with the reference profiles to generate the error states. Using the method proposed in Section 3.3, the search boundaries of the drive forces and steering angles are obtained. Meanwhile, the error states are transfered to the offset model. Virtual inputs calculated by the MPC algorithm are delivered to the control allocation modual. Then, the actual control inputs including drive forces and steering angles are generated and substituted into the dynamic model for the next iteration. The dashed box at the bottom represents the main controller. PSO controller can be substituted in place of the MPC controller.

The PSO and MPC controllers are applied to drive the simulated vehicle to track the reference path shown in Figure 7a. In the simulation, the varying terrain conditions are considered to verify the validity and robustness of each controller. As presented in Figure 7b, the reference path is divided into ten sections, which have different tire stiffnesses k_{li} and k_{Li} acting on each wheel. To simulate the terrain disturbances, the disturbances F_{gi} applied to all four wheels were always in the vehicle longitudinal direction. As shown in Figure 8, the disturbances are modeled as step signals and their magnitudes are generated randomly to show the uncertainties.

As a comparison, a kinematic model based control method is applied to drive the same vehicle. According to relative kinematic path tracking research [8,19], the side slip is the main disturbance that leads to the unpredicted tracking errors. Using the dynamic model based observer, the side slip of the vehicle can be predicted with an error of 10% to 30%. In this simulation, a random side slip velocity that is less than 10% of its longitudinal velocity is added in the kinematic model.

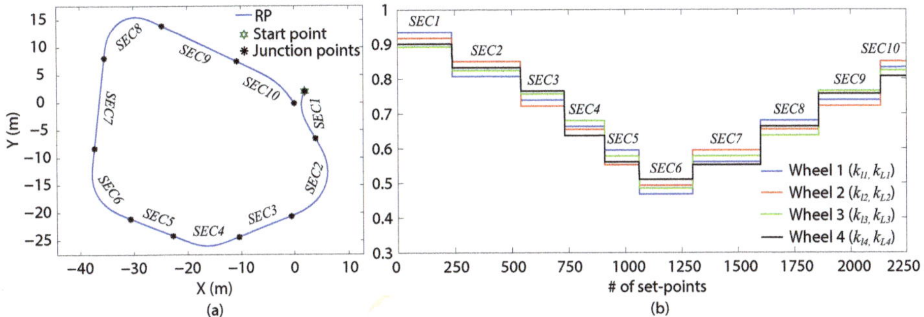

Figure 7. Reference path (RP) and terrain coefficients along the RP. (**a**) Reference path; (**b**) terrain coefficients.

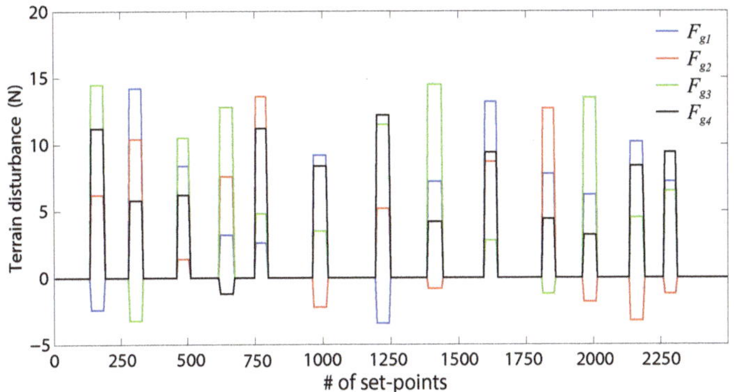

Figure 8. Terrain disturbance.

4.2. Parameters of Simulation

The first two sections of Table 1 list the parameters of vehicles used in simulation. The common parameters in the objective functions are listed in the third section of Table 1, in which the vehicle mass and inertia values used in the objective function are different from actual vehicle values to simulate parametric uncertainties. This helps to verify the robustness of the vehicle control method in dealing with variations in the system. In order to run in real time, both optimization methods are limited to 15 ms of search time per iteration, less than the system sample time $T_s = 20$ ms. In optimization, the optimal solution is searched within the boundaries until time reaches TI or the error gradient of 0.001 is achieved by the objective function. The last two sections present the particular parameters for each method.

Table 1. Parameters used in the simulation.

Para	Value	Unit	Para	Value	Unit
Vehicle model					
M	200.0	kg	m_d	15.0	kg
L_f	0.85	m	L_r	0.85	m
L_h	0.5	m	R_w	0.25	m
J_z	45.0	kg·m^2	J_w	0.8	kg·m^2
System constraints					
$F_{d\,\max}$	250.0	N	$\delta_{\max}$	40.0	$^\circ$
$F_{d\,\min}$	-250.0	N	$\delta_{\min}$	-40.0	$^\circ$
$\Delta F_{d\,\max}$	0.8	N	$\Delta\delta_{\max}$	0.35	$^\circ$
Common parameters in objective function					
M_c	205.0	kg	J_{zc}	40.0	kg·m^2
λ_l	2.0	-	λ_r	1.5	-
λ_a	2.8	-	TI	15	ms
MPC and SQP parameters					
$\bar{R}$	$0.1 \times I^{3\times3}$	-	Q_f	$1e^{-4} \times I^{4\times4}$	-
Q_δ	$0.1 \times I^{2\times2}$	-	Q_s	$I^{3\times3}$	-
PSO parameters					
SS	24	-	ζ	0.9	-
ϕ_1	1.85	-	ϕ_2	1.85	-
σ_F	0.4	-	σ_{δ_f}	0.5	-
σ_{δ_r}	0.5	-	C_l	0.35	-
C_r	0.35	-	C_a	0.3	-

$I^{i\times i}$ represents the i-th-order identity matrix. MPC: model predictive control; SQP: sequential quadratic programming; PSO: particle swarm optimization.

4.3. Simulation Results

Figure 9 shows path tracking results by different controllers. The MPC-SQP controller has superior performance with offset error less than 3.2 cm, and heading error less than 2°. The PSO controller has maximum offset error of 6.1 cm and heading error of 3.2°. Both dynamic model based control methods perform better than the kinematic model based one in constraining the offset errors.

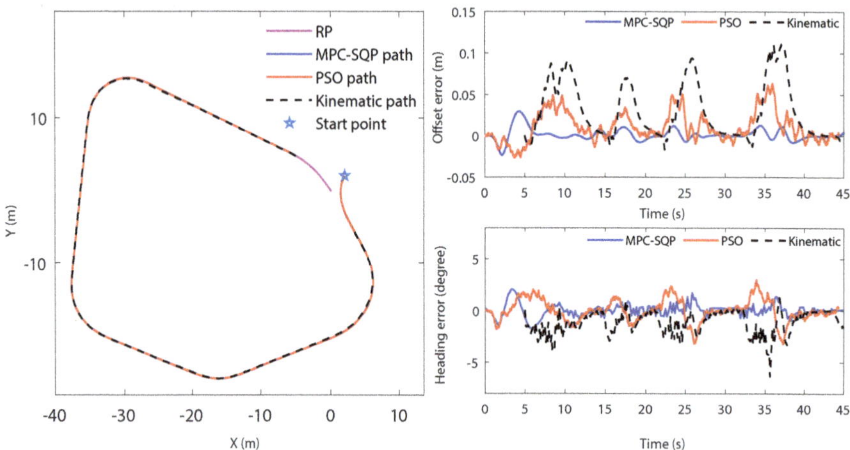

Figure 9. Path following performance.

The drive torques and steering angles applied on four wheels are presented in Figure 10. At each cornering, the torques applied at the two outside wheels (i.e., T_1 and T_4) firstly increase and then decrease, which are contrary to that of the two inside wheels (i.e., T_2 and T_3). The difference in torques is used to compensate the insufficient angular accelerations of vehicle in corners. The steering angles of MPC-SQP and PSO controller follow the same varying trend in which steering curves of MPC-SQP show smoother variation. Thus, the MPC-SQP controller can provide more stable steerings compared with the PSO controller.

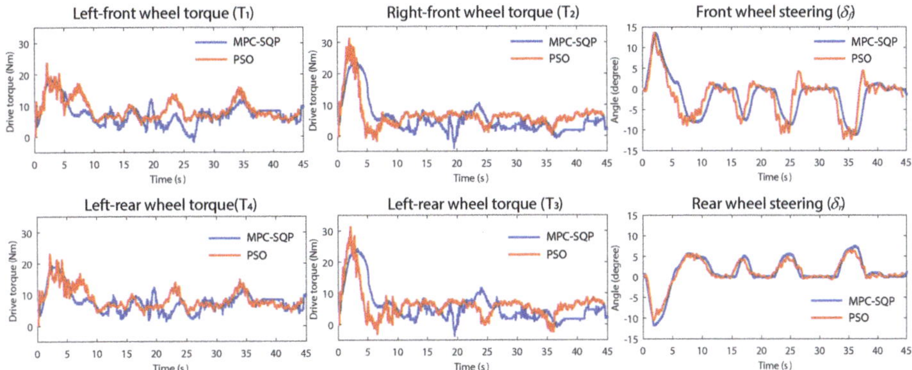

Figure 10. Actual inputs.

In this work, another aim of control is to maintain high vehicle motion quality including its velocity and acceleration performances. Figure 11a presents the longitudinal velocity curves achieved by MPC-SQP and PSO controllers. Starting at 0.5 m/s, the two curves reach 3 m/s at 6.5 s and 4.8 s , respectively. Thus, the trajectory and error curves of the MPC controller responses are slower than the PSO controller in Figure 9. Both controllers are capable of maintaining the longitudinal velocity around 3 m/s in the rest process. In Figure 11b, the longitudinal acceleration curve of PSO increases faster than MPC-SQP and has a peak value of 1 m/s^2. The curve of MPC-SQP controller reaches 0.6 m/s^2 at a maximum and has smoother variation during the whole process.

From the offset model, lateral velocity is undesirable as it causes tracking error. As shown in Figure 11c, both controllers maintain the lateral velocities fluctuating around zero. The MPC-SQP controller has a maximum lateral velocity of 0.08 m/s, whereas PSO has a maximum value of 0.17 m/s. The lateral velocities are well constrained, which provides higher accuracy of path tracking. The lateral acceleration curves follow the same varying trend in Figure 11d. It can be seen the PSO controller causes a lateral acceleration oscillation of 0.6 m/s^2 relative to the MPC-SQP. The smoother change of lateral acceleration achieved by MPC-SQP controller reduces the jerk effect, decreasing the deviation from the reference path. The angular velocities are demonstrated in Figure 11e, in which the MPC-SQP has higher accuracy and a smoother manner, compared with that of the PSO. In Figure 11f, the angular acceleration curves are presented. MPC-SQP and PSO curves vary with oscillations of 12 °/s^2 and 20 °/s^2, respectively. The undesirable oscillations are eliminated in the process of the integration, which can be demonstrated in Figure 11e. From all the acceleration plots given in Figure 11, the MPC-SQP controller performs better in constraining oscillations of accelerations, which provides better stability in vehicle motion control.

Considering that the optimization based method may lead to an expensive computation, it is essential to analyze the computing efficiency for each controller. As shown in Figure 12, the computing efficiency of each controller is compared. Figure 12a shows the computing time used by the controllers in each sampling time T_s. It can be seen that both controllers achieve finishing computing within T_s, which validates the feasibility of proposed controllers. In the box plot in

Figure 12b, both controllers have an average computing time of 8 ms. The PSO controller gives a stable variation range from 4 ms to 10 ms, as the particle velocity vector is decided by the limitation of the sampling time, which guarantees relatively high quality solutions within a short time. The computing time of MPC-SQP controller substantially increases and reaches 15 ms to 20 ms during each cornering because the control allocation may encounter a complex optimization problem when the lateral forces start to vary. It indicates that the PSO controller performs better in computing efficiency.

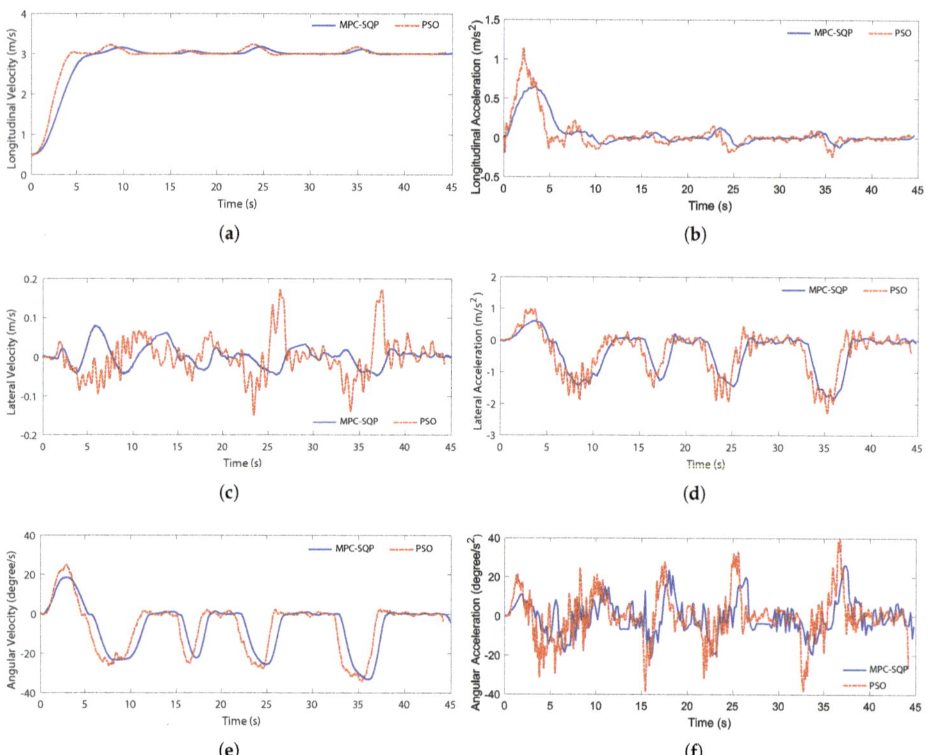

Figure 11. Acceleration and velocity performances. (**a**) longitudinal velocity; (**b**) longitudinal acceleration; (**c**) lateral velocity; (**d**) lateral acceleration; (**e**) angular velocity; (**f**) angular acceleration.

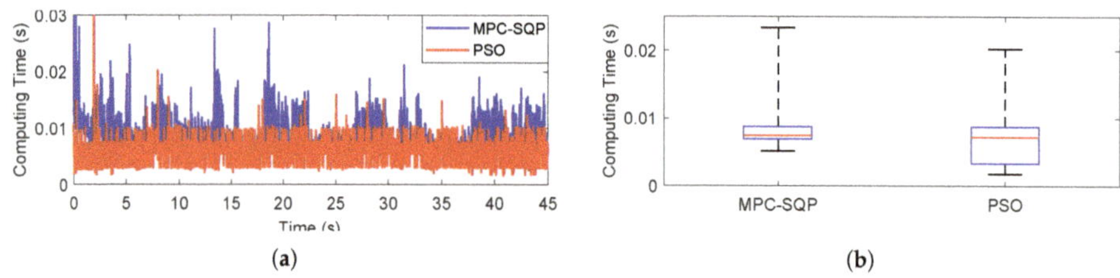

Figure 12. Computing efficiency comparison. (**a**) computing time; (**b**) box plot of computing time.

5. Discussion

Based on the theoretical analysis and simulation results provided above, the discussion between MPC based method and PSO method are given as follows:

D1   Comparing the simulations, it is obvious that dynamic-based methods proposed in this paper perform better than kinematic model based methods. In the kinematic model based methods, it is difficult to measure the side slip directly. The majority of research works are trying to design observers to predict its side slip. However, this kind of method can only get an approximate estimate. Thus, it is not feasible to use a kinematic model based controller to completely eliminate the error due to side slip. Both MPC-SQP and PSO methods are proposed based on dynamic models and controlled by drive forces. In the dynamic model, the lateral forces can be obtained easily using force sensors. This kind of method gives a practical way to avoid the side slip estimation and achieve precise tracking along curved paths.

D2   Both MPC-SQP and PSO methods are proposed partly or completely based on optimizations. As is compared in the simulation, the PSO controller achieves obtaining the solution with a stable computing time. The algorithm can optimize the quality of solutions and the calculation times at the same time, which guarantees the feasibility and capability of the controller. The MPC-SQP controller may encounter the calculation timeout in the simulation results. This is because the SQP solver spends more time when calculating a complex Hessian matrix. Thus, it is necessary to set a sufficient sampling time when applying the MPC-SQP method.

D3   The MPC-SQP method gives a stability proof of the control system while the PSO method has difficulty with the mathematical proof due to the limitation of intelligent optimization. The stability analysis of MPC-SQP method provides a possibility to analyze the stable range of its control parameters and margin of errors.

D4   Both control algorithms are reduced to constrained optimization problems with six input variables. PSO is a global algorithm that performs better with searching for a global optimal solution while SQP is a reliable solver but may be held in local optimal solutions. Thus, for each method, it is important to define a proper search range. In this paper, the boundaries are calculated based on the vector s, which improves the efficiency and stability of the optimization process.

6. Conclusions

This paper presented two control methodologies applicable to four wheel steering and four wheel drive vehicle systems to track paths accurately. The system model is a nonlinear coupled dynamic model that is over-actuated. The first methodology determines the drive force inputs and steer inputs using an MPC based method coupled with a control allocation based on SQP. The second methodology is proposed as a fully optimization based method that is developed by refining a previously proposed PSO based method. The performance of the MPC-SQP method has been compared with the PSO based control method and a kinematic model based control method. In the simulation, the path tracking results have proved the superior performance of the MPC-SQP based controller. In the cases when computing times are considered, the PSO based controller offers more efficiency and better stability in computing compared with MPC-SQP based controller.

Author Contributions: Conceptualization, Q.T. and J.K.; Methodology, Q.T. and P.D.; Writing—Original Draft Preparation, Q.T. and Z.Z.; Writing—Review & Editing, J.K.; Supervision, J.K.; Project Administration, J.K.; Funding Acquisition, Q.T.

Funding: This research was funded by the Fundamental Research Funds for the Central Universities, Grant No. 2017JBM051 and 2015JBC022, and the China Postdoctoral Science Foundation, Grant No. 2016M600910.

Acknowledgments: This work was carried out at the School of Mechanical and Manufacturing Engineering, the University of New South Wales, Sydney, Australia.

References

1. Oppenheimer, M.W.; Doman, D.B.; Bolender, M.A. Control Allocation for Over-actuated Systems. In Proceedings of the 14th Mediterranean Conference on Control and Automation, Ancona, Italy, 28–30 June 2006; pp. 1–6.
2. Ackermann, J.; Sienel, W. Robust yaw damping of cars with front and rear wheel steering. *IEEE Trans. Control Syst. Technol.* **1993**, *1*, 15–20. [CrossRef]
3. Itoh, H.; Oida, A.; Yamazaki, M. Numerical simulation of a 4WD-4WS tractor turning in a rice field. *J. Terramech.* **1999**, *36*, 91–115. [CrossRef]
4. Wu, J.; Wang, Q.; Wei, X.; Tang, H. Studies on improving vehicle handling and lane keeping performance of closed-loop driver-vehicle system with integrated chassis control. *Math. Comput. Simul.* **2010**, *80*, 2297–2308. [CrossRef]
5. Wang, W.; Song, Y. A new high dimension nonlinear dynamics simulation model for four-wheel-steering vehicle. *J. Mech. Eng. Sci.* **2013**, *227*, 29–37. [CrossRef]
6. Udengaard, M.; Iagnemma, K. Kinematic analysis and control of an omnidirectional mobile robot in rough terrain. In Proceedings of the IEEE/RSJ International Conference on Intelligent Robots and Systems, IROS, San Diego, CA, USA, 29 October–2 November 2007; pp. 795–800.
7. Grand, C.; Amar, F.B.; Plumet, F. Motion Kinematic analysis of wheeled-legged rover over 3D surface with posture adaptation. *Mech. Mach. Theory* **2010**, *45*, 477–495. [CrossRef]
8. Han, G.; Fu, W.; Wang, W.; Wu, Z. The Lateral Tracking Control for the Intelligent Vehicle Based on Adaptive PID Neural Network. *Sensors* **2017**, *17*, 1244. [CrossRef] [PubMed]
9. Qian, H.; Lam, T.; Li, W.; Xia, C.; Xu, Y. System and design of an Omni-directional vehicle. In Proceedings of the IEEE International Conference on Robotics and Biomimetics (ROBIO), Bangkok, Thailand, 22–25 February 2009; pp. 389–394.
10. Grepl, R.; Vejlupek, J.; Lambersky, V.; Jasansky, M.; Vadlejch, F.; Coupek, P. Development of 4WS/4WD Experimental Vehicle: Platform for research and education in mechatronics. In Proceedings of the IEEE International Conference on Mechatronics (ICM), Istanbul, Turkey, 13–15 April 2011; pp. 893–898.
11. Ramaswamy, S.A.P.; Balakrishnan, S.N. Formation control of car-like mobile robots: A Lyapunov function based approach. In Proceedings of the American Control Conference, Seattle, WA, USA, 11–13 June 2008; pp. 657–662.
12. Camacho, E.; Alba, C. *Model Predictive Control*; Advanced Textbooks in Control and Signal Processing; Springer: London, UK, 2013.
13. Lenain, R.; Thuilot, B.; Cariou, C.; Martinet, P. Model predictive control for vehicle guidance in presence of sliding: Application to farm vehicles path tracking. In Proceedings of the 2005 IEEE International Conference on Robotics and Automation (ICRA 2005), Barcelona, Spain, 18–22 April 2005; IEEE: Piscataway, NJ, USA, 2005; pp. 885–890.
14. Lenain, R.; Thuilot, B.; Cariou, C.; Martinet, P. High accuracy path tracking for vehicles in presence of sliding: Application to farm vehicle automatic guidance for agricultural tasks. *Auton. Robots* **2006**, *21*, 79–97. [CrossRef]
15. Prasetya, D.A.; Yasuno, T. Cooperative control of multiple mobile robot using particle swarm optimization for tracking two passive target. In Proceedings of the SICE Annual Conference (SICE), Akita, Japan, 20–23 August 2012; pp. 1751–1754.
16. Lin, L.; Sun, Q.; Wang, S.; Yang, F. Research on PSO based multiple UAVs real-time task assignment. In Proceedings of the Chinese Control and Decision Conference (CCDC), Guiyang, China, 25–27 May 2013; pp. 1530–1536.
17. Thomas, J. Integrating Particle Swarm Optimization with Analytical Nonlinear Model Predictive Control for nonlinear hybrid systems. In Proceedings of the International Conference on Informatics in Control, Automation and Robotics (ICINCO), Colmar, France, 21–23 July 2015; pp. 294–301.
18. Dai, P.; Katupitiya, J. Force control for path following of a 4WS4WD vehicle by the integration of PSO and SMC. *Veh. Syst. Dyn.* **2018**, 1–35. [CrossRef]
19. Wang, X.; Taghia, J.; Katupitiya, J. Robust Model Predictive Control for Path Tracking of a Tracked Vehicle with a Steerable Trailer in the Presence of Slip. *IFAC-PapersOnLine* **2016**, *49*, 469–474. [CrossRef]
20. Rajamani, R. *Vehicle Dynamics and Control*; Springer Science: Berlin, Germany, 2006.

21. Smith, D.; Starkey, J. Effects of Model Complexity on the Performance of Automated Vehicle Steering Controllers: Model Development, Validation and Comparison. *Veh. Syst. Dyn.* **1995**, *24*, 163–181. [CrossRef]
22. Brach, R.; Brach, R. *Tire Models for Vehicle Dynamic Simulation and Accident Reconstruction*; SAE Technical Paper; Society of Automotive Engineers: Warrendale, PA, USA, 2009; Volume 1.
23. Jazar, R.N. *Vehicle Dynamics: Theory and Applications*; Springer: New York, NY, USA; London, UK, 2008.
24. Milliken, W.F.; Milliken, D.L. *Race Car Vehicle Dynamics*; Society of Automotive Engineers: Warrendale, PA, USA, 1995.
25. Dai, P.; Katupitiya, J. Force Control of a 4WS4WD Vehicle for Path Tracking. In Proceedings of the IEEE International Conference on Advanced Intelligent Mechatronics (AIM), Busan, Korea, 7–11 July 2015; pp. 238–243.
26. Johansen, T.A.; Fossen, T.I. Control allocation—A survey. *Automatica* **2013**, *49*, 1087–1103. [CrossRef]
27. Slotine, J.J.E.; Li, W. *Applied Nonlinear Control*; Prentice Hall: Upper Saddle River, NJ, USA, 1991.
28. Eberhart, R.C.; Shi, Y. Particle swarm optimization: Developments, applications and resources. *Evol. Comput.* **2001**, *1*, 81–86.

Article

The Dynamic Coupling Analysis for All-Wheel-Drive Climbing Robot Based on Safety Recovery Mechanism Model

Fengyu Xu [1,2], Quansheng Jiang [3,*], Fan Lv [1], Mingliang Wu [4] and Laixi Zhang [4]

[1] College of Automation, Nanjing University of Posts and Telecommunications, Nanjing 210003, China; xufengyu598@163.com (F.X.); lvfan6514@163.com (F.L.)

[2] State Key Laboratory of Robotics and System (HIT), Harbin Institute of Technology, Harbin 150001, China

[3] School of Mechanical Engineering, Suzhou University of Science and Technology, Suzhou 215009, China

[4] School of Mechanical & Electronical Engineering, Lanzhou University of Technology, Lanzhou 730050, China; wml10757@163.com (M.W.); laixi_zh@163.com (L.Z.)

* Correspondence: qschiang@163.com; Tel.: +86-512-6832-0622

Received: 9 September 2018; Accepted: 30 October 2018; Published: 1 November 2018

Abstract: Cable is one of the most important parts on cable-stayed bridges. Its safety is very important. The aim of this study is to design an all-wheel-drive climbing robot based on safety recovery mechanism model for automatic inspection of bridge cables. For this purpose, a model of a three-wheel-drive climbing robot with high-altitude safety recovery mechanism is constructed and the basic performances such as climbing ability and anti-skidding properties are analyzed. Secondly, by employing the finite element method, natural frequency of the robot is calculated and that of a cable with concentrated masses is obtained through use of the Rayleigh quotient. Based on the mentioned quantities, the dynamic characteristics of the robot–cable system are further analyzed. In order to verify the climbing ability of the designed robot, a prototype of the robot is made, a robot testing platform is established and the climbing & loading experiments of the robot are carried out. The experiment results illustrated that the robot can carry a payload of 10 kg and safely return along the cable under the influences of inertial force.

Keywords: climbing robot; safety recovery mechanism; cable detection; dynamic coupling analysis

---

1. Introduction

A cable-stayed bridge is a common structural form. The cables which are as important bearing parts of the bridge, need to be inspected regularly due to the influences of extreme conditions, such as wind- and rain-induced vibration. At present, manual detection methods are widely used. This study designs a robot to replace workers for cable defect detection tasks.

Robots for cable detection are non-standard automated items of equipment and many researchers have proposed various innovative mechanisms therein: for example, Cho et al. used a six-wheel robot for bridge cable inspection [1,2]. Such a robot can adjust its clamping forces according to the situation encountered on the surface of a cable, so as to adapt to cables of different diameters. Moreover, they designed a cable visual detection system and took pictures of the cable surface from different directions with three cameras. The pictures were sent to the ground-based monitoring station in real-time. Park et al. designed a cable crawling robot [3,4]. The cables are automatically monitored by using non-destructive testing (NDT) technology. With the development of an eight-channel sensor for magnetic flux leakage inspection, the measured magnetic field signals were transformed into 3-d graphics to allow intuitive monitoring. Ho et al. proposed a robot for cable detection [5]. By combining it with an efficient image-based damage detection system, the damage to the cable surface was automatically recognized through image processing and pattern recognition technology. Type-I and -II robots for cable maintenance are capable of carrying 250 kg payloads and designed by Luo Jun et al., which have been tested on Xupu Bridge in Shanghai, China for maintenance issues such as the painting of bridge cables [6].

There are a lot of researches on bar and pipeline climbing robots. For instance, Mahmoud et al. proposed a 3D CLIMBER bar climbing robot that can climb along a spatial tubular structure [7]. The 3D CLIMBER can measure position deviations against three different classifications by using a sensor and then the errors therein are corrected one-by-one. Zhu et al. proposed a method of computing operational regions for transition in global path searching for biped climbing robots [8]. Furthermore, Mahmoud et al. designed an oil and gas pipeline climbing robot to detect bent pipelines and T-shaped oil and gas pipelines [9]. Allan et al. developed a wheeled bar climbing robot which is capable of climbing along a straight bar with heavy payloads [10]. Li Nan et al. designed a climbing robot that can move along the bar and rotate around it [11]. In addition, Lam et al. investigated a Treebot climbing robot with two claws [12]: when the robot climbs a tree, it detects the position of the trunk based on the information provided by an encoder, a contact sensor and an inclinometer to deduce the optimal climbing path. Schempf proposed a series of robots for pipeline detection and designed corresponding NDT methods [13]. Guan proposed the use of a climbing robot with two claws [14]. By using three sensors (laser range scanner, color camera and ultrasonic ranging module) the deviation of different poses of the holder at the end is detected and then corrected step-by-step, thus realizing automatic gripping and grasping. A grasping module of cross-arranged claw is designed for the inspection of rough stone and concrete wall surfaces, which performs well in grasping vibrating walls with certain anti-rollover capability [15]. Yanagida et al. designed a shape shifting rolling–crawling–wall-climbing robot with wall-climbing locomotion abilities [16]. Xu et al. designed a robot for cable inspection with double wheels on both sides [17,18]. Such a robot can overcome the influences of high-altitude wind loads and conduct NDT by entering the high-altitude inclined cable environment well beyond human reach. An energy saving and recovery method based on the theory of counter electromotive force was proposed to realize the energy-saving recovery of robots at an altitude of hundreds of meters. The author also analyzed the anti-dynamic ability of the climbing robot based on the claws [19].

Other representative climbing robots for low-altitude operation are described as follows: one is a robot with claws, designed by Moon, that can span between interlaced bars [20]. The other is a humanoid wheeled climbing robot designed by Ahmadabadi, that moves under joint control of two motors and the clamping principle lies between spring-loaded-type and squirmy-type-designs [21]. Such a robot is used for cleaning street lamps on high-speed roads. In addition, for some wall climbing robots, their adhesion principle provides some inspirations for cable detection robots. For example, the robot with vacuum suckers designed by Lee [22] and a wheeled robot designed by Espinoza adopt vacuum adhesion and magnetic adhesion devices [23]. They show simple principles and could be easily controlled. Furthermore, the robots can move freely on arc-shaped walls and operate in various environments, such as pipelines and walls of pipelines.

In conclusion, researchers have studied the use of robots for cable inspection, while the dynamic characteristics of such robots are rarely touched. The reasons for this include the following facts: robot climbing can change the vibration characteristics of cables themselves and cable vibration influences the climbing ability of the robot. Therefore, we proposed a new all-wheel-drive climbing robot and solved the dynamic problems in robot–cable coupling. The research is structured as follows: Section 2 summarizes models of all-wheel-drive climbing robot based on safety recovery mechanism according to the needs of cable inspection at high altitude. Section 3 covers the analysis of its basic climbing performance. Coupling dynamic characteristics of the robot–cable system are studied in Section 4. Section 5 covers the fabrication of the prototype climbing robot and its testing platform, with which the testing and analysis of climbing performance are conducted. Finally, Section 6 summarizes the research and outlines the future research work.

2. Structural Model of the Climbing Robot

Firstly, the mechanical structure including climbing principle and specific driving structure of the robot was introduced. Secondly, a centrifugal–friction recovery mechanism was introduced and the key factors influencing braking ability were analyzed through simulation.

2.1. Overall Scheme for the Climbing Robot

The proposed robot involved the clamping of a cable clamped with three wheels. According to the principle illustrated in Figure 1a, wheels A and C were set on the left and right support plates which are opposite to each other, while wheel B was set on the left and right swing plates in an opposed arrangement. Swing plates and support plates were hinged and rotated freely around the hinge, while the spring mechanism pushed the upper and lower support bars and thus the robot clamping cables. The shapes of support plates and swing plates are shown in Figure 1b. Motors 1 and 2 directly drove wheels A and B through the synchronous belt drive and wheel C was driven by motor 1 through another synchronous belt, thus forming the all-wheel-drive climbing robot (Figure 2).

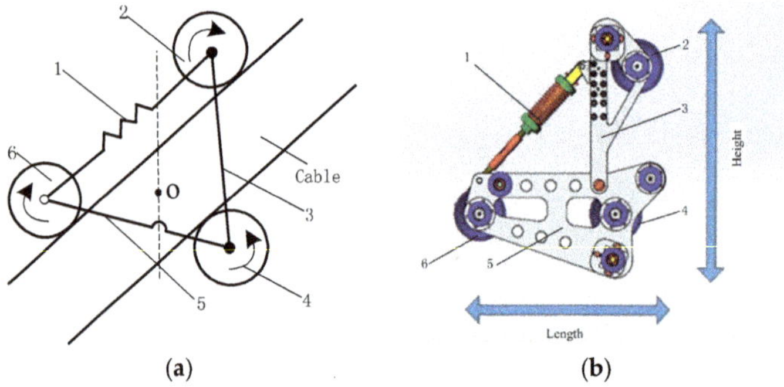

Figure 1. The model of climbing robot. 1—Spring mechanism; 2—Wheel B; 3—Swing plate; 4—Wheel A; 5—Support plate; 6—Wheel C. (**a**) The principle of the climbing robot; (**b**) Left view of the robot.

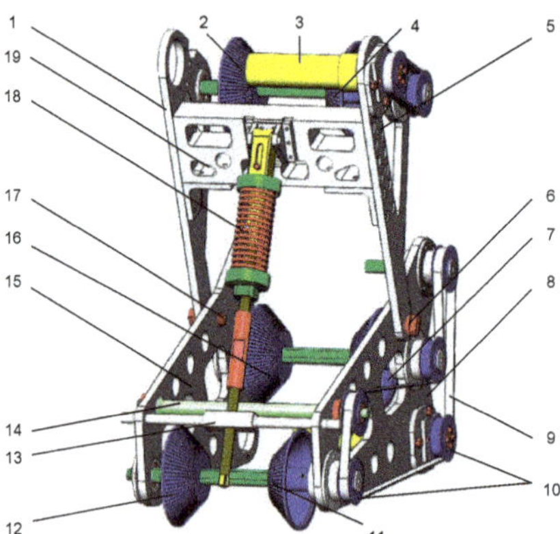

Figure 2. The structure of the climbing robot. 1—Left swing plate; 2—Left half-wheel B; 3—Motor 2; 4—Right half-wheel B; 5—Right swing plate; 6—Right hinge; 7—Right half-wheel A; 8—Right support plate; 9—synchronous belts; 10—Synchronous pulley; 11—Right half-wheel C; 12—Left half-wheel C; 13—Lower support bars; 14—stop lever; 15—Left support plate; 16—Left half-wheel A; 17—Left hinge; 18—Spring mechanism; 19—Upper support bars.

Each of the three rolling wheels were comprised of two half-wheels on the left and right (Figure 2), which allowed easy adjustment of the distance between the two half-wheels and the clamping force. All parts of the robot were set between the left and right support plates.

Based on the joint effects of the relative rotation of swing plates and support plates, scaling of the spring and adjustment of spacing of rolling wheels, the robot clamped the cable. In practical applications, it is more convenient to adjust clamping forces by adjusting the spacing of the rolling wheels, so the design allowed for a mass of 11 kg and a length ranging from 370 mm to 400 mm. Moreover, the width was 236 mm and the height was able to be adjusted within the range from 410 to 440 mm.

2.2. The Safety Recovery Mechanism Model of the Climbing Robot

When electrical failure occurs, the robot freely falls to the ground from high altitude and the fast rate of descent and impact velocity damages the equipment and even threatens the safety of workers. In order to avoid fast gliding speed of the robot after electrical failure, a centrifugal–friction recovery mechanism was designed. The braking force was dynamically adjusted in accordance with different inclination angles of cables as shown in Figure 3a. The recovery mechanism consisted of mass blocks, a brake disc, a friction plate, a gearbox and a rotary shaft (Figure 3b–d) and each part was set on the rotary shaft. The chute was processed on one side of the brake disc and mass blocks slid in the chute to push the brake disc to overcome the spring force and then move along the rotary shaft. The brake disc contacted with the friction plate to generate braking torque, while the gearbox generated an acceleration to increase the braking torque of the friction plate on the robot.

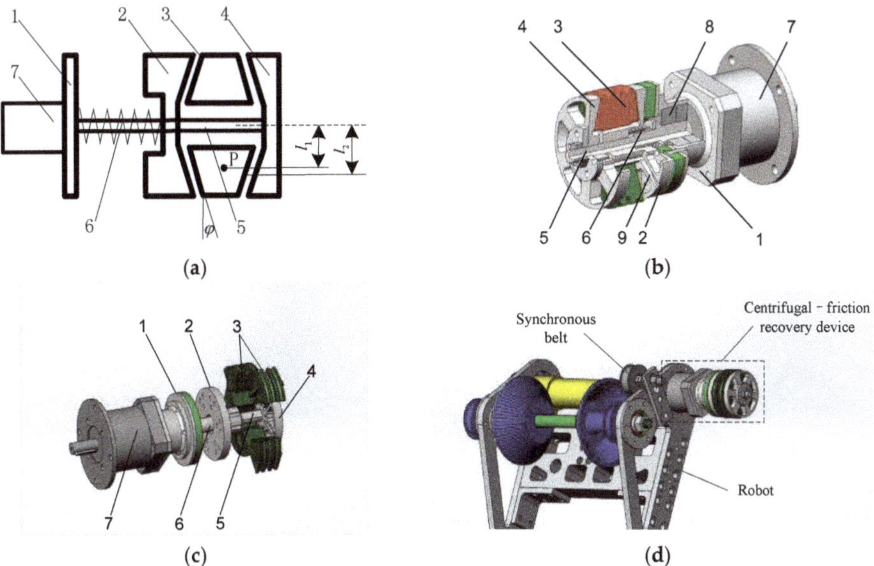

Figure 3. Safety recovery structure of the robot. 1—Main case; 2—Brake disc; 3—Mass blocks; 4—Base; 5—Rotary shaft; 6—Spring mechanism; 7—Gearbox; 8—Ball bearing; 9—slideway. (**a**) Schematic diagram of recovery mechanism; (**b**) The inner structure of the recovery device; (**c**) Breakdown drawing of recovery mechanism; (**d**) Centrifugal–friction recovery mechanism.

When the descent speed of the robot is low, the brake disc is separated from the friction plate under the effects of the spring (Figure 4a). If the robot drops quickly after a power outage, the mass blocks slide outside under the effects of centripetal forces and the axial component thereof overcomes the spring forces, so that the brake disc makes complete contact with the friction plate and braking forces are dynamically transferred to the robot (Figure 4b).

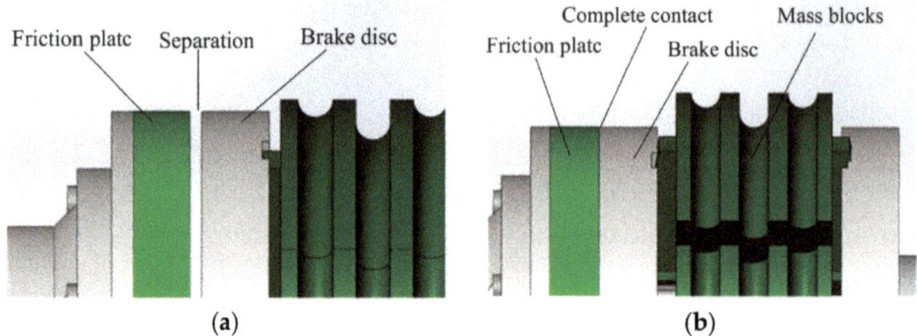

Figure 4. Movement state of the recovery mechanism. (a) Low-speed movement of the robot; (b) High-speed movement of the robot.

With the changes of descent speed of the robot, contact forces between the friction plate and brake disc changes constantly. In theory, the braking torque can automatically reach a stable value according to the inclination angle of cables and loads on the robot.

Both sides of mass blocks are designed for a certain inclination angle φ so as to transform the centripetal force into an axial force. The component of this force in the axial direction of the sum of centripetal forces of six mass blocks can be expressed as:

$$F = 3 \cdot m \cdot w^2 \cdot l_1 \cdot \cot \varphi \tag{1}$$

where m, w and l_1 represent the mass of mass blocks, the angular velocity of the rotary shaft and the distance from the center (Point P) of gravity of the mass blocks to the axis of the rotary shaft, respectively. If small changes in the spring forces of the brake in working are ignored, axial force can be expressed as:

$$\begin{cases} F' = 0 & (F_{t1} > F) \\ F' = F - F_{t1} = 3 \cdot m \cdot w^2 \cdot l_1 \cdot \cot \varphi - F_{t1} & (F_{t1} < F) \end{cases} \tag{2}$$

where F_{t1} denote the spring force. The braking torque thus generated is:

$$M' = F' \cdot \mu_f \cdot l_2 = (3 \cdot m \cdot w^2 \cdot l_1 \cdot \cot \varphi - F_{t1}) \cdot \mu_f \cdot l_2 \tag{3}$$

where μ_f denote the coefficient of friction. l_2 represents the distance from the geometrical center of the mass blocks to the axis of the rotary shaft. When the robot runs at a high speed, it needs to satisfy the following equation in order to limit the speed and move more uniformly in a certain range:

$$i \cdot M' = M \tag{4}$$

where i is the gear ratio and M is the driving torque during the descent of the robot, thus giving the angular speed of the rotary shaft:

$$w = \sqrt{\left(\frac{M}{\mu_f l_2 \cdot i} + F_{t1} \right) \Big/ (3 \cdot m \cdot l_1 \cdot \cot \varphi)} \tag{5}$$

The descent speed of the robot is:

$$v = \frac{1}{i} w \cdot d \tag{6}$$

where d represents the diameter of rolling wheels of the robot and the falling torque M can be expressed as:

$$M = G \cdot \cos \beta \cdot d \tag{7}$$

where G indicates the self-weight of the robot and β is the angle of inclination of cables, so the descent speed of the robot can be rewritten as:

$$w = \sqrt{\left(\frac{M}{\mu_f l_2 \cdot i} + F_{t1}\right) \bigg/ (3 \cdot m \cdot l_1 \cdot \cot \varphi)} \tag{8}$$

According to Equation (8), by adjusting gear ratio and spring force, the descent speed of the robot can be changed. Assuming that the friction coefficient is $\mu = 0.5$, the mass of the sliding block is $m = 0.2$ kg, the inclination angle at the both ends of sliding block is $\varphi = 10°$, the distance from the center of gravity of the sliding block to the axis is $l_1 = (16–20)$ mm and $l_2 = (18–22)$ mm. Furthermore, all the values of l_1 and l_2 are changeable in the process of the robot landing. To simplify the process of calculation, the maximum values are used in the simulation. At last, the relationship linking the gear ratio, spring force and descent speed is shown in the simulation diagram.

As shown in Figure 5a, the spring force does affect braking ability (although only slightly), while the influence of the gear ratio is greater. Furthermore, by setting the spring force is 30 N and setting gear ratio is 10, the effects of the mass of the robot and the inclination angle of cables on the rate of descent were studied. Let the mass changes between 10 kg and 80 kg and the inclination angle of cables vary from 0° to 80°, Figure 5b shows the influences of inclination angle of the cables on braking force is far greater than that of the mass of the robot. Therefore, the designed recovery mechanism has more obvious effects on cables with larger inclination angles.

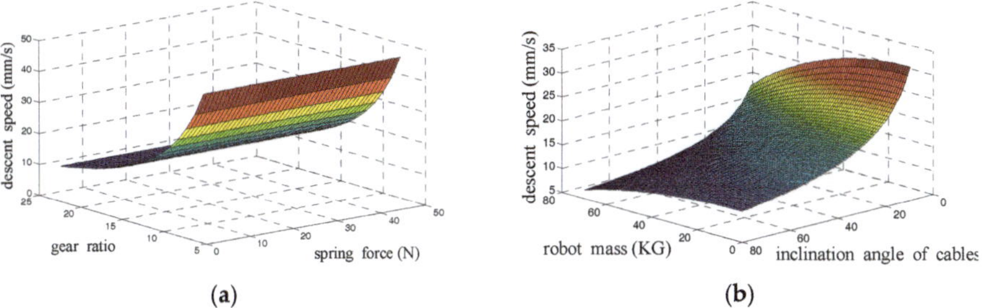

Figure 5. Simulation diagram of descent speed. (**a**) Relationship of descent speed, gear ratio and spring force. (**b**) Relationship of descent speed, robot mass and inclination angle of cables.

3. Analysis of Climbing Performance for the Climbing Robot

This section mainly analyzed the basic conditions for robot climbing and basic mechanical performance, such as the minimum output torque of the motors and the minimum friction force on the rolling wheels to prevent skidding. The robot was driven by two motors. In the schematic diagram, the driving wheels (A and B) are represented by solid circles, while the center of wheel C is indicated by a hollow circle. Moreover, wheel C is jointly driven by the motor installed on wheel A through a synchronous belt.

3.1. Analysis of Climbing Ability

The climbing robot utilizes the spring mechanism to provide force to clasp the cable and the driving wheel is actuated by DC motor to roll along the cable. The friction force must be greater than the gravity of the robot and payloads, so that it can climb along the cables inclined at any angle. The climbing principle, coordinate system and external force analysis are shown in Figure 6. For the whole robot mechanism, if the motor with enough input torque is used, we can establish the equilibrium equation:

$$\begin{cases} F_{fa} + F_{fb} + F_{fc} = Mg\cos\beta \\ N_b + N_c = N_a + Mg\sin\beta \end{cases} \tag{9}$$

where $F_{fa} = N_a\mu$, $F_{fb} = N_b\mu$, $F_{fc} = N_c\mu$, denote the maximum static friction force of the three wheels. N_a, N_b, N_c represent the supporting force of the three wheels, μ is static friction coefficient. M, g, β and O represent the mass of the robot including external payloads, the gravitational acceleration, the inclination angle of the cable and the center of gravity of the robot, respectively.

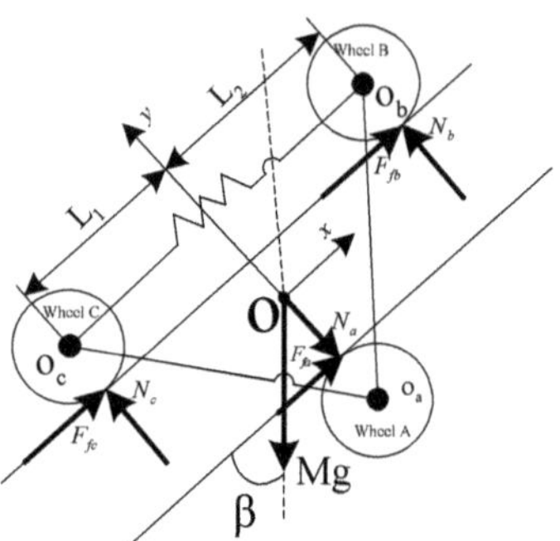

Figure 6. Climbing principle.

In this paper, various obstacles on cable surfaces are neglected. Meanwhile, by ignoring small losses during the drive and regarding the sum of output torques of driving wheels as τ, the following balance equation can be established:

$$\tau = (F_{fa} + F_{fb} + F_{fc})\cdot r = Mg\cdot\cos\beta\cdot r \tag{10}$$

where r represents the equivalent radius of the rolling wheels.

As the two motors drive three groups of rolling wheels, in ideal conditions, the single actual friction torque t' of rolling wheels is as follows in ideal conditions:

$$\tau' = \frac{1}{3}Mg\cdot\cos\beta\cdot r \tag{11}$$

In addition, in order to prevent skid in climbing, the sum F_f of the minimum sliding frictions of the rolling wheels must satisfy: $F_f > Mg\cdot\cos\beta$.

Suppose the velocity of the robot is v and the efficiency of the motor is η. In the design process, we should choose a driving wheel that possesses a rather large friction coefficient with the cable to reduce the clasped force. Here, ignore the inner friction of the robot, which the required whole power of DC motor P is:

$$P = F_f v/\eta = \frac{Mgv\cdot\cos\beta}{\eta} \tag{12}$$

3.2. Anti-Skid Analysis in Operating Conditions

Skid results from the lack of friction force between rolling wheels and cables, while friction is mainly transformed from the force generated by the spring. To study the relationship between spring force and friction force, some assumptions were made as follows: (1) The robot is climbing on a smooth cable without any obstacles in a constant velocity. For the wheel with the smallest support force, the driving torque of the motor equals the torque produced by the tangential friction forces. (2) All the driving motors are the same, therefore, we suppose the robot reached the maximum load when the wheel with smallest support force is in the slipping state. (3) To simply the calculation, the self-weight of the swing plate is ignored as it has less effect on the robot climbing ability. That is to say, the two sets of robot swing plates are considered as two two-force bars. (4) All the inner frictions of the robot mechanism are ignored. According to Figures 6 and 7, which shows the schematic diagram of the movement of the robot, the balance equations of the three groups of rolling wheels are established.

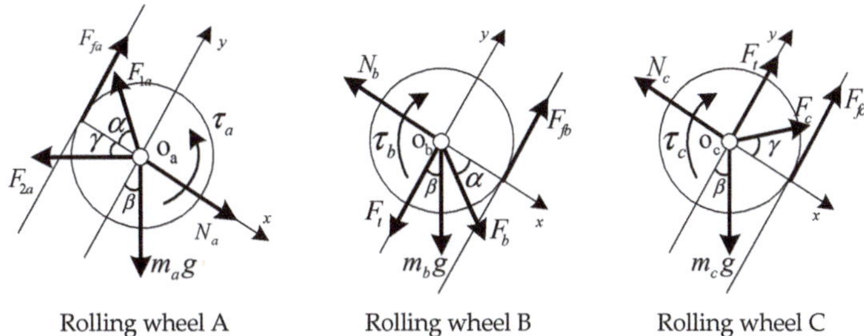

Rolling wheel A   Rolling wheel B   Rolling wheel C

Figure 7. Force balance relationships of rolling wheels.

In Figure 7, α and γ indicate the angles of two-force bars with the horizontal direction respectively and F_t represents the spring force. N_a, N_b and N_c are the support force, which denote the constraint reactions of cables on rolling wheels, while F_{1a}, F_b, F_{2a} and F_c denote forces acting on the bars of the rolling wheels. According to the nature of two-force bars, it is obvious that $F_{1a} = F_b$, $F_{2a} = F_c$:

As presented in Figure 7, the forces on rolling wheels A, B and C are presented, so the force balance equation for rolling wheel A is

$$\begin{cases} N_a + m_a g \sin \beta = F_{1a} \cos \alpha + F_{2a} \cos \gamma \\ F_{fa} + F_{1a} \sin \alpha = F_{2a} \sin \gamma + m_a g \cos \beta \\ \tau_a = F_{fa} r \end{cases} \tag{13}$$

For the rolling wheel B we have

$$\begin{cases} N_b = F_b \cos \alpha + m_b g \sin \beta \\ F_{fb} = F_b \sin \alpha + F_t + m_b g \cos \beta \\ \tau_b = F_{fb} r \end{cases} \tag{14}$$

Therefore we can get: $N_b = \frac{F_t - m_b g \sin \beta tg\alpha + m_b g \cos \beta}{\mu - tg\alpha}$, For the rolling wheel C

$$\begin{cases} N_c = F_c \cos \gamma + m_c g \sin \beta \\ F_{fc} + F_t + F_c \sin \gamma = m_c g \cos \beta \\ \tau_c = F_{fc} r \end{cases} \tag{15}$$

Therefore we can deduce: $N_c = \frac{-F_t + m_c g \sin \beta tg\gamma + m_c g \cos \beta}{\mu + tg\gamma}$, For the whole robot (Figure 6):

$$\begin{cases} N_a + M g \sin \beta = N_b + N_c \\ F_{fa} + F_{fb} + F_{fc} \geq M g \cos \beta \\ F_{fa} r + N_b L_2 = N_c L_1 + F_{fb} r + F_{fc} r \end{cases} \tag{16}$$

In the above equations, F_{fa}, F_{fb} and F_{fc} represents the friction force of three rolling wheels respectively. m_a, m_b and m_c represents the mass of three rolling wheels. τ_a, τ_b and τ_c represents the torques acted on three rolling wheels. μ denotes the friction coefficient between the rolling wheels and the cable surface.

Therefore we can get: $N_a = N_b + N_c - Mg \sin \beta$. Combining Equations (13)–(16), the following constraint equation is obtained:

$$\begin{cases} N_a = N_b + N_c - Mg \sin \beta \\ N_b = \frac{F_t - m_b g \sin \beta tg\alpha + m_b g \cos \beta}{\mu - tg\alpha} \\ N_c = \frac{-F_t + m_c g \sin \beta tg\gamma + m_c g \cos \beta}{\mu + tg\gamma} \end{cases} \tag{17}$$

For the three rolling wheels, we conclude the supporting force of the wheel C is the smallest one from the results the numerical analysis. Therefore, wheel C slides first. For wheel C climbing on the cable vertical cable ($\beta = 0°$), we suppose it is the critical state when $F_{fc} = N_c \cdot \mu$. The motor whose rated torque $\tau_c = F_{fc} r$ is chosen as the source of power. Because all the driving motors are the same, that means $\tau_a' = \tau_b' = \tau_c'$, so $F_{fa} = \frac{\tau_a'}{r}$, $F_{fb} = \frac{\tau_b'}{r}$ is deduced. Here, τ_a', τ_b', τ_c' represents the rated torque of the driving motors. Therefore, the maximum friction of the robot meet $F_f = 3F_{fc} \geq Mg \cdot \cos \beta$.

To study climbing performance of the robot under the condition of cable vibration, it is assumed that the robot is affected by inertial force during the vibration and simple harmonic vibration is applied to the cables with a vibration amplitude and frequency of 0.01 m and 10 Hz. Moreover, $\alpha = \gamma = 45°$, $\beta = 0°$ and friction coefficient $\mu = 0.5$, $m_a = m_b = m_c = 0.5$ kg. On this basis, the changes in friction force provided by the robot under cable vibration are simulated, as shown in Figure 8. The blue curves represent the maximum static friction force the cable exerted on the wheel, while red curves (Figure 8a–c) indicate the maximum (critical) friction force the motor exerted on the wheel. From the figures, the maximum static friction force is larger than the climbing force generated by the motor. Therefore, all the three wheels did not slide. Figure 8d shows the sum of the three driving force. The robot can take 15 kg payload besides self-weight. In accordance with Formula (17) and simulation results, the minimum spring force F_t is 130 N when the wheel C did not slide. To guarantee the robot climbs in a stable manner on the cable, the suitable scope of spring force is 130 N–150 N.

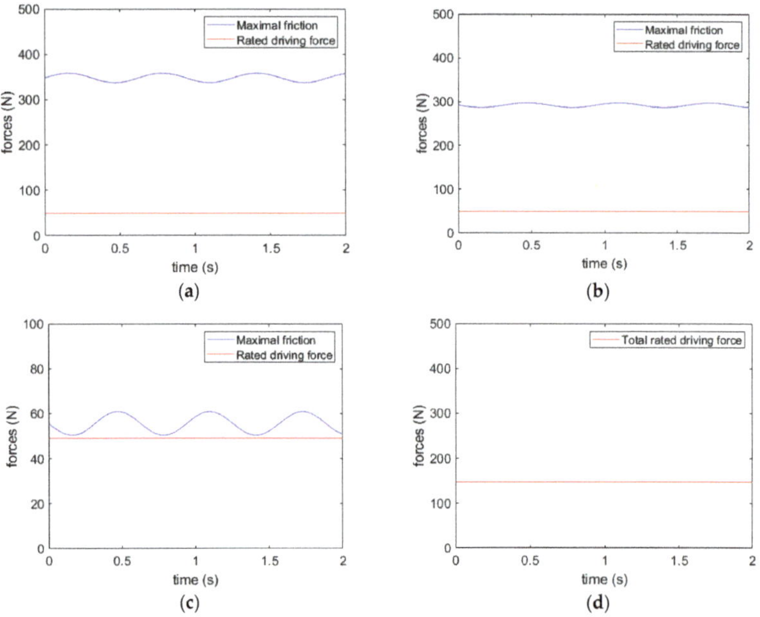

Figure 8. Changes in friction forces on the rolling wheels. (**a**) Rolling wheel A; (**b**) Rolling wheel B; (**c**) Rolling wheel C; (**d**) The payload of the robot.

4. Dynamic Characteristics of the Robot–Cable System

When the robot climbs the cable, once resonance occurs, the safety of the cable is affected. By using the finite element method, the natural frequency of the robot rack is estimated and then natural frequency of the cable is obtained with Rayleigh quotient. Finally we analyze the dynamic characteristics of the robot–cable system.

4.1. Natural Frequency of the Climbing Robot

By modeling the robot rack, the robot includes the following parts, such as an aluminum alloy rack, steel wheel shafts, rolling wheels and a spring. Owing to the rack contributing most to the weight of the robot and as the elastic modulus of steel is much larger than that of aluminum alloy, the aluminum alloy rack is the key component here. The parameters of the robot are illustrated in Table 1. By removing some holes and slots that slightly affect the overall structure and replacing the motor with a lumped mass of 0.7 kg, the finite element model of the robot is obtained (Figure 9).

Table 1. The parameters of the robot.

Item	Value
Length	370~400 mm
Width	236 mm
Height	410~440 mm
Material	Aluminum Alloy
Gravity	11.7 kg

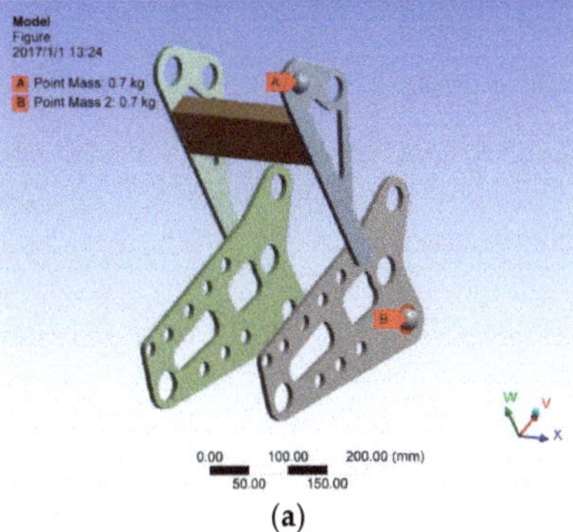
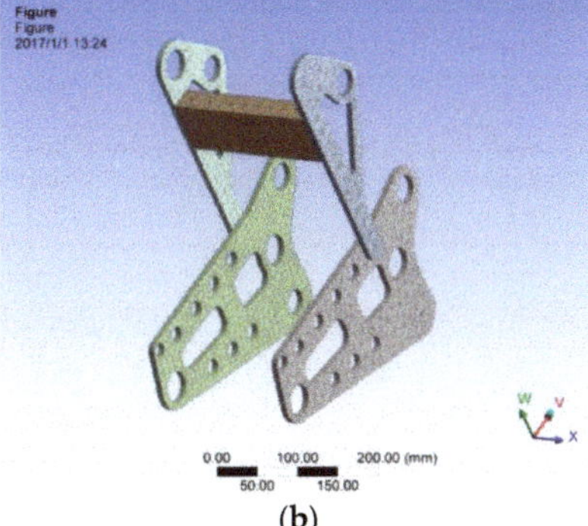

(a)
(b)

Figure 9. Finite element analysis of the robot. (**a**) Finite element model of the robot; (**b**) Finite element mesh.

The fixed constraints were applied on wheel shafts and the spring and the model was divided into grids measuring 10 mm. Through modal analysis, the first five orders of natural frequencies of the robot rack are 322 Hz, 463 Hz, 606 Hz, 723 Hz and 911 Hz.

4.2. Analysis of the Natural Frequency of the Cable

The string vibration equation is used to describe cable vibration (Figure 10). Owing to the robot climbing on a long cable, which has certain influences on cable vibration, the robot is regarded as a point mass to establish a string vibration equation with this lumped mass:

$$\begin{cases} u_{tt} - a u_{xx} = 0 & x \neq x_0 \\ u_{tt} - a_{x0} u_{xx} = 0 & x = x_0 \end{cases} \tag{18}$$

where x_0, l and u represent the position of the robot, the length of the cable and the normal displacement of the cable from the initial position, respectively. In addition, a denotes the ratio of tension T of the cable to the linear density ρ of the cable, namely $\frac{T}{\rho}$, while a_{x0} represents the ratio of T to the linear density ρ_{x0} of the cable at the position of the robot, namely, $\frac{T}{\rho_{x0}}$.

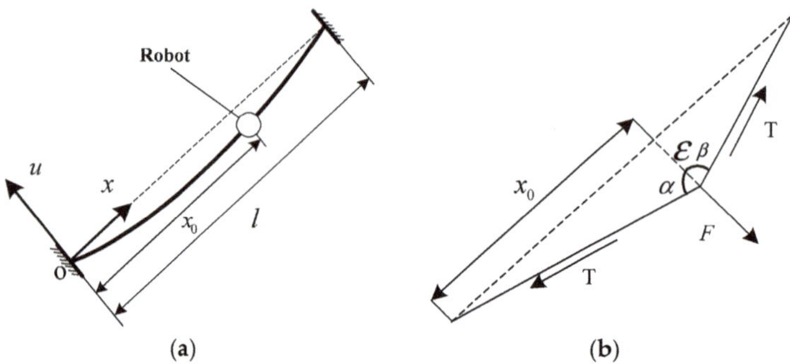

Figure 10. Robot–cable dynamic analysis. (**a**) Schematic diagram of cable vibration; (**b**) Force applied on the robot.

It is difficult to solve string vibration problems when they include a lumped mass. This section adopts an approximate method, namely, the use of the Rayleigh quotient, which can directly estimate the natural frequency of the system without solving its vibration equation. The Rayleigh quotient is expressed in the following form:

$$R = \frac{V_{\max}}{T_{ref}} = \omega_n^2 \tag{19}$$

$V_{\max}$ and T_{ref} represent the maximum potential energy and kinetic energy for reference of the cable respectively, which indicate quantities relating to potential energy and kinetic energy. ω_n denotes the nth-order natural frequency. $V_{\max}$ and T_{ref} can be written as:

$$\begin{cases} V_{\max} = \frac{1}{2}\int_0^l T(W_n'(x))^2 dx \\ T_{ref} = \frac{1}{2}\int_0^l \rho W_n^2(x) dx \end{cases} \tag{20}$$

$W_n(x)$ denotes the nth-order vibration mode function relating to x.

Free vibration of a tensile string is a linear combination of infinite multi-order vibrations. In general, the higher the order is, the smaller the influence on the spectrum, so it is important to study the first-order vibration. Experience shows that the first-order vibration mode of the cable is similar to static deformation under inertial loads. The first-order natural frequency can represent the frequency of the cable, so it is selected as the static deformation of the cable under the self-weight of the robot.

Figure 10b shows that the following equation can be established through the force balance and geometric relationships:

$$\cos\left(\arctan\frac{x_0}{\varepsilon}\right) + \cos\left(\arctan\frac{l - x_0}{\varepsilon}\right) = \frac{F}{T} \tag{21}$$

where F and ε represent the component of the self-weight of the robot in the vertical direction of the cable and the maximum static deformation, respectively. Because the distance from the position of maximum deformation to the end on the cable changes linearly, the vibration mode function can be obtained as long as ε is known from the following equation:

$$W(x) = \begin{cases} \varepsilon\frac{x}{x_0} & ,0 < x < x_0 \\ \varepsilon\frac{l-x}{l-x_0} & ,x_0 < x < l \end{cases} \tag{22}$$

Therefore, the relationship between x_0 and ε is established according to Equation (21).

$$\cos\left(\arctan\frac{x_0}{\varepsilon}\right) + \cos\left(\arctan\frac{115 - x_0}{\varepsilon}\right) = \frac{1}{45,300} \tag{23}$$

Based on this, an expression describing the deformation ε through the position of the robot is established, which is to say, $\varepsilon = f(x_0)$. Through data fitting, the samples are obtained such that $x_0 = [0, 20, 40, 60, 80, 100, 115]$. At two ends (0 m and 115 m), the deformation is zero. In view of the sample point being in the middle, the dichotomy is utilized for calculation purposes. In accordance with the zero point theorem, the search interval always shrunk. In this way, the root of the equation can be approached as close as possible. The flowchart of this dichotomy is shown in Figure 11. It is assumed that the self-weight of the robot, tensile force, length and linear density of the cable are $G - 200N$ (including loads), $T = 4530$ kN, $l = 115$ m and $\rho = 65.6$ kg/m, respectively.

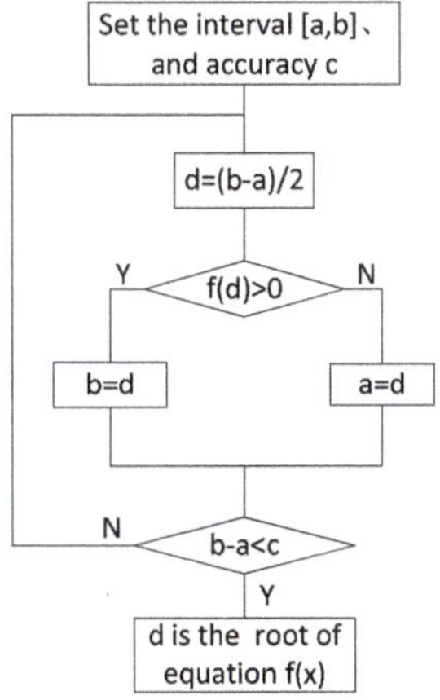

Figure 11. Flow chart of the dichotomy.

The search interval is set to $[0.0001, 0.01]$. It is assumed that $f(\varepsilon) = 45,300[\cos(\arctan\frac{x_0}{\varepsilon}) + \cos(\arctan\frac{115 - x_0}{\varepsilon})] - 1$ and $f(\varepsilon)$ is the continuous elementary function in the interval. Moreover, for $x_0 = [20, 40, 60, 80, 100]$, there exists $f(0.0001) < 0$ and $f(0.001) > 0$. According to the zero point theorem, there are solutions in the interval.

Through calculation, the solutions correspond to $x_0 = [20, 40, 60, 80, 100]$ are $[0.0003653, 0.0005755, 0.0006336, 0.0005381, 0.0002879]$. Therefore, it is obvious that the weight of the robot affects the cable, albeit only slightly.

As shown in Figure 12a, the distribution of sample points is parabolic. We can get the following equation with quadratic polynomial fitting technique:

$$\varepsilon = 10^{-4}(-0.0019x_0^2 + 0.2208x_0 + 0.0008) \tag{24}$$

Combining the above equation with Equations (22) and (23), the vibration mode function of the robot is:

$$W(x) = \begin{cases} 10^{-4}(-0.0019x_0^2 + 0.2208x_0 + 0.0008)\frac{x}{x_0} & ,0 < x < x_0 \\ 10^{-4}(-0.0019x_0^2 + 0.2208x_0 + 0.0008)\frac{l-x}{l-x_0} & ,x_0 < x < l \end{cases} \tag{25}$$

The first-order frequency is calculated according to Equations (19) and (20). For the cable with its lumped mass, the kinetic energy for reference is obtained:

$$T_{ref} = \frac{1}{2}\int_0^l \rho W^2(x)dx + \frac{1}{2}m[W(x_0)]^2 \tag{26}$$

where m represents the mass of the robot.

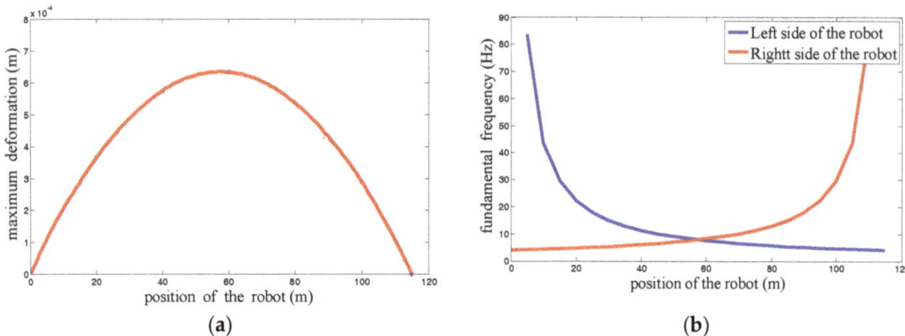

Figure 12. Simulation results: cable deformation and frequency. (**a**) The relationship between the position of the robot and the maximum deformation; (**b**) The change in fundamental frequency of the cable.

As illustrated in Figure 12b, the range of changes in the fundamental frequency of the cable is always less than 90 Hz, showing large differences across the first five fundamental frequencies of the robot rack, so resonance does not occur.

5. Climbing Experiment of the Robot Prototype

To verify the climbing ability of the robot, we fabricated a prototype of the robot, set-up a test platform for the robot and completed climbing & loading experiment of the robot.

5.1. Preparation of the Robot Prototype

According to the structural model of the robot described in Section 2, a prototype of the climbing robot was designed (Figure 13). A current (DC) motor system with high volume and torque was selected as power producer, which include a DC motor, a reducer and a rotary encoder. Based on the model in Figure 1, multiple synchronous pulleys and two synchronous belts were added. Motor 1 drove rolling wheels A and C through synchronous pulleys and belts, while rolling wheel B was driven by motor 2 through synchronous pulleys and belts, to form the all-wheel-drive mechanism climbing along the cable.

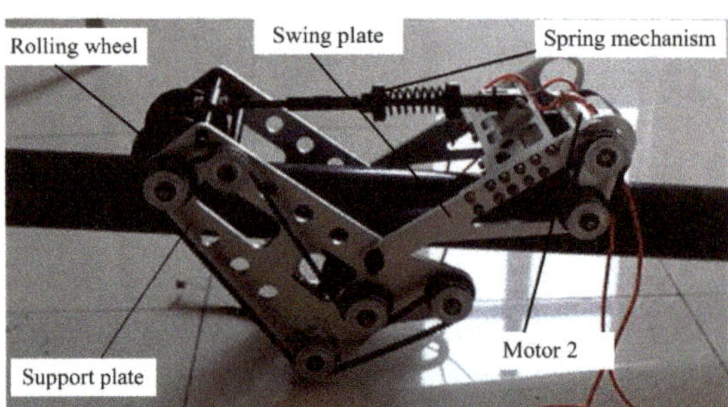

Figure 13. Three-wheel climbing robot.

Through the relative rotation of the swing plates and support plates, scaling of spring and adjustment of spacing between the left and right rolling wheels, the robot could fit around cables of different diameters.

5.2. Vibration Test Platform

By establishing a set of rigid and flexible hybrid robot test platforms, climbing tests using the robot in static and vibration environments was simulated. The vibration test platform is comprised of a simulated cable, fixtures, a vibration table, a lower rigidly-fixed fixture and an upper flexibly-fixed fixture (Figure 14). A hard plastic tube with a diameter of 90 mm was selected to replace the cable and the surface friction coefficient of the tube is similar to that of a typical cable.

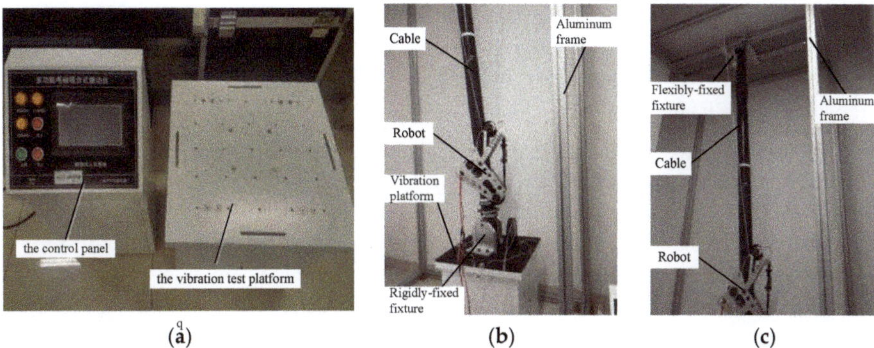

Figure 14. Vibration test platform. (**a**) Electromagnetic vibrator; (**b**) Lower rigidly-fixed fixture; (**c**) Upper flexibly-fixed fixture.

The lower rigidly-fixed fixture was mounted on the vibration table and connected to the lower end of the cable with a fixed hinge, which accurately transferred vibration and adjusted the inclination angle of the cable according to the need. A flexible connection was applied to the upper end of the cable and the upper fixture and ropes and spring were used for fixing. The cable and the upper fixture did not make complete contact and vibrated (within a fixed range of amplitudes) to simulate the vibration of real bridge cables.

By using an electromagnetic absorption type of vibration table to drive the cable vibration, the actual vibration of a cable on a bridge was simulated. The vibration table can bear 100 kg and the maximum vibration amplitude was 5 mm. Moreover, the frequency was adjustable in the range from 1 Hz to 400 Hz. The parameters of the electromagnetic vibrator are shown in Table 2.

Table 2. Parameters of the electromagnetic vibrators.

Maximum load: 100 kg	**Power**: 0.75 KW~2.2 KW
Maximum acceleration: <20 g	**The vibration direction**: X + Y + Z
Maximum amplitude: 0~5 mm	**Precision**: 0.01 HZ
Frequency range: 1~400 HZ	**Frequency sweep**: 1~400 HZ

By setting the low and high frequencies used, as well as the low and high vibration intensities of the vibrator, the frequency and amplitude of the cable, as well as the dead time at various frequencies and amplitudes, could be controlled. The main steps involved are as follows:

(1) First, the regions on the cable where the robot can stably grasp onto were explored and marked. The positions of the robot were adjusted so that all the wheels can stably grasp the cable under the influence of the driving forces.

(2) Then, the high-speed camera system and the vibrator were started and debugged.

(3) Main test started. The high-speed camera was used to record various parameters (output frequency and amplitude of the vibrators) and the output frequency and intensity of the test platform when the robot is climbing on the cable, before and after imposing the driving forces.

5.3. Testing Experiment

According to the analysis in Section 3 and for the given diameter of the cable together with those parameters, such as the clamping force of the robot, spacing of the rolling wheels and pre-tightening force of the spring were adjusted. The space between the left and right half-wheels was adjusted to be 42 mm.

As shown in Figure 15, a climbing test was conducted on the robot. Although the surface was smooth and there were obstacles (less than 10 mm in height), the robot could climb in a stable manner. By selecting heavy objects (mass, 10 kg), the climbing ability of the robot with loads was verified. The robot could still climb normally under such loads. The test proved that the climbing speed of the robot met the required performance for cable inspection. The following conclusions were drawn by combining simulation and test, data:

(1)　When masses of 10 kg were added to the robot, the robot could still climb along cables with inclination angle between 20° to 90°.
(2)　The robot could be installed and maintained by only one worker.
(3)　The climbing speed could be adjusted within the range from 0 m/s to 0.164 m/s.
(4)　The robot could pass across obstacles over a maximum length of 10 mm.

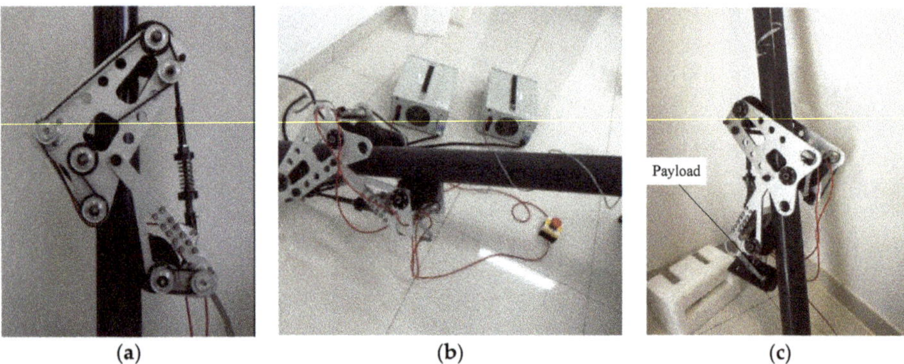

(a)　　　　　　　　　　(b)　　　　　　　　　　(c)

Figure 15. Climbing test of the robot. (**a**) Climbing on the smooth surface; (**b**) Climbing on the surface with obstacles; (**c**) Load test.

The climbing ability of the robot under the low-frequency and high-frequency vibration was further investigated and parameters, such as vibration direction, frequency and intensity of the vibration table could be adjusted through a controller. Based on the combined vibration in horizontal and vertical directions, the test was conducted using the combination of frequencies of 10 Hz and 100 Hz and vibration intensities of 20%, 40% and 60% (namely, 1 mm, 2 mm and 3 mm) of the maximum amplitude. The test shows that the robot could climb normally and its driving wheels did not skid. As a result, it did not leave the cable under the influence of inertial force during the vibration testing. Meanwhile, when the vibration table worked at a low frequency, the whole test system vibrated strongly, which exceeded the bearing limit of the floor of the building and limited the applicable test range. This should be addressed before in the future research.

6. Conclusions

In view of automatic inspection demands for bridge cables at high altitude, the all-wheel-drive climbing robot based on safety recovery mechanism model was developed. On this basis, this work analyzed the climbing performance and fabricated a prototype on which verification testing was conducted. The main conclusions are as follows:

(1) In order to limit the excessive speed of the robot when it is in electrical fault, the centrifugal–friction recovery mechanism model of the all-wheel-drive climbing robot is designed.

(2) The climbing performances of the robot are analyzed, including basic climbing conditions under minimum output motor torque and minimum friction force on the rolling wheels.

(3) By using the finite element method, the natural frequency of the robot was found and the first five modal frequencies were 322 Hz, 463 Hz, 606 Hz, 723 Hz and 911 Hz. Furthermore, the natural frequency of the cable was obtained by using the Rayleigh quotient. The results show that the change in fundamental frequency of the cable was always less than 90 Hz, showing significant differences with the first five modal frequencies of the robot, so resonance did not occur, thus ensuring the safety of the robot–cable system.

(4) The prototype of the robot is made and complete climbing & loading testing experiments is conducted. The testing results proved that the robot can carry a payload of 10 kg and safely return the cable under the influences of inertial force and vibration at the combination of frequencies of 10 Hz and 100 Hz and intensities accounting for 20%, 40% and 60% (namely, 1 mm, 2 mm and 3 mm) of the maximum vibration amplitude. This proved the feasibility of the robot.

In future research, it is necessary to optimize the mechanical structure of the robot, propose more accurate control methods and study inspection methods which are better suited for detection of defects in different cables.

Author Contributions: Data curation, F.X. and L.Z.; Formal analysis, F.X. and M.W.; Investigation, F.X.; Methodology, Q.J. and F.L.; Writing—original draft, F.X.; Writing—review & editing, Q.J.

Funding: This research was funded by [National Natural Science Foundation of China] grant number [51775284], [Primary Research & Development Plan of Jiangsu Province] grant number [BE2018734], [State Key Laboratory of Robotics and System (HIT)] grant number [SKLRS-2017-KF-10], [Jiangsu Six Talent Peaks Program] grant number [JY-081] and [Joint Research Fund for Overseas Chinese, Hong Kong and Macao Young Scholars] grant number [61728302].

Conflicts of Interest: The authors declare no conflict of interest.

References

1. Cho, K.H.; Jin, Y.H.; Kim, H.M.; Moon, H.; Koo, J.C.; Choi, H.R. Multifunctional Robotic Crawler for Inspection of Suspension Bridge Hanger Cables: Mechanism Design and Performance Validation. *IEEE/ASME Trans. Mechatron.* **2017**, *22*, 236–246. [CrossRef]

2. Cho, K.H.; Kim, H.M.; Jin, Y.H.; Liu, F.; Moon, H.; Koo, J.C.; Choi, H.R. Inspection Robot for Hanger Cable of Suspension Bridge: Mechanism Design and Analysis. *IEEE/ASME Trans. Mechatron.* **2013**, *18*, 1665–1674. [CrossRef]

3. Kim, J.W.; Lee, C.; Park, S.; Lee, J.J. Magnetic flux leakage-based steel cable NDE and damage visualization on a cable climbing robot. *Proc. SPIE* **2012**, *8345*, 46.

4. Park, S.; Kim, J.W.; Lee, C.; Lee, J.J. Magnetic Flux Leakage Sensing-Based Steel Cable NDE Technique. *Shock. Vib.* **2014**, *5*, 1–8. [CrossRef]

5. Ho, H.-N.; Kim, K.-D.; Park, Y.-S.; Lee, J.-J. An efficient image-based damage detection for cable surface in cable-stayed bridges. *NDT & E Int.* **2013**, *58*, 18–23.

6. Luo, J.; Xie, S.R.; Gong, Z.B. Development of cable maintenance robot for cable-stayed bridges. *Ind. Robot. Int. J.* **2007**, *34*, 303–309. [CrossRef]

7. Tavakoli, M.; Marques, L.; Almeida, A. A low-cost approach for self-calibration of climbing robots. *Robotica* **2011**, *29*, 23–34. [CrossRef]

8. Zhu, H.; Gu, S.; He, L.; Guan, Y.; Zhang, H. Transition Analysis and Its Application to Global Path Determination for a Biped Climbing Robot. *Appl. Sci.* **2018**, *8*, 122. [CrossRef]

9. Tavakoli, M.; Cabrita, G.; Faria, R.; Marques, L.; de Almeida, A.T. Cooperative multi-agent mapping of three-dimensional structures for pipeline inspection applications. *Int. J. Robot. Res.* **2012**, *31*, 1489–1503. [CrossRef]

10. Allan, J.F.; Lavoie, S.; Reiher, S.; Lambert, G. Climbing and pole line hardware installation robot for construction of distribution lines. In Proceedings of the 2010 1st International Conference on Applied Robotics for the Power Industry (CARPI), Montreal, QC, Canada, 5–7 October 2010; IEEE: Piscataway, NJ, USA, 2010; pp. 1–5.

11. Li, N.; Ji, X.G.; Xu, K.; Cui, A.Q. A Multi-pose pole-climbing Robot to Adapt Guidebar with Different Diameter. *Mod. Manuf. Eng.* **2011**, *9*, 60–61.

12. Lam, T.L.; Xu, Y. Biologically inspired tree-climbing robot with continuum maneuvering mechanism. *J. Field Robot.* **2012**, *29*, 843–860. [CrossRef]

13. Schempf, H.; Mutschler, E.; Gavaert, A.; Skoptsov, G.; Crowley, W. Visual and Nondestructive Evaluation Inspection of Live Gas Mains Using the ExplorerTM Family of Pipe Robots. *J. Field Robot.* **2010**, *27*, 217–249.

14. Xiao, Z.; Wu, W.; Wu, J.; Zhu, H.; Su, M.; Li, H.; Guan, Y. Gripper Self-Alignment for Autonomous Pole-Grasping with a Biped Climbing Robot. In Proceedings of the IEEE International Conference on Robotics and Biomimetics, Guangzhou, China, 11–14 December 2012; pp. 181–186.

15. Jiang, Q.; Xu, F. Grasping Claws of Bionic Climbing Robot for Rough Wall Surface: Modeling and Analysis. *Appl. Sci.* **2018**, *8*, 14. [CrossRef]

16. Yanagida, T.; Elara Mohan, R.; Pathmakumar, T.; Elangovan, K.; Iwase, M. Design and Implementation of a Shape Shifting Rolling–Crawling–Wall-Climbing Robot. *Appl. Sci.* **2017**, *7*, 342. [CrossRef]

17. Xu, F.; Wang, X. Cable Inspection Robot for Cable-Stayed Bridges: Design, Analysis, and Application. *J. Field Robot.* **2011**, *28*, 441–459. [CrossRef]

18. Xu, F.; Hu, J.L.; Wang, X.; Jiang, G. Helix Cable-Detecting Robot for Cable-Stayed Bridge: Design and Analysis. *Int. J. Robot. Autom.* **2014**, *29*, 406–414. [CrossRef]

19. Xu, F.; Wang, B.; Shen, J.; Hu, J.; Jiang, G. Design and realization of the claw gripper system of a climbing robot. *J. Intell. Robot. Syst.* **2018**, *89*, 301–317. [CrossRef]

20. Han, S.; Ahn, J.; Moon, H. Remotely controlled prehensile locomotion of a two-module 3D pipe-climbing robot. *J. Mech. Sci. Technol.* **2016**, *30*, 1875–1882. [CrossRef]

21. Sadeghi, A.; Moradi, H.; Nil Ahmadabadi, M. Analysis, simulation, and implementation of a human-inspired pole climbing robot. *Robotica* **2012**, *30*, 279–287. [CrossRef]

22. Lee, G.; Kim, H.; Seo, K.; Kim, J.; Kim, H.S. MultiTrack: A multi-linked track robot with suction adhesion for climbing and transition. *Robot. Auton. Syst.* **2015**, *72*, 207–216. [CrossRef]

23. Espinoza, R.V.; Oliveira, A.S.; Arruda, L.V.R.; Junior, F.N. Navigation's stabilization system of a magnetic adherence-based climbing robot. *J. Intell. Robot. Syst.* **2015**, *78*, 65–81. [CrossRef]

Article

Step-Climbing Tactics Using a Mobile Robot Pushing a Hand Cart

Hidetoshi Ikeda *, Takuya Kawabe, Ryousuke Wada and Keisuke Sato

National Institute of Technology, Toyama College, 13 Hongo-Chou, Toyama 939-8630, Japan;
toyama_nct_ikedah_lab@yahoo.co.jp (T.K.); ikedalabo.tnct@gmail.com (R.W.); sato@nc-toyama.ac.jp (K.S.)
* Correspondence: ikedah@nc-toyama.ac.jp; Tel.: +81-764-93-5444

Received: 3 September 2018; Accepted: 29 October 2018; Published: 1 November 2018

Abstract: The present paper describes step-climbing tactics using a wheeled robot and a hand cart that has a hand brake. The robot has two arms that are used to hold or push the handle of the cart and a lower extendable wheel mechanism that can push against the bottom of the cart. Some of the manipulator joints are controlled passively when moving over the step. To lift the front wheels of the cart, the robot holds the handle steady and pushes against the bottom of the cart using the extendable wheel mechanism. This action is similar to that performed by a human. The robot then pushes the entire cart forward so that the front wheels of the cart are above the step. When the rear wheels of the cart have climbed the step, the upper-arm links of the manipulators are pressed against the robot chest to allow the robot to push the cart. When the cart has fully climbed the step, the robot then uses the cart to climb the step. The present paper describes the details of the robot system, and theoretical analyses were performed to determine the requirement of masses and the centers of gravity of both vehicles to lift the cart. Experiments were also carried out in which the robot was controlled using an intranet connection, and the results demonstrated the effectiveness of the proposed method.

Keywords: cart; robot; step climbing; transportation; stopper

---

1. Introduction

Hand carts and shopping carts are used all over the world to help people convey baggage. In recent years, several robotic carts have been investigated, including an autonomous cart that follows the user [1,2], a robotic cart that helps the operator when transporting heavy baggage [3], and transportation using a cart connected to an omnidirectional mobile robot [4,5]. Iwanuma et al. studied two shopping assistant robots that can be used in supermarkets [6]. They proposed not only a cart robot, which has a shopping basket attached to the robot body for transportation of an item, but also a humanoid with dual manipulators, which can bring the shopping basket. If the partner robot can operate a cart freely, the possibilities of transportation using a cart for improvement are endless. Research on such carts that can be manipulated by a partner robot has been carried out. Ohno proposed a method of transportation of a hand cart using ASIMO (HONDA Co., Ltd., Tokyo, Japan), a legged locomotion robot with dual manipulators [7]. Similarly, Inaba et al. studied the locomotion of a wheelchair, rather than a cart, using a legged locomotion robot [8]. However, they showed how to push and transport a wheelchair using a personal robot with dual manipulators.

A hand cart has a wheel mechanism that it uses to move. Excellent energy efficiency is one of the advantages of a wheel mechanism. It is easy for people to move a hand cart on a flat road. However, it is difficult for a vehicle that has a wheel mechanism to move up and down steps. Numerous research programs aimed at improving the mobility of wheel mechanisms on steps have been conducted, based on the following strategies: additional legs [9,10], multiple wheels and variable wheelbases [11], and a combination of an adjustable center of gravity (COG) and multiple wheels [12]. Reports of multiple vehicles cooperating for crossing irregular terrain include those of Asama et al. [13], who considered a forklift system. Transporting a cart using a mobile robot that has numerous manipulators, similar to the human arm, is not an easy task. When the cart is heavier than the robot, controlling a robot to push or pull the cart is very difficult because the reaction force from the cart strongly influences the robot. In particular, one weakness for a wheeled vehicle is step climbing [14]. In transporting heavy baggage using a push-cart, it is very difficult to control the cart at a step because the incline of the vehicle is always changing, and the reaction force from the cart is also changing. As such, there have been no studies of a cart climbing a step using a partner robot.

The research group of the present study has achieved cooperative step climbing using a wheelchair and a partner robot with manipulators [15] and have proposed a cooperative step-climbing system using a hand cart and a robot [16]. In the present paper, we intend to describe in detail the concept of step climbing of a hand cart and present the results of both theoretical analysis and experiments.

The proposed step-climbing method uses the difference in velocity between connected vehicles, which considers not only the robot driving force, but also the wheelchair driving force. However, since a typical hand cart does not generate a driving force, the step-climbing method that we have proposed for a wheelchair cannot be applied to a hand cart. The present paper proposes a step-climbing method for a hand cart and a partner robot that uses an action that is similar to that performed by a human.

Based on preliminary measurements of friction coefficients using the vehicles in wet and dry conditions on asphalt, concrete, wood, and interior flooring, the ground surface considered in the present study was assumed to have a friction coefficient in the range 0.6 to 0.9, which satisfies all of the above conditions. The heights of steps at the entrances of typical buildings and other structures were measured, and the target step height was set to 120 mm, which accounts for more than 80% of the observed heights.

The remainder of the present paper is organized as follows. Section 2 describes the step-climbing system using a robot and a hand cart. Section 3 describes the process of climbing a step. Section 4 provides a theoretical analysis. Section 5 presents the experimental results. Section 6 concludes the paper.

2. Hand Cart and Robot

This system consists of a wheeled robot and a hand cart (Figure 1). The robot used in the present study was the wheeled "Tateyama", which was developed in our laboratory. This robot has three sets of wheels (front, middle, and rear) paired on the left and right. The front pair of wheels are casters, and the middle and rear pairs of wheels are driving wheels. The rear pair of wheels can be shifted in position, and this mechanism is used by the robot to climb steps (Figure 2, see Section 3). The hand cart and robot are deployed in a forward-and-aft configuration (Figure 3). The specifications of the robot and the hand cart are shown in Tables A1 and A2 (Appendix A).

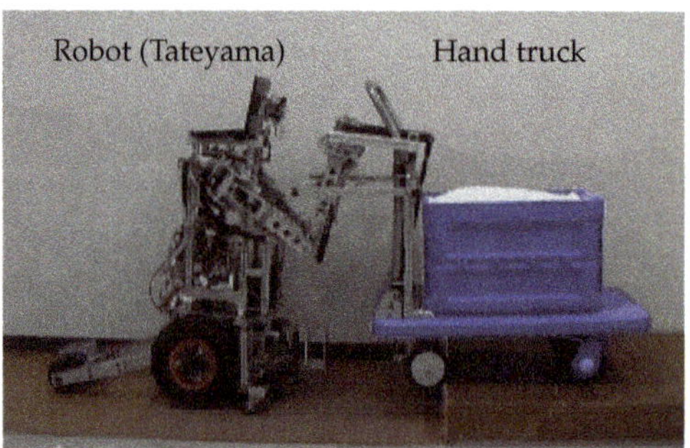

Figure 1. Photograph of the robot and the hand cart.

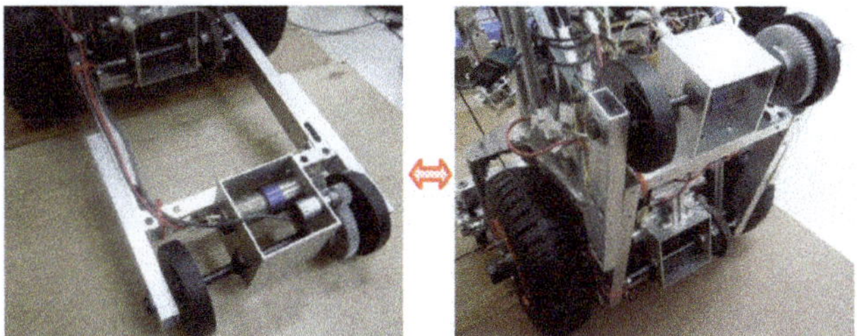

Figure 2. Robot rear wheel mechanism.

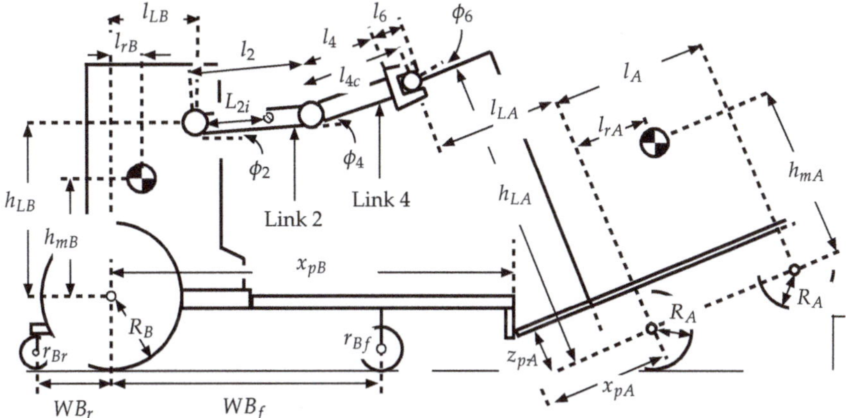

Figure 3. Model of the robot and the hand cart.

Figure 4 shows the process in which a human moves a hand cart over a step. Most people push the rear bottom of the cart with their foot when lifting the front wheels of the hand cart during step climbing. When some people push and lift the rear wheels of a heavy cart, they limit the passive rotation about the shoulder joints as the upper arms are pushed into their chest. In the present paper, this motion is performed using the front-wheel mechanism of the robot (Figures 5 and 6) and the robot stopper (Figure 7).

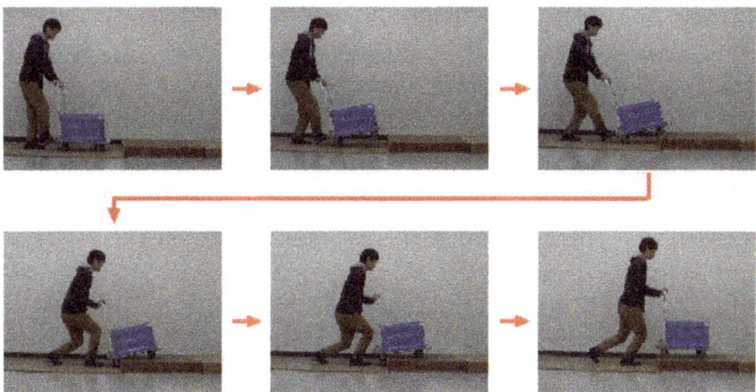

Figure 4. Human method of pushing a hand cart up a step.

Figure 5. Extendable robot front-wheel mechanism.

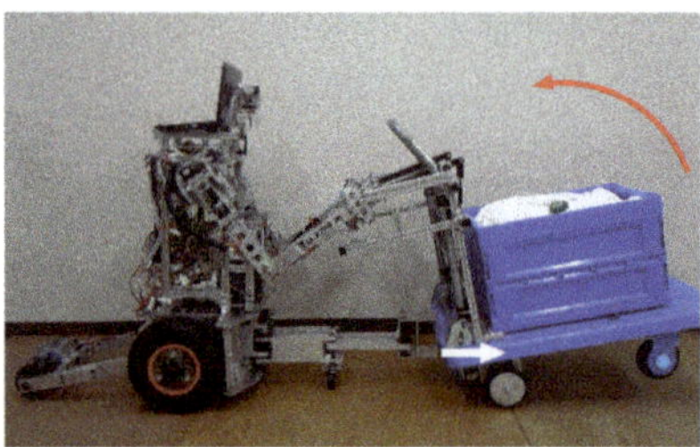

Figure 6. Robot pushing against the bottom of a hand cart using the front-wheel mechanism.

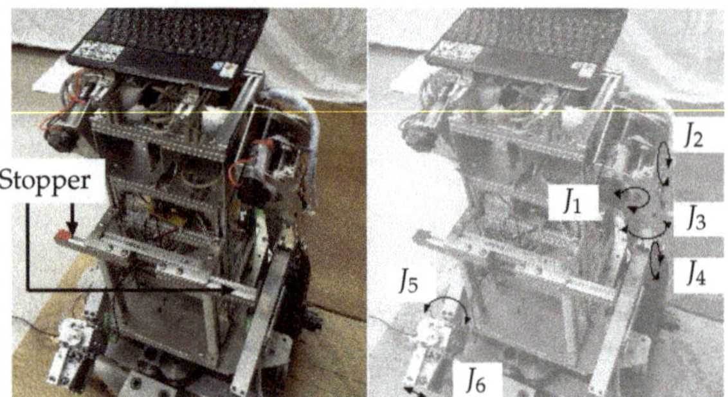

Figure 7. Stopper on the robot and manipulator joints.

The robot has manipulators attached to the left and right sides of its upper half. In addition to the five degrees of freedom (DOFs) of the arm, the hand has one DOF, for a total of six DOFs (Figure 7). The manipulator joint angles are -90 [deg] $\leq \phi_2 \leq +90$ [deg] and 0 [deg] $\leq \phi_4 \leq +100$ [deg]. The robot has a stopper mounted on the front of its body (Figure 7). First, the upper links of the manipulators are located behind the stopper when the front wheels of the cart are lifted (Figure 8). As described below (Section 3), the manipulators are pulled when lifting the front wheels of the cart so that the front-wheel mechanism of the robot can reach the rear bottom of the cart when lifting the front wheels of the cart.

The manipulators are pulled by the cart when the front wheels of the cart ascend the step, and the stopper limits the passive rotational travel around the shoulder axes of the manipulators (Figure 9, Section 3). After climbing of the front wheels of the cart, the upper links of the manipulators are located in the front of the stopper (Figure 10). When the rear wheels of the cart climb a step, the manipulators of the robot are pushed by the force from the cart. However, the stopper of the robot limits the passive rotational travel of the manipulator (Figure 10, Section 3). This enables the robot to imitate a human pushing an object by limiting the passive rotation about the shoulder joints as the upper arms are pushed into the chest (Figure 4).

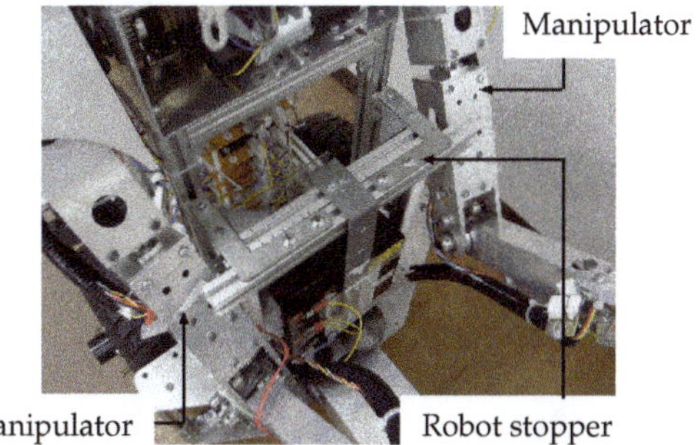

Figure 8. Locations of manipulators for lifting the front wheels of the cart.

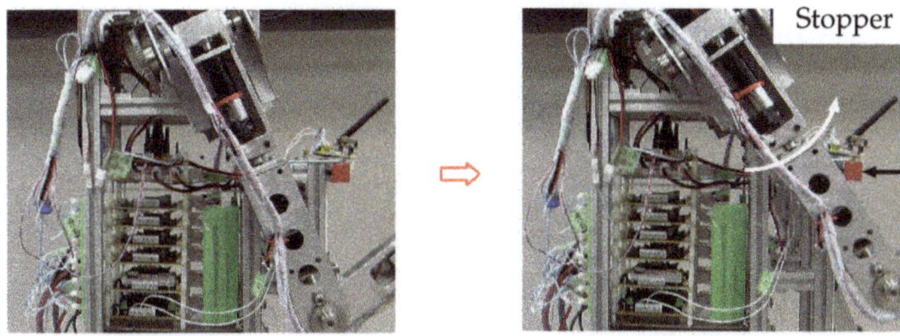

Figure 9. Limit of manipulator rotation when the robot is being pulled.

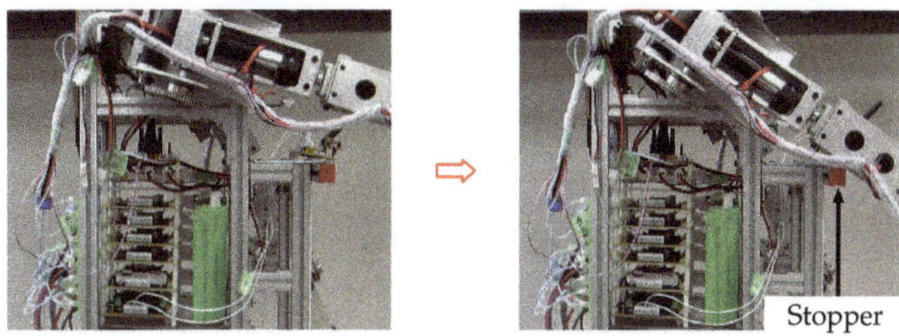

Figure 10. Limit of manipulator rotation when the robot is being pushed.

The hand cart (PLA250-DB, Kanatsu, Chiyoda-ku, Tokyo, Japan) used in the present study is a commercially available hand cart (Figure 1, Table A2). The cart has a hand brake mechanism, which was remodeled to allow grasping (Figure 11). This hand brake mechanism has two shafts to allow holding, which are used to apply or release the cart brakes. When the robot hand grasps the hand brake system, the brake system is releases and the cart is able to move. When the robot hand releases the hand brake system, the brake is applied and the cart is not able to move. The robot and the cart are connected by the robot hands throughout the step-climbing process. Releasing or applying the hand brake is controlled by the angle of the forearm links of the manipulators because the hands have spaces between fingers (Figure 11). When the angle of forearm links is small, the brake is released, and when the angle of the forearm links is large, the brake is applied.

This cart also has a stopper that is composed of front and rear bars (Figure 12), and the stopper is mounted on the rear side. The stopper is not used for cart climbing, but rather for robot climbing (Figure 13). The details of the use of the cart stopper are presented in Section 3.

Figure 14 is a system configuration diagram. The robot has a notebook PC (OS: Windows Xp, msi U100, New Taipei City, Taiwan) on its body. The motors, encoders, and touch sensors were connected to motion controllers (MCDC3006-S, MCDC3003-S, Faulhaber Co. Ltd, Baden-Wuerttemberg, Germany).

These motion controllers were connected to the notebook PC of the robot. The motors were controlled via commands issued using the Motion Manager 4 software package (Faulhaber). The robot used a camera built into the PC, and moving images from the camera and the Motion Manager 4 operating window were displayed on the notebook PC mounted on the robot. The screen of this notebook PC used Real VNC software (version 4.1.2) and was transmitted as-is over an intranet to the display of a PC used by an operator at a different location. The operator of the robot was able to control the robot by operating Motion Manager 4 from their PC. Commands for Motion Manager 4 were issued using "JoyToKey" software. The keyboard commands, which were activated by pushing the buttons of a "Joypad" (PL-USGP12, Planex Communications Inc., Shibuya-ku, Tokyo, Japan), corresponded to manipulation of the controller to operate the robot. The controller is a commercially available game pad. The operator controls the robot by watching moving images returned from the camera on the robot. In the present study, the robot was operated at a constant speed (0.76 [km/h]).

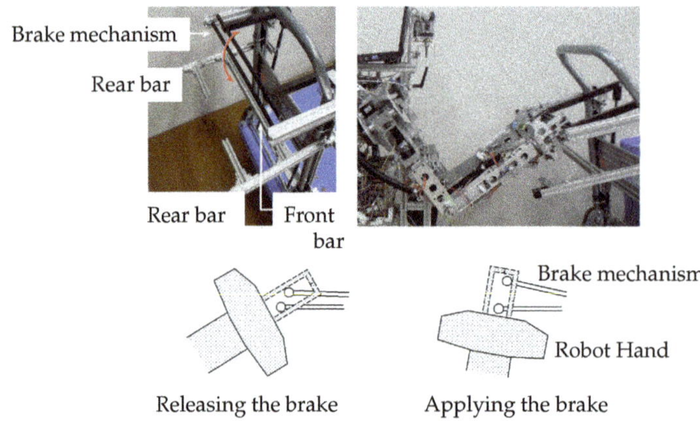

Figure 11. Brake mechanism on the hand cart.

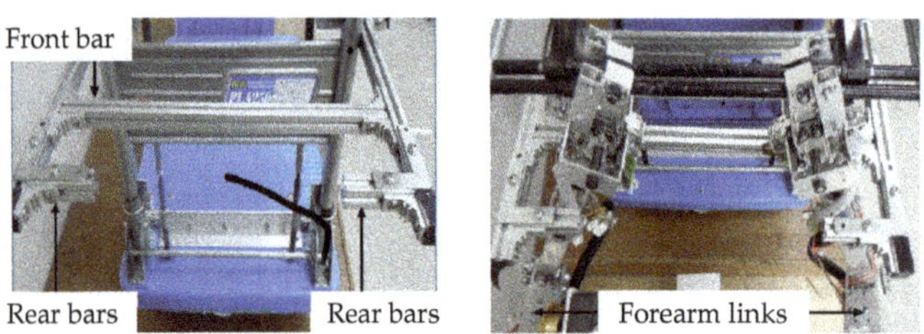

Figure 12. Stopper on the hand cart.

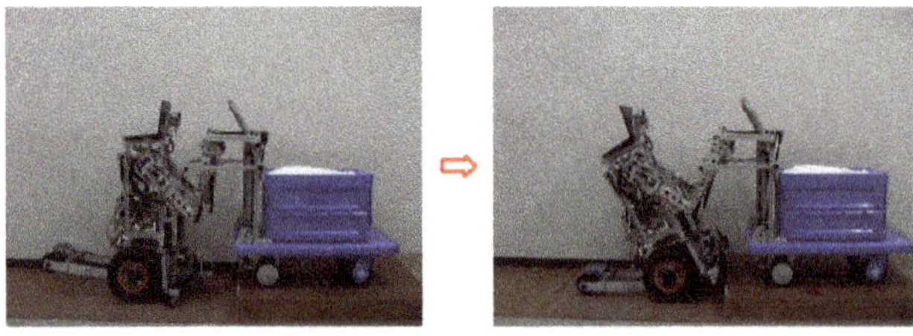

Figure 13. Robot lifting its front wheels.

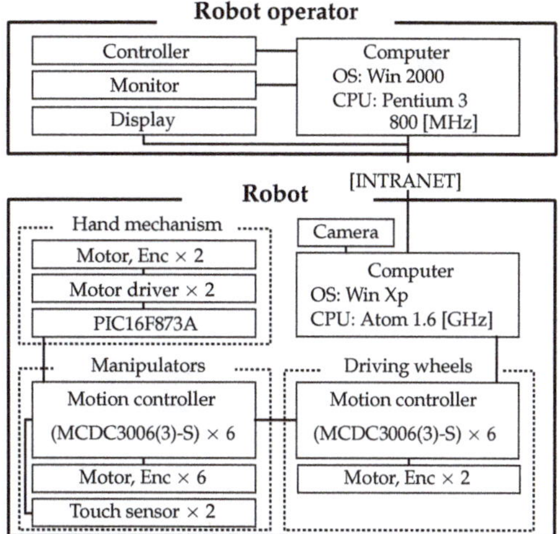

Figure 14. Block diagram of the robot system.

3. Process of Step Climbing

In the proposed step-climbing method, the hand cart first climbs a step, and then the robot climbs the step. In the present study, Stages 1 and 2, respectively, describe the processes by which the front and rear wheels of the cart ascend the step. Stage 3 describes the process by which the front wheels of the robot ascend the step. Stage 4 describes the processes in which the middle and rear wheels of the robot ascend the step. Numbers (1)–(16) in Figures 15 and 16 correspond to the states described below.

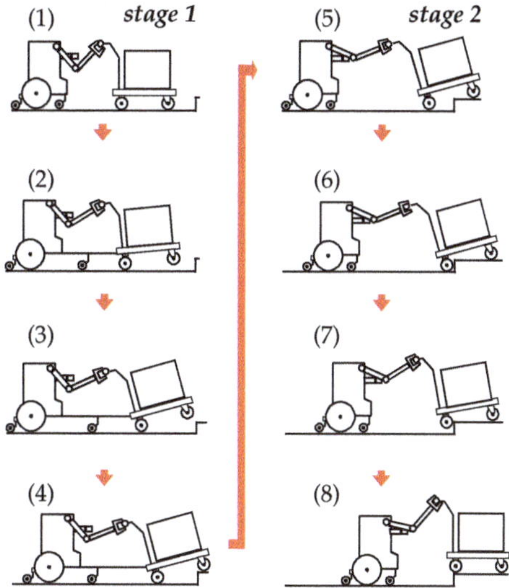

Figure 15. Step-climbing process of the hand cart.

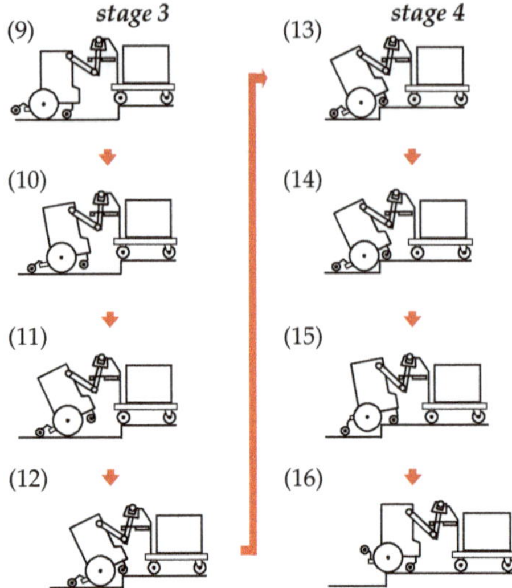

Figure 16. Step-climbing process of the robot.

[Stage 1]

(1) The upper links of manipulators are located behind the robot stopper (Figures 8 and 9). Joints 2, 4, and 6 are allowed to rotate passively until the ascent of the cart has been completed (Figure 7).

(2) The robot stops and pushes against the rear bottom part of the cart using the robot front-wheel mechanism (Figure 6). The front wheels of the cart start to be lifted.

(3) The robot continues to push the rear bottom part of the cart, and the cart tilt increases. The action for lifting the front wheels of the cart exerts forces on the manipulators, causing passive rotation about Joint 2. However, the upper arm link of the manipulator comes into contact with the robot stopper, limiting the extent of rotation (Figure 9).

(4) The robot moves forward and the front wheels of the cart are positioned over the edge of the step. As the operator closes the robot front-wheel mechanism, the incline of the cart becomes small. The front wheels of the cart are located on the upper level of the step.

[Stage 2]

(5) First, the sides of the robot manipulators are opened, and the upper links of the manipulators are located in front of the robot stopper (Figure 10). This limits the passive rotational travel of the manipulators when the robot is pushed. The robot then moves forward, pushing the cart from behind.

(6) The rear wheels of the cart come into contact with the step.

(7) The robot pushes the cart, and the rear wheels of the cart begin to climb the step. The robot imitates the operation of a human pushing an heavy object by limiting the passive rotation about the shoulder joints as the upper arms are pushed into the chest (Figure 4).

(8) Once the rear wheels of the cart have reached the upper level of the step, the robot stops.

[Stage 3]

(9) The sides of the robot are opened, and the two manipulators are inserted into the stopper of the cart (Figure 12). The rear wheels of the robot are folded upward (Figure 13). The robot moves forward and the manipulator forearm links come into contact with the stopper of the cart. The hand brake is applied. The cart maintains its position (Figure 11).

(10) The robot continues to push on the cart, and the front wheels of the robot are lifted (Figure 13).

(11) When the location of the robot center of mass shifts behind the contact point between the robot middle wheels and the ground as the robot tilt increases, the robot begins to tip over backward. However, the robot is supported by the middle and rear wheels, and the rotational travel of the manipulators are stopped by the rear bars of cart's stopper. The brake of cart is then released (Figure 11).

(12) The robot moves forward using the middle and rear wheels, and the front wheels of the robot are placed on the upper level of the step.

[Stage 4]

(13) The robot drives and both vehicles move forward.

(14) The middle wheels of the robot come into contact with the step.

(15) The middle wheels of the robot continue to drive, and both vehicles continue to move forward. The rear-wheel mechanism of the robot is lowered and pushes up the robot body. The middle wheels of the robot start to climb the step.

(16) The middle wheels of the robot are able to climb the step. After the middle wheels of the robot have reached the upper level of the step, the vehicles are stopped. The rear-wheel mechanism of the robot is then folded upward.

4. Theoretical Analysis

The proposed step-climbing method is sensitive to the relationship between the masses of the connected vehicles. In this section, we discuss the requirement for lifting the front wheels of the hand cart. The vehicles slowly climb a step and maintain their balance in the proposed method, which is analyzed by considering statics.

The basic coordinate system of the robot if denoted Σ_B, where contact point B between the robot middle (driving) wheels and the ground is the origin (Figure 17). In Stage 1, the inclination of the robot is zero ($\phi_0 = 0$). In Figure 3, the position vectors for these joints in system Σ_B are expressed as

$$^B\boldsymbol{p}_{2i} = [x_{2i}\ z_{2i}]^T \quad (i = 1 - 3), \tag{1}$$

where

$$^B\boldsymbol{p}_2 = [x_2\ z_2]^T = [l_{LB}\ R_B + h_{LB}]^T \tag{2}$$

$$^B\boldsymbol{p}_4 = [x_4\ z_4]^T = [l_{LB} + l_2 \cos\phi_2\ R_B + h_{LB} + l_2 \sin\phi_2]^T \tag{3}$$

$$^B\boldsymbol{p}_6 = [x_6\ z_6]^T = [l_{LB} + l_2 \cos\phi_2 + l_{4c} \cos(\phi_2 + \phi_4)\ R_B + h_{LB} + l_2 \sin\phi_2 + l_{4c} \sin(\phi_2 + \phi_4)]^T \tag{4}$$

In the same way, the position vectors for the contact points between the robot front and rear wheels and the ground are expressed as

$$^B\boldsymbol{p}_{fwb} = [WB_f\ 0]^T \tag{5}$$

and

$$^B\boldsymbol{p}_{rwB} = [-WB_r\ 0]^T \tag{6}$$

The position vectors for the contact points between the robot front-wheel mechanism and the rear of the hand cart are expressed as

$$^B\boldsymbol{p}_{fp} = [x_{pB}\ z_{pB}]^T. \tag{7}$$

The body of the robot, neglecting the manipulators, is Link 0 with mass m_0. If the centers of gravity of the robot body and each manipulator link (Links 2, 4, and 6) are denoted by

$$^B\boldsymbol{p}_{gj} = [x_{g2j}\ z_{g2j}]^T \quad (j = 0 - 3) \tag{8}$$

the vector of the driving force for the robot middle wheels and the resistance force from the ground surface is given by

$$f_m = [f_{1x} \ f_{2z}]^T. \tag{9}$$

In addition, the resistance force vector at the robot front wheels is

$$f_{fwB} = [0 \ f_{3z}]^T \tag{10}$$

and that at the rear wheels is

$$f_{rwB} = [f_{4x} \ f_{5z}]^T. \tag{11}$$

The reaction force vector from the linked cart is

$$f_L = [f_l \cos(\phi_2 + \phi_4) \ \ f_l \sin(\phi_2 + \phi_4)]^T \tag{12}$$

and the reaction force vector from the cart to the robot front-wheel mechanism is given by

$$f'_p = [-f_p \ 0]^T. \tag{13}$$

The coordinate system fixed at the point of contact, A, between the cart rear wheels and the ground is given by Σ_A. The incline of the cart body with respect to the road surface is $\phi_2 + \phi_4 + \phi_6 = \phi_{246}$ (in Stages 1 and 2, the robot incline is zero $\phi_0 = 0$). In Σ_A, the center of gravity of the cart is located at

$$^A p_{GA} = [x_{GA} \ z_{GA}]^T = [l_{rA} \cos \phi_{246} - h_{mA} \sin \phi_{246} \ \ \ l_{rA} \sin \phi_{246} + h_{mA} \cos \phi_{246} + R_A]^T \tag{14}$$

and the push handle location (which is held by the robot hand), P_c (P_6), is

$$^A p_c = [x_c \ z_c]^T = [-l_{LA} \cos \phi_{246} - h_{LA} \sin \phi_{246} \ \ \ - l_{LA} \sin \phi_{246} + h_{LA} \cos \phi_{246} + R_A]^T. \tag{15}$$

The contact position between the body of cart and the robot front-wheel mechanism is located at

$$^A p_{fp} = [-l_{pA} \cos \phi_{246} - h_{pA} \sin \phi_{246} \ \ \ R_A - l_{pA} \sin \phi_{246} + h_{pA} \cos \phi_{246}]. \tag{16}$$

The resistance force vector of the rear wheels of the cart from the ground surface is

$$f_{rwA} = [0 \ f_{6z}]^T. \tag{17}$$

In addition, the reaction force from the linked robot is given by

$$f'_L = [-f_l \cos(\phi_2 + \phi_4) \ \ - f_l \sin(\phi_2 + \phi_4)]^T. \tag{18}$$

The reaction force vector from the robot front-wheel mechanism to the cart is given by

$$f_p = [f_p \ 0]^T. \tag{19}$$

In moving statically, the equilibrium of the robot along the x and z axes is given by the following equation:

$$f_{\Sigma B} + M_B g = 0, \tag{20}$$

where $f_{\Sigma B} \in R^2$ is the sum of forces on the robot due to the driving force and the resistances of the ground surface and the linked cart, and is given as

$$f_{\Sigma B} = [f_{1x} + f_{4x} + f_l \cos(\phi_2 + \phi_4) - f_p \ \ \ f_{2z} + f_{3z} + f_{5z} + f_l \sin(\phi_2 + \phi_4)]^T \tag{21}$$

and

$$g = [0 \ -g]^T \tag{22}$$

is the gravitational acceleration vector.

The following equation is obtained from the equilibrium of moments about the point of contact between the robot middle wheels and the ground:

$$^B\boldsymbol{p}_{GB} \times M_B\boldsymbol{g} + {}^B\boldsymbol{p}_4 \times \boldsymbol{f}_L + {}^B\boldsymbol{p}_{fwB} \times \boldsymbol{f}_{fwB} + {}^B\boldsymbol{p}_{rwB} \times \boldsymbol{f}_{rwB} + {}^B\boldsymbol{p}_{fp} \times \boldsymbol{f}'_p = 0. \tag{23}$$

Equation (24) is obtained from Equation (23), as follows:

$$-x_{GB} \cdot M_B g + x_4 \cdot f_l \sin(\phi_2 + \phi_4) - z_4 \cdot f_l \cos(\phi_2 + \phi_4) + WB_f \cdot f_{3z} - WB_r \cdot f_{5z} + z_{pB} \cdot f_p = 0. \tag{24}$$

Summing the total forces on the hand cart exerted by the ground surface and the linked robot for $f_{\Sigma A} \in \boldsymbol{R}^2$, we find that

$$f_{\Sigma A} = [f_p - f_l \cos(\phi_2 + \phi_4) \quad f_{6z} - f_l \sin(\phi_2 + \phi_4)]^T \tag{25}$$

When the linked vehicles move together in static equilibrium, the equilibrium for both the x and z axes yields

$$f_{\Sigma A} + M_A g = 0, \tag{26}$$

whereas the equilibrium of moments about the point of contact between the rear wheels of the hand cart and the ground yields

$$^A\boldsymbol{p}_{GA} \times M_A\boldsymbol{g} + {}^A\boldsymbol{p}_c \times \boldsymbol{f}'_L + {}^A\boldsymbol{p}_{fp} \times \boldsymbol{f}_p = 0. \tag{27}$$

We obtain the following equation from Equation (26):

$$f_p = f_l \cos(\phi_2 + \phi_4). \tag{28}$$

The requirement for lifting the front wheels of the hand cart is obtained from Equation (27) as follows:

$$-x_{GA} \cdot M_A g - x_c \cdot f_l \sin(\phi_2 + \phi_4) + z_c \cdot f_l \cos(\phi_2 + \phi_4) - z_{pA} \cdot f_p > 0. \tag{29}$$

Substituting Equation (28) for Equation (29), we obtain

$$x_{GA} \cdot M_A g + f_l \cdot \{x_c \sin(\phi_2 + \phi_4) + (z_{pA} - z_c) \cos(\phi_2 + \phi_4)\} < 0, \tag{30}$$

and substituting Equation (28) for Equation (24), we obtain

$$f_l = \frac{x_{GB} \cdot M_B g - WB_f \cdot f_{3z} - WB_r \cdot f_{5z}}{x_4 \sin(\phi_2 + \phi_4) + (z_{pB} - z_4) \cos(\phi_2 + \phi_4)}. \tag{31}$$

We assume the most difficult situation is to lift the cart's front wheels, in which the robot does not use its rear wheels (Figure 17, $f_{5z} = 0$) and the robot is pushed by the reaction force from the hand cart and the normal force of the front wheels of the robot ($f_{3z} = 0$). In this case, by substituting Equation (31) for Equation (30), we obtain the requirement for lifting the front wheels of the hand cart as follows:

$$\frac{M_A}{M_B} < \frac{(z_{pA} - z_c) \cos(\phi_2 + \phi_4) + x_c \sin(\phi_2 + \phi_4)}{(z_4 - z_{pB}) \cos(\phi_2 + \phi_4) - x_4 \sin(\phi_2 + \phi_4)} \cdot \frac{x_{GB}}{x_{GA}}. \tag{32}$$

When Equation (32) is satisfied, the front wheels of the hand cart can be lifted by the robot. Tables A1 and A2 indicate that the robot is able to lift the hand cart in theory. In the next section, we present the results of the experiment.

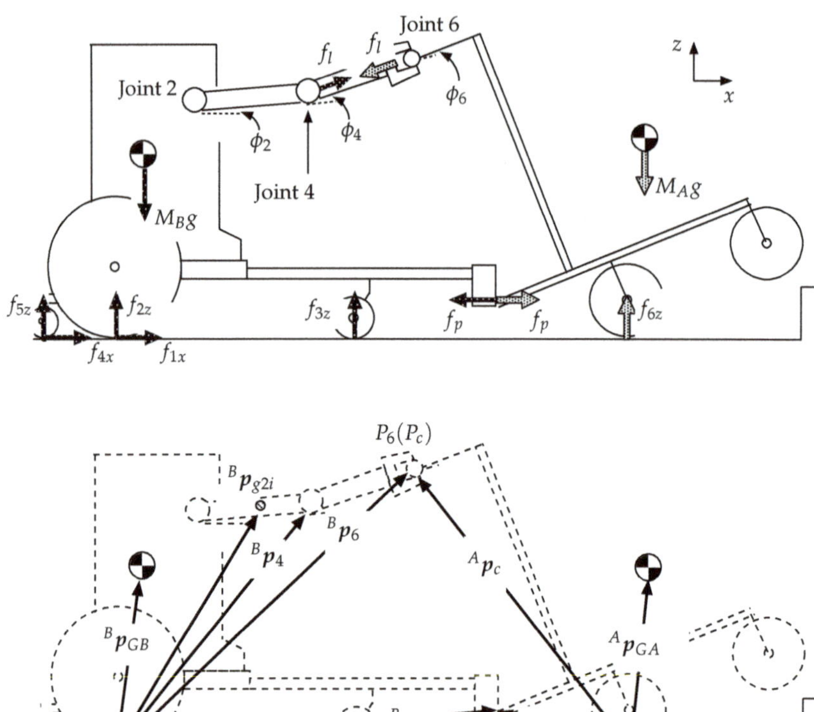

Figure 17. Model showing lifting of the front wheels of the hand cart.

5. Experiment

Experiments were carried out under an environment with a step of 120 mm in height, and the mass of the hand cart was 75 kg (body of the cart: 25 kg, baggage: 50 kg).

The speed of the robot was constant (0.76 km/h), and the friction coefficient between the tires and the road surface was $\mu = 0.72$ (Figure 18). The robot and the hand cart were located on one floor of the National Institute of Technology of Toyama College, and the operator of the robot (adult male) was located on another floor. The robot operator performed his task over an intranet while observing the video from the camera on the robot.

In Stage 1, the robot operator was able to lift the front wheels of the cart and position them on the step by watching video captured by the robot camera. After climbing of the front wheels of the hand cart, the operator was able to change the manipulators' upper link positions from behind the robot stopper to in front of the robot stopper (Section 3, Figures 9 and 10). However, the operator needed approximately 90 s to change these positions. In the present paper, the robot and the hand cart are connected, and the trajectories of these vehicles are limited. Teleoperation of the robot using a game pad controller when the positions of the manipulators change was difficult, indicating that the robot should have an autonomous system for positioning the manipulators.

In Stage 2, it was possible for the rear wheels of the hand cart to climb the step with ease.

In Stages 3 and 4, the step-climbing process of the robot was performed. Although the robot was able to climb the step, it was difficult to operate the robot while viewing video captured by the camera on the robot because the robot is inclined during Stages 3 and 4 and the operator was not able to view the state of the vehicles. Another experiment was carried out using an additional camera on the chest of the robot (same operator), and the operability of robot was improved. The climbing task was also performed successfully by another operator (adult male).

The results of these experiments clarified that the proposed step-climbing method is useful and can be improved by adding an autonomous system to change the manipulator position.

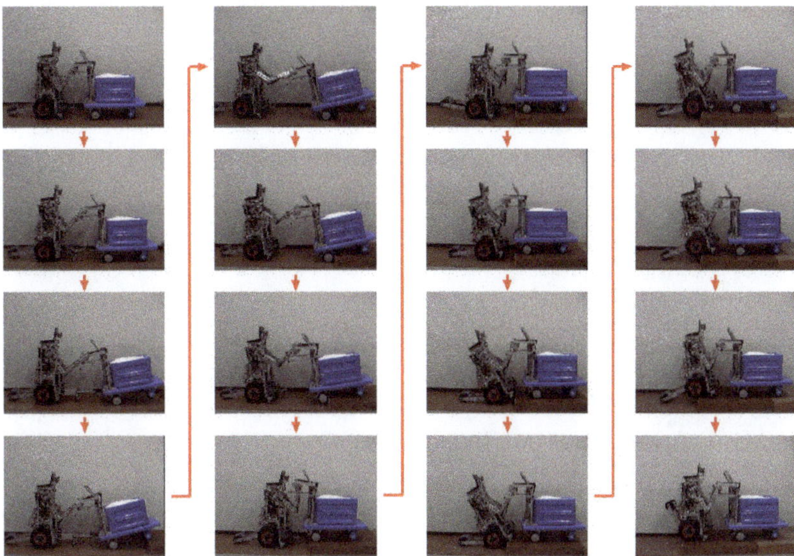

Figure 18. Step-climbing experiment using the robot and the hand cart.

6. Conclusions

The present paper describes the proposed step-climbing method of the hand cart using a partner robot that imitates human motion when operating a hand cart. The system, which consists of a mobile robot and a hand cart, was constructed. Numerical calculations clarified the requirements of masses and centers of gravity for both vehicles to lift the cart. Experiments were carried out incorporating teleoperation of the robot over an intranet, in which the robot operator controlled the robot while viewing video captured by the robot camera.

The results of the experiment are listed below.

(1) The teleoperated robot and the hand cart are able to climb a step using the proposed method.
(2) The operator requires some time (approximately 90 s) to change the position of the manipulators (after Stage 1). Both vehicles have individually driven wheels and are connected, and thus the trajectories of the vehicles are limited. Moreover, the teleoperated robot was controlled using a game pad, making operation of the robot difficult and indicating that system for changing the positions of the manipulators should be autonomous.
(3) The camera on the robot cannot capture the entire situation of the vehicles. However, the additional camera on the robot chest was able to improve operability. The climbing task was also performed successfully by another operator.

Nevertheless, a mobile robot with manipulators can make a heavy hand cart climb a step. The proposed method is relatively simple and can be applied to other partner robots, thereby improving their abilities.

In the future, we intend to improve the maneuverability of the proposed step-climbing method and evaluate the operability of the system. We are going to construct an autonomous system for positioning the manipulators.

Author Contributions: Conceptualization, H.I.; methodology, H.I.; software, T.K., R.W., K.S.; experiment; T.K., R.W., H.I., K.S. formal analysis, H.I.; investigation, H.I.; writing–original draft preparation, H.I.; writing–review and editing, H.I.; supervision, H.I., K.S.; project administration, H.I.

Funding: This research was funded by a grant from the Mazda Foundation (2011KK-240).

Acknowledgments: The writing of the present paper was made possible largely through a grant from the Mazda Foundation (2011KK-240), and we would like to acknowledge here the generosity of this organization.

Conflicts of Interest: The funder had no role in the design of the study.

Appendix A

Table A1. Specifications of the robot.

Overall length	230–800 mm
Overall height	747 mm
Radius of front wheels (r_{Bf})	25 mm
Radius of center wheels (R_B)	145 mm
Radius of rear wheels (r_{Br})	19 mm
Length of the front-wheel mechanism (l_{pB})	250–750 mm
Height of the front-wheel mechanism (h_{pB})	250–70 mm (when the incline of the cart is 0)
Wheelbase (WB_f)	190–440 mm
Wheelbase (WB_r)	270 mm
Mass position from the rear axis (l_{rB})	93 mm
Height of the mass from the rear axis (h_{mB})	286 mm
Position of Joint 2 from the rear axis (l_{LB})	90 mm
Height of Joint 2 from the rear axis (h_{LB})	532 mm
Mass of the robot's body	56.2 kg
Mass of Link 2 (from Joint 2 to Joint 4)	2.55 × 2 kg
Mass of Link 4 (form Joint 4 to the human hand)	0.8 × 2 kg
Length of Link 2 (l_2)	330 mm
Length of Link 4 (l_4)	300 mm
Length of the hand (l_6)	105 mm
Length from Joint 4 to the connecting position (l_{4c})	370 mm
Mass position of Link 2 (L_2)	67 mm
Mass position of Link 4 (L_4)	169 mm
Mass position of Link 6 (L_6)	35 mm

Table A2. Specifications of the hand cart.

Overall length	1020 mm
Overall height	900 mm
Radius of the front and rear wheels (R_A)	65 mm
Wheelbase (l_A)	470 mm
Connecting height (h_{LA})	695 mm
Connecting position (l_{LA})	270 mm
Stopper position (front bar) from the axis of the rear wheels	210 mm
Stopper position (rear bars) from the axis of the rear wheels	360 mm
Stopper Height from the axis of the rear wheels	500 mm
Length of contact position (l_{pA})	160 mm
Height of contact position (h_{pA})	150 mm
Mass position (l_{rA})	235 mm
Mass height (h_{mA})	350 mm
Mass (M_A)	75 kg

References

1. Onozato, T.; Tamura, H.; Kambayashi, Y.; Katayama, S. A Control System for the Robot Shopping Cart. In Proceedings of the 2010 ERAST International Congress on Computer Applications and Computational Science, Singapore, 4–6 December 2010; pp. 907–910.

2. Sales, J.; Mart, J.V.; Marn, R.; Cervera, E.; Sanz, P.J. CompaRob: The Shopping Cart Assistance Robot. *Int. J. Distrib. Sens. Netw.* **2015**, 1–15. [CrossRef]

3. Lee, H.; Lee, G.; Kwon, C.; Noguchi, N.; Chong, N.Y. Swithched Observer Based Impedance Control for an Assistive Robotic Cart unbder Unknown Parameters. In Proceedings of the IEEE RO-MAN: The 21st IEEE International Symposium on Robot and Human Interactive Communication, Paris, France, 9–13 September 2012; pp. 101–106.

4. Takahashi, T.; Suzuki, T.; Shitamoto, H.; Moriguchi, M.; Yoshida, K. Developing a mobile robot for transport applications in the hospital domain. *Robot. Auton. Syst.* **2010**, *58*, 889–899. [CrossRef]

5. Scholz, J.; Chitta, S.; Marthi, B.; Likhachev, M. Cart pushing with a mobile manipulation system: Towards navigation with moveable objects. In Proceedings of the 2011 IEEE International Conference on Robotics and Automation, Shanghai, China, 9–13 May 2011; pp. 6115–6120.

6. Iwamura, Y.; Shiomi, M.; Kanda, T.; Ishiguro, H.; Hagita, N. Do elderly people prefer a conversational humanoid as a shopping assistant partner in supermarkets? In Proceedings of the 6th International Conference on Human-Robot Interaction, Lausanne, Switzerland, 6–9 March 2011; pp. 449–456.

7. Ohno, N.; Hasegawa, T. Controller of Mobile Robot. U.S. Patent US8340823 B2, 25 December 2012.

8. Nozawa, S.; Maki, T.; Kojima, M.; Kanzaki, S.; Okada, K.; Inaba, M. Wheelchair support by a humanoid through integrating environment recognition, whole-body control and human-interface behind the user. In Proceedings of the 2008 IEEE/RSJ International Conference on Intelligent Robots and Systems, Nice, France, 22–26 Septmber 2008; pp. 1558–1563.

9. Nakajima, S.; Nakano, S.; Takahashi, T. Free gait algorithm with two returning legs of a leg-wheel robot. *J. Robot. Mechatron.* **2008**, *20*, 661–668. [CrossRef]

10. Kumar, V.; Krovi, V. Optimal Traction Control In A Wheelchair With legs And Wheels. In Proceedings of the 4th National Applied Mechanisms and Robotics Conference, Cincinnati, OH, USA, 10–13 December 1995.

11. Gonzalez, A.; Ottaviano, E.; Ceccarelli, M. On the Kinematic Functionality of s Four-bar Based Mechanism for Guiding Wheels in Climbing Steps and Obstacles. *Mech. Mach. Theory* **2009**, *44*, 1507–1523. [CrossRef]

12. Independence Technology, L.L.C., iBOT. Available online: http://www.hizook.com/blog/2009/02/11/ibot-discontinued-unfortunate-disabled-perhaps-budding-robotics-opportunity (accessed on 11 February 2009).

13. Asama, H.; Sato, M.; Goto, N.; Kaetsu, H.; Matsumoto, A.; Endo, I. Mutual transportation of cooperative mobile robots using forklift mechanisms. In Proceedings of the 1996 IEEE International Conference on Robotics and Automation, Minneapolis, MN, USA, 22–28 April 1996.

14. Ikeda, H. Step climbing strategy for a wheelchair. In *Advances in Intelligent Systems: Reviews*; Reviews Book Series; Yurish, S., Ed.; IFSA Publishing, S.L.: Barcelona, Spain, 2017; Volume 1, Chapter 10, pp. 249–288.

15. Ikeda, H.; Hashimoto, K.; Murayama, D.; Yamazaki, R.; Nakano, E. Collision avoidance between a wheelchair front wheels and a step wall during step climbing using a care robot. *Adv. Sci. Technol. Eng. Syst. J.* **2017**, *2*, 732–740. [CrossRef]

16. Ikeda, H. Establishment of step climbing technology for a heavy carrier using a robot which is driven by low power motors. *Rep. Matsuda Found.* **2014**, *26*, 87–94. (In Japanese)

 *applied sciences*

Article

Effective Behavioural Dynamic Coupling through Echo State Network

Christos Melidis [1,*] and Davide Marocco [2]

[1] School of Computing, Electronics and Mathematics, Plymouth University, Drake Circus, Plymouth PL4 8AA, UK

[2] Dipartimento di Studi Umanistici, University of Naples "Federico II", via Porta di Massa 1, 80138 Naples, Italy; davide.marocco@unina.it

* Correspondence: christos.melidis@plymouth.ac.uk

Received: 13 December 2018; Accepted: 6 March 2019; Published: 28 March 2019

Abstract: This work presents a novel approach and paradigm for the coupling of human and robot dynamics with respect to control. We present an adaptive system based on Reservoir Computing and Recurrent Neural Networks able to couple control signals and robotic behaviours. A supervised method is utilised for the training of the network together with an unsupervised method for the adaptation of the reservoir. The proposed method is tested and analysed using a public dataset, a set of dynamic gestures and a group of users under a scenario of robot navigation. First, the architecture is benchmarked and placed among the state of the art. Second, based on our dataset we provide an analysis for key properties of the architecture. We test and provide analysis on the variability of the lengths of the trained patterns, propagation of geometrical properties of the input signal, handling of transitions by the architecture and recognition of partial input signals. Based on the user testing scenarios, we test how the architecture responds to real scenarios and users. In conclusion, the synergistic approach that we follow shows a way forward towards human in-the-loop systems and the evidence provided establish its competitiveness with available methods, while the key properties analysed the merits of the approach to the commonly used ones. Finally, reflective remarks on the applicability and usage in other fields are discussed.

Keywords: dynamic neural networks; mobile robot navigation; gesture recognition; behaviour dynamics; real-time action recognition

---

1. Introduction

Coupling the dynamics of humans' movements and the dynamics of a machine in order to control and direct the machine dynamics is a complex task. Mapping signals from one to the other in a continuous manner, in such a way that human users find intuitive and capable of expressing their own wishes and intentions, implies both detection and recognition of the input signals, as well as the full exploitation of their temporal aspects. Moreover, such detection and classification of sequences should be performed on the fly, in order to make the user in full control of the machine, therefore, a computational system that performs both tasks in real-time is of crucial importance in the field of human machine interaction.

Whether the machine is a computer, a robot, or an integrated system in which both human's and machine's autonomy are involved (i.e., a self-driving car), being able to provide a direct and natural way of interaction between the human to the machines which are in control can ease the usage of such systems, and also bring them 'closer' to the operator. 'Closer' in the sense that users do not perceive the machine as an external entity, but a continuation and expansion of their own body. At the same time, the emergence of adaptive computational techniques allows for systems that seamlessly adapt to user preferences. Indeed, being able to connect humans and machines in such a way that the machine adapts to the user willing and intentions, rather than forcing the user to learn how to use and forcefully direct a given machine, has the potential to produce an easier, more comfortable and, above all, "natural" usage of the system [1,2]. Therefore, the approach towards a system that can adapt to the users, in order to detect and classify their actions, has an unquestioned importance in the advancement of action recognition systems, and ultimately in human-machine and human-robot interaction.

Adaptation towards the user is important [3], as it allows for personalised patterns of communication between the user and the machine. Indeed, it is shown to improve the user experience, personalised controls can also enhance the usability of the system itself, making its usage easier and more intuitive [3]. At the same time, to provide a natural way of communication between the user and the machine, the system must be able to recognise a specific sequence in a timely manner from a stream of data, effectively placing the human user in the interaction loop. The latter enables coupling the user with the robot (i.e., performing regression), rather than working under the more typical paradigm of classification.

Imagine the action of driving a car: Not only a consistency and accuracy in recognising the driving commands is needed from the on-board mechanics and electronics of the car, but also the driver's ability to perform adjustments on the steering wheel—the input for the car control system—based on the car's behaviour is equally important. Likewise, when a human and a robot are coupled in their actions and behaviour, the extent at which the user performs an input command depends on how the robot implements the corresponding behaviour. Modulating the behaviour of the robot requires that the corresponding user behaviour is effectively recognised and propagated to the control system of the robot. In the particular case of continuous interaction, being able to inform the robot of the input magnitude or intensity is also fundamental. Thus, having user and robot behaviours coupled requires the following characteristics: (i) partial input observations to yield partial output results; (ii) the input signal's intensity to be propagated to the output; and (iii) smooth transitions in the recognition of different input signals.

Another important aspect of such interaction is time. Specifically, the time required for the computations of the recognition model and for handling the dynamics of the input signals. In this context, three are the main aspects that require attention: (i) the recognition model should be able to accommodate input patterns of different lengths; (ii) it should be trained and able to adapt to different users needs and preferences in a short time, so that the user does not disengage; (iii) the recognition should be implemented with a low complexity of computation [3].

Adaptive methodologies that present useful features like the ones above have only started to appear, most of them working under a classification paradigm [4,5]. In this context, the challenges presented are mainly two: (a) detecting that a sequence is actually present in the data stream received from the input and (b) correctly classifying it. Most research features these two aspects with independent mechanisms [6–10], however, having a unified mechanism for the two tasks, saves computational resources overall and, at the same time, the recognition process becomes faster.

Moreover, the task of dynamic sequence recognition becomes especially complicated when working with real and continuous streams of data and the complexity increases when the sequences have different lengths. Methods used for the classification span from distance measures (e.g., Dynamic Time Warping) [11,12] and statistical models (e.g., Hidden Markov Models) [13,14], to artificial neural architectures (e.g., Recurrent Neural Networks) [15–19] and hybrid solutions [20]. These methods vary in complexity and adaptability, with Recurrent Neural Networks being one of the most promising direction in the field [21]. Adaptation of RNNs though, is known to have high computational complexity. In addition, the training procedure is shown to have difficulties in finding good solutions, usually referred to as a gradient vanish problem [22–24].

Given the inner complexity of the recognition task itself, working in real world environments is particularly difficult and demanding for adaptive models. Performance degrades rapidly when working directly with noisy user data taken from real input devices, making most methods not applicable in real world situations. Cleaning and pre-processing input data, as it is often required for model to work, is not a viable option when the fundamental demand is for a method that should be readily available to the user and work reliably in real-time. The task becomes even more difficult when the input is sampled in real time and is treated continuously. Not having the ability to segment the input data, i.e., not having a starting and stopping point, makes the usage of recurrent methods necessary, as they can integrate the signal continuously in time. On the other hand, training such models requires clean data to perform well, making them difficult to train with data obtained from real users. A potential solution in this case is a computational model that is able to capture the internal dynamics of a behaviour, such as an action performed by the user on a given input device, and thus provide a robust recognition [3].

A recurrent architecture that is shown to work well with noisy data under the restrictions mentioned above is the Echo State Network approach. ESNs seems to perform surprisingly well with noisy data directly taken from a user actions and can also adapt rapidly, making their usage for user oriented systems particularly appealing [19,25–29]. In the present paper, since we are interested in behaviour recognition, data comes directly from the user manipulations of an input device. Data can be noisy and the user repetition is not always perfect, resulting to training sets of data with high degrees of noise and variation between samples (e.g., gestures, behaviours). The ESN approach followed here provides a stable and robust mapping of the input commands for user behaviour recognition.

For the investigation and validation of the method, we have followed a methodology that encompasses three stages. Firstly, we establish the validity of the proposed setup and neural architecture by benchmarking and reporting its accuracy on the recognition of actions obtained from a publicly available dataset. This allows to compare the proposed neural architecture against alternative state-of-the-art methods and also against baseline methods. Secondly, we investigate the properties of the architecture on a dataset of sequences created in house with actions recorded by the experimenter with a Leap Motion device and made on purpose to better resemble the real ones that might be obtained by casual users. The intention is to get more detailed information about the property of the system on realistic sequences before exposing it to real users. Finally, we perform a user testing of the system on a small group of people, asking them to control a simulated robot. Characteristics of the neural architecture employed, as well as methodology and results of such investigations are described in the following sections.

2. Material and Methods

2.1. Echo State Network

Echo State Networks (ESN), as seen in Figure 1, provide an architecture for efficient training of Recurrent Neural Networks (RNN) in a supervised manner [30,31]. One can distinguish two main components in an ESN. Firstly, the Dynamic Reservoir (DR), a large, random, recurrent neural network with fixed weights. These weights get initialised once and are not adapted through the training procedure. The DR is activated by the input and the feedback from the output providing a non-linear response to the input signal. The neurons of the DR usually have sigmoidal activation functions, with hyperbolic tangents to be the prevailing choice. The second part of the ESN is the output, resulting from a linear combination of the reservoir's activations. Only these weights connecting the reservoir with the output are adapted through the training procedure.

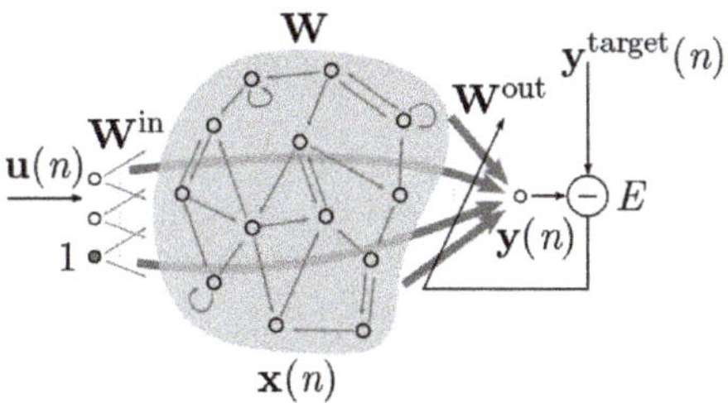

Figure 1. The Echo State Network architecture [32].

For an ESN to function properly, the *echo state property* (ESP) is essential. ESP states that the dynamics of the DR will asymptotically washout, from the initial conditions. It has been observed, that this can be achieved by scaling the *spectral radius* of the DR weights W to be less than unity [33]. That is the largest eigenvalue of the weight matrix for the DR weights should be less that unity. This condition states that the dynamics of the ESN is uniquely controlled by the input, and the effect of the initial states vanishes.

The setting of spectral radius is also associated with the *memory* of the DR [33,34]. That is the time steps it takes for the dynamics of the reservoir to washout and thus the past time steps for which information is incorporated to produce the output.

Echo State Network's Dynamics Formalisation

Assuming an ESN consisting of N units in the DR, K input units and L output units. A matrix W_{in} of size $[K \times N]$ connecting the input to the DR, a matrix W of size $[N \times N]$ describing the connections amongst the DR units and a matrix W_{out} of size $[N \times L]$ connecting the DR to the output, and finally a matrix W_{outFb} of size $[L \times N]$ connecting the output to the DR establishing the feedback connections from the output to the DR.

Assuming time n, the input signal driving the reservoir is $u(n) = [u_1(n) \cdots u_K(n)]$, the state of the DR neurons is $x(n) = [x_1(n) \cdots x_N(n)]$ and the output signal is $y(n) = [y_1(n) \cdots y_L(n)]$. The state of the reservoir is updated according to

$$x(n+1) = (1-\alpha)x(n)+$$
$$\alpha f(\mathbf{W}x(n) + \mathbf{W}_{in}\mathbf{u}(n+1) + \mathbf{W}_{outFb}\mathbf{y}(n)) \tag{1}$$

where f is a sigmoid function usually the logistic sigmoid or the *tanh* function, in our case selected to be a hyperbolic tangent. The parameter α (referred to as leaking rate) regulates the percentage of the contribution of the state's previous time step to the current one. For small α continuous time dynamics can be approximated [35]. Setting the leaking rate to small values forces the reservoir's dynamics to a slower adaptation, in cases increasing the short term memory of the reservoir. Generally the α parameter can be understood as the speed of the reservoir's update dynamics discretised in time. Thus, it provides an approximation of the time interval between to consecutive -discrete- samples in the continuous -real- world. In our case the leaking rate allows for and explicit control over the memory of the reservoir, by effectively re-sampling part of the reservoir's state x every time step.

The extended system state $\mathbf{z}(n+1) = [\mathbf{x}(n+1); \mathbf{u}(n+1)]$ at time n is the concatenation of the reservoir and input states. The extended system state, depending on the particulars of the implementation can also include the output of the reservoir, if the output connections of the reservoir are recurrent. Here there is no recurrency in the output and thus the extended system state is as shown above.

The output signal is obtained from the network, given the extended system state by,

$$y(n+1) = g(\mathbf{W}_{out}\mathbf{z}(n+1)), \tag{2}$$

where g is an output activation function typically the identity or a sigmoid, in our case the identity.

2.2. Training Procedure

During training the only weights adapted are the ones connecting the DR to the output, W_{out}. Let us assume a driving signal $u = [u(1), \ldots, u(n_{max})]$ and a desired output signal $d = [d(1), \ldots, d(n_{max})]$. The training procedure of the ESN involves two stages: (a) sampling and (b) weight computation.

Sampling

In this stage the output is 'written' in output units, a procedure referred to as teacher forcing, and the input is provided through the input units. The network is initialised using a zero initial state x.

The network is driven by the input and output signals for n times $n = 0, \cdots n_{max}$, at each time step having as input $u(n)$ and teacher signal $d(n-1)$, this since there exists the feedback from the output. For the first time step where d does not exist, it is set to zero.

For each time step, after the washout period, the extended system states $z(n)$ and the teacher signal $d(n)$ are collected. The washout period includes those time steps just after the presentation of an input signal to the network where the systems extended states are discarded and not used in the training. This is to wait for the network to settle and the internal dynamics to stabilise and the network to settle to the input provided.

The extended states are collected in a matrix S of size $[n_{max} \times (N+K)]$ and the desired outputs $d(n)$ in a matrix D of size $[n_{max} \times L]$.

Now, the desired output weights W^{out} can be calculated as follows. First, the correlation matrix of the extended system states is calculated, $R = S'S$. Then, the cross-correlation matrix of the extended states against the desired outputs d, $P = S'D$. Finally, the calculation of the output weights of the network W_{out} is done by calculating the pseudoinverse of S, $S^\dagger$,

$$W^{out} = (S^\dagger D)'$$

(3)

2.3. Intrinsic Plasticity

Selecting the spectral radius of the reservoirs weight matrix is one of the most important parameters while using dynamic reservoirs. Intrinsic Plasticity (IP) provides an unsupervised method for the adaptation of the Dynamical Reservoir [36,37]. The idea is that the activation functions of the neurons are adapted to fire under a certain, usually exponential, distribution. This results to sparse activations of the reservoir neurons, with each one capturing only important features of the input signal. The IP rule is local in space and time and aims at maximizing input to output information transmission for each neuron.

In our case, where the training data is noisy, IP is show to alleviate the overall performance helping in the decorrelation of the noisy input signals in the training procedure.

Using a hyperbolic tangent as an activation function for the reservoir's neurons, the intrinsic parametrisation can be derived by adding a gain a and a bias b, to the activation function $f'(x) = f(ax + b)$ and now working with f' as the activation function of the reservoir's neurons. Then, the online adaptation rule of IP according to [38] is derived to be,

$$\Delta b = -\zeta(-\mu\sigma^{-2} + y\sigma^{-2}(2\sigma^2 + 1 - y^2 + \mu y))$$

(4)

$$\Delta a = \zeta a^{-1} + x\Delta b$$

(5)

where ζ is the learning rate for the IP, μ the mean of desired activation distribution and σ^2 it's variance. All signals, x, y and parameters a, b are of the same time step n.

2.4. Parametrisation of the System

The matrices W_{in} and W_{outFb} have 10% connectivity and are initialised in ranges $[-0.9, 0.9]$ and $[-10^{-4}, 10^{-4}]$ respectively. The DR matrix W has a 20% connectivity and is adapted through the IP rule, needing no explicit *spectral radius* setting. The parameter α in state calculation is set to 0.5 for both training and usage of the network. The size of the DR was chosen to be $N = 128$.

For the IP learning rule, the learning rate is $\zeta = 0.0001$, the mean $\mu = 0.0$ and the variance $\sigma^2 = 0.8$. The gain parameter a is initialised to unity, while the bias parameter b to zero.

For the adaptation of the ESN the training sequence is presented to the network and the IP rule is applied according to Equations (4) and (5). Then the sequence is presented once more and the collection matrices S and D are created and the output weights W_{out} are calculated as described above in Equation (3).

The optimal configuration was achieved by repeated experiments, although there exist methods for automatic or semi-automatic fixing of the parameters. The autonomous adaptation of the reservoir through the IP rule allows for a variability in the setting of the parameters, since the reservoir neurons are adjusted to have a maximal information transfer for the given input signal.

3. Experimental Setup

3.1. Technical Details

For the testing of the system a Leap Motion sensor was used, as seen in Figure 2. The system is initialised as described above for the input device. The ESN architecture described above was coded in Python using Theano [39]. A client-server model was implemented to provide the connection between the input device and the learning algorithm (i.e., ESN).

Figure 2. The Leap Motion input device used in the experimental setup.

3.2. Input Signal

The Leap Motion device is a sensory device providing tracking and skeleton data for hand and fingers positions in space. Using the *JavaScript* library provided by the manufacturer, and the client-server setup described above, we recorded six, 6, values to describe the hand position at each frame. The values recorded represent the 3 rotational and 3 translational DoF of the centre of the palm of the hand. The setup allows us to stream the input signal through the network at the sampling rate of the device i.e., >30 fps. A frame rate as high as 100 fps was able to be produced, just for testing purposes of the setup.

There is no sub-sampling performed, nor for the training set acquisition nor for the testing phase. The device is sampled at each time step, and the sample is directed to the server side where it is fed to the ESN. The ESN provides an output for each time step, recorded and used for the analysis of the performance of the system in the results section.

3.3. Testing Cases

For the testing of the proposed system we have worked as follows:

- We tested the validity of the proposed method on a publicly available dataset, the Cornell Activity Dataset (CAD-120) [40]. This, in order to assess the quality of the work presented and to provide evidence of the generalisation capability and flexibility of the proposed method. Although CAD is rather distant to the application field of the proposed method, it allows for the comparison of our method against a baseline, while also shows the applicability of the setup regardless the type of input signal. The results of our method are reported and compared to alternative state-of-the-art computational methods.
- In order to investigate the system in more detail, an extra set of sequences was recorded by the experimenter using the Leap Motion as input device. This has produced a new dataset on which the proposed system has been further tested, labelled in this work as (*Dataset Testing*). I this way, we are able to test the system with input sequences more applicable to our specific interest of robot control and also highlight general characteristics of the system.
- A small number of users were asked to control a simulated robot, visible on the screen of a computer, using the proposed system and the Leap Motion as input device. We refer to this test within this work as *User Testing*. In this phase of the testing, the users were asked to perform gestures using the Leap Motion device, in relation to behaviours of the simulated robot shown by the experimenter on the computer screen. It is important to mention here that the users were not instructed on the kind of gestures they should use in order to control the robot, allowing them to freely manipulate the input device at their own preferences. This resulted to different gestures being used by the users in relation to the same robot behaviour shown to them. This fact indicates the flexibility of the proposed system in personalising the control sequences and the associations between the user's gestures and the robot behaviours. Since the gestures performed by the users were different for each one of them, based on their preference, only the accuracy of the system is reported under this setup. Once the ESN was trained with the gestures performed by the users, they were asked to control the robot using their own provided gestures.

3.3.1. CAD-120 Testing

The CAD-120 is a publicly available dataset with the recording of skeleton data of 10 daily activities: *making cereal, taking medicine, stacking objects, unstacking objects, microwaving food, picking objects, cleaning objects, taking food, arranging objects, having a meal.* These activities are performed by 4 different people and each is repeated 3 or 4 times. For each person and repetition, a time series of the skeleton data is used as input for the network. Although the dataset offers a confidence value for the skeletal data at any point, all have been used regardless, since ESN are known to work well with noisy data. For the reporting of the accuracy of the method on the CAD-120 dataset, a leave-one-person-out cross-validation scheme is used as found in the literature [40,41]. Given the application domain—iterative robot control—from the dataset only the skeletal data were used as input (i.e., avoiding ground truth labels and objects in the scene), including erroneous entries. The accuracy is reported as the mean of the respective accuracies for each person. Following the accuracy reporting scheme found in literature, for each activity presented to the network the readouts were averaged for the whole length of the sequence.

3.3.2. Dataset Testing

In order to test the system in a setup more similar to its intended functionality the *Dataset Testing* was created in house by the experimenter. This dataset consists of 7 generic dynamic hand gestures performed using the Leap Motion device. In this case a different measure is used to calculate the accuracy, in order to highlight the mapping paradigm under which the method is used. We report the percentage of time the network output is indicating the correct input sequence presented, since we assume no segmentation of any sort of the input sequences, neither logical (e.g., by performing a moving average of the output), nor physical (e.g., by removing the hand from the Leap Motion's recording area before and after the performance of the gesture). The Leap Motion was selected for the testing as its larger input size is more demanding for the system. Each input gesture was repeated 3 times, the system was trained using two sequences out of the three and tested on the third, unseen, one. Thus, the accuracy is reported based on a 3-fold cross validation methodology. Each gesture of the training set includes the preparation, the nucleus, and the retraction of the gesture without tagging any of those moments [42,43]. That is, each gesture includes the positioning of the hand within the device's receptive field and its removal. There was no care whether each execution of gestures was starting from the same point, nor that it had the same time span, nor that it was performed in the same manner, so to follow the exact same shape every time (e.g., performing a clockwise rotation of the same radius for the 3 times). This has been done in order to account for spatial and time variability between the input sequences.

The actions performed by the experimenter with the right hand within the range of the Leap Motion are the following:

Push The hand moves forward from the centre of the receptive field in the horizontal plane;
Pull The hand moves backwards from the centre of the receptive field in the horizontal plane;
Swipe-right Repeated swipe movements from the centre to the right of the receptive field in the horizontal plane;
Swipe-left Repeated swipe movements from the centre to the left of the receptive field in the horizontal plane;
Clockwise Circle the hand moves repeatedly clockwise in a circle within the receptive field in the vertical plane;
Anti-Clockwise Circle the hand moves repeatedly anticlockwise in a circle within the receptive field in the vertical plane;
Up-Down the hand moves up and down within the receptive field in the vertical plane.

It is important to note that sequences varies in length, as seen in Table 1, which presents the 7 gestures recorded for the testing together with their length. Furthermore, it is also interesting to note that for the system to be able to discriminate between sequences 4, 5, and 6 it should be able to follow their ongoing dynamics. In fact, the higher and lower hand position in gesture 6 can also be found in gesture 4 and 5, as they are part of the circle described by the hand on the vertical plane. Similarly, gestures 2 and 3 share some of the hand positions with gestures 4 and 5, since the leftmost and rightmost points also belong to the circle described by the hand on the latter gestures.

Table 1. The 7 gestures recorded for testing. Their description is provided in the text. The sequence length is given in frames captured by the input device.

Description	ID	Sequence Length
Push	0	150
Pull	1	195
Swipe right	2	144
Swipe left	3	129
Clockwise Circle	4	225
Anti-Clockwise Circle	5	147
Up-Down	6	147

3.3.3. User Testing

In this final stage of the testing, eight participants were asked to perform gestures that they would deem appropriate in order to control 4 simple robotic behaviours shown to them with a simulated mobile robot. The robot selected for this test was a simple 2 D.o.F. differential drive mobile robot. That is, 2 drive wheels are mounted on a common axis and each wheel can independently be driven either forward or backward. A set of 4 behaviours were implemented on the robot: Forward, backward, clockwise rotation and anticlockwise rotation. The users were asked to perform their own set of input signals for these behaviours and then control the robot in a continuous fashion using their own generated signals. The system does not require that the users segment their input gestures, with transition between the input sequences being handled by the network's dynamics autonomously. The recording of the gestures was done in the same fashion as described in the previous section. The only relevant difference was that each gesture was performed only once for the training of the ESN. The participants were also asked, at the end of the testing, whether they realised any lags in the executions of the commands they sent to the robot.

In all cases where the system was used, each time point of a gesture performed is recorded, placed in a bucket and labelled with an index at the allowed frame rate of the Leap Motion. Once all gestures are performed, the network is trained, following the procedure described in Section 2.2. The machine used for the training and testing of the system, in both test cases, was a mid-range laptop with an Intel Core i5-3340M CPU @ 2.70GHz $\times$ 4 (2 cores, 4 threads), with 3.7GB of RAM and without the use of any GPU acceleration methods. The training procedure took less than a second <1 s in all cases, even for the larger testing set.

4. Results

Results are split in three sections for the three test cases used. First the results from the *CAD-120 Testing* are reported, followed by the *Dataset Testing* and finally the results from the *User Testing*.

4.1. CAD-120 Testing

In Table 2 the accuracy of the proposed method is reported. The skeletal recordings for each activity are used as input, while at the same time, all the available data regardless of their corresponding confidence value are used. The confidence value is available in the dataset and reports whether the given skeletal pose at a given frame is valid or not. For comparison the best results found in bibliography are reported for which the same input was used, that is, skeletal data without ground truth labels, or objects in the scene.

Table 2. Accuracy of the method on the Cornell Activity Dataset (CAD-120). * Koppula et.al reports on results with information about the objects in a scene.

Method	Accuracy (%)
[44]	70.2
[40] *	75.0
ESN	73.5

Although not many research reports classification results excluding objects in the scene, we can observe that the method presented here is able to achieve comparable performance.

At the same time, results show how the ESN architecture is able to handle the vastly different lengths of the recorded activities in the CAD-120, which varies from 150 to 900 frames.

In addition, the unsupervised adaptation of the reservoir through the IP rule, allows for a parametrisation of the network specific to the input sequences. Indeed, we observe that, because of the IP rule, a much smaller reservoir of only 128 neurons can be used, compared to the 300 units reservoir reported in [41]. We believe this is possible thanks to the IP rule, by which the activation function for each neuron is adjusted to maximise the information transfer. Furthermore, locality in time and space makes the adaptation computationally efficient [37].

4.2. Dataset Testing

After running the ESN according to the training procedure described, the system was always able in every case to converge and to find the right set of output weights for the task. Once the system is trained, sequences are then presented in random order to test the accuracy of the training. The network provides a response (output) for each time step an input is provided. Comparing the output with the gesture performed, we measured an accuracy of 87.8% for all the gestures performed. That is, 87.8% of the time steps an output was generated, it was indicating the correct gesture. It is worth to note here that this measure cannot reach 100% accuracy, since the ESN needs some time to stabilise its output for the input signal. For a more stable measure, the output of the network should have been segmented and observed only after the stabilisation. However, since we do not want to use any arbitrary set of parameters to judge the stabilisation point, we proceed with this holistic measure in the reporting of the results. Comparable performances, using less gestures, have been reported by Weber [45]. It is to be noted, however, that in Weber's work gestures have a starting and ending points, that we have not included in our work, to avoid any arbitrary interpretation of the gestures.

In Table 3 a more detailed representation of individual results obtained for each sequence are presented. During testing, each gesture is recognised during the exhibition. What we present in table is the average of correct recognitions for all time steps each pattern is presented to the network. Since the system is meant to provide a continuous output for every point of the sequence provided in input, we measure the percentage of correct recognition in time, as the input sequence is presented to the ESN. It can be seen as a measure of the correct mapping between input and output in time.

Table 3. Training and testing accuracy scores for the 7 gestures. Score is measured as the mean of recognised time steps for each gesture.

Description	Training	Testing
Push	0.99	0.99
Pull	0.99	0.99
Swipe right	0.99	0.98
Swipe left	0.99	0.72
Clockwise Circle	0.99	0.89
Anti-Clockwise Circle	0.99	0.70
Up-Down	0.96	0.88
Mean	0.98	0.87

4.3. User Testing

As a final step of the testing phase, the system was finally exposed to users. That is, eight people were asked to use the system and control the simulated wheeled robot without a specific task. Their only goal was to control the robot in the way they wanted.

Results in this case are very similar to the ones observed with the *Dataset Testing* condition. The ESN was able to find the right set of output weights for all sequences provided by all height users every time. Although the input sequences recorded by the users where completely arbitrary and very different in terms of overall length and gesture patterns, the proposed architecture was able to cope with the incoming signal and mapping it to the output. Notably, the overall training performance was significantly increased with respect to the previous testing, since the network had only to distinguish between four input patterns, i.e., the four gestures associated to the four pre-coded movement of the robot.

Table 4 shows the lengths of the 4 recorded input patterns (gestures) from the participants. From the table we can observe the high variability of the gestures in term of length of the input and to also, therefore, highlight the capability of the setup to deal with different lengths. The input device was sampled at the maximum allowed frame rate (i.e., 100 fps) with the length of each sequence being the number of frames recorded from the user.

Table 4. The length of each of the 4 gestures used by each participant are presented. The table displays the variability of the length of the gestures. Fluctuations of the gesture lengths are observed between participants and between gestures. We can clearly observe that there is no particular tendency in terms of the lengths of the selected input sequences. The length is measured by the frames recorded for each sequence, with the average frame rate of the device being 100 frames per second.

Participant	G1	G2	G3	G4
P1	2613	841	975	1142
P2	210	180	192	121
P3	721	619	360	701
P4	205	409	384	602
P5	187	155	68	101
P6	207	128	203	266
P7	604	614	436	596
P8	241	521	522	492

Furthermore, in order to provide a qualitative appreciation of the variability observed between user gestures, Figure 3 shows the training set (i.e., the recorded sequences made by the Leap Motion device) of 3 users depicted within the 3 respective graphs. Each figure shows 6 lines representing the 6 D.o.F of the input device during the recording of the user. Those recorded values are then fed to the ESN as input. In each figure, the separation of the four sequences, representing the four gestures made by the user, is indicated above the graphs with the label G1, G2, G3, and G4 respectively, referring to the forward, backward, clockwise, and anticlockwise movements of the robot. By visually comparing the patterns for the three users it is possible to appreciate their differences, both *within* the same user and *between* users. As already seen in Table 4, the differences in length of the sequences are noticeable. Moreover, it is also possible to appreciate the difference in which users have decided to associate their gestures to the four robot behaviours. Some users preferred periodic movements for all their input sequences (e.g., *P6*), while others chose more stable and non-periodic movements (e.g., *P8*). At the same time, as shown in *P7*, the system was also able to handle cases were periodic and non-periodic input behaviours were mixed by the user.

After the test, each user was asked to respond to a questionnaire, in order to investigate the quality of the interaction and the feasibility of the methodology proposed. It is to be noted, in fact, that given the characteristic of the task presented the subjects, it is not possible to disentangle the input sequences performed by the users at run-time with the corresponding output and isolate the single gestures recorded during the training phase by the users itself, in order to make a comparison. This makes impossible to assess the accuracy of the network in the same way as it was done for the *Dataset Testing* condition. This is also the reason behind the creation of the *Dataset Test*, i.e., to have tangible and quantitative proof of the actual works of the ESN.

Besides the specific analysis of the responses, which is not central for this work, all of the users did not report any delay in the systems response. Seven subjects out of the eight reported that they felt in control of the robot by using the Leap Motion device. This indicates that the ESN was able to map their input signals in the corresponding robot behaviours, as they were expecting. Also, all users reported that the network's training time was short, most of them having not noticed it, and the training procedure short enough, having to perform only one repetition for each control signal.

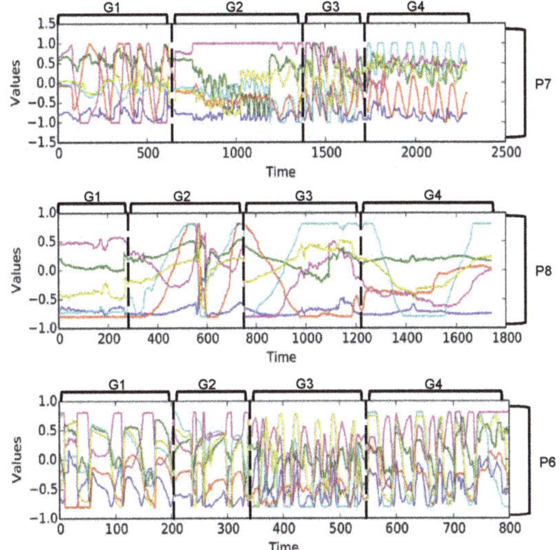

Figure 3. Three examples of four gestures, input behaviours, from participants P6, P7, and P8 used for the training of the Echo State Networks (ESN). The four different gestures are labelled with G1, G2, G3 and G4 at the top of each figure. Lines represent the 6 dimensions of the input signal of the Leap Motion device, plotted against time. It is possible to appreciate: (a) the visible differences in the quality of the input sequences and (b) the different lengths in time. That is, the different span along the x-axis.

5. Properties of the Echo-State Network and Human-Machine Interface

During the running of the ESN and the tests that have been performed, a number of observations have led to a more detailed investigation of some aspects that represents particularly interesting features for the field of human-machine and human-robot interaction. Such interesting features that the proposed ESN shows regards the way in which it solves problems concerning the variability of the length of the patterns to be classified, the complexity posed by the real-time processing of the input streams and the huge amount of noise, which is typical of the raw data that we use as input for the system.

From the same perspective, the next sections present some of the properties that we have discovered by analysing the trained ESN. Those properties, together with the above features, we believe can have an interesting impact in the way in which a system like the one presented here can shed new lights on the construction of flexible interfaces between human and machines.

5.1. Variability in Pattern Length

As mentioned in Section 4 the training patterns varied in length. This is a characteristic of all actions and behaviours performed by humans in real life and is also evident both in the CAD-120 dataset and in the dataset created by ourselves. Therefore, this is a fundamental problem that a human-machine interface has to face. Indeed, not confining the patterns to be of equal time scales, which will be artificial, allows for greater freedom for the user and enhances the robustness of a system trained on raw user data. The immediate benefit is that there is no need for explaining to the user the way in which the interface works. At the same time, the user will be free to behave in natural and intuitive way.

The high degree of recurrency within the DR allows for temporal dynamics of different time scales to be recorded and retained, without any explicit specification on the duration of the patterns. Both the number of neurons and the spectral radius of their connecting weights accounts for the memory size of the DR. In this work we have shown that it is possible to adapt those parameters in an unsupervised manner using the IP rule, allowing for the DR to adapt to the input behaviours. In this way we can obtain at the same time a general architecture capable of recognising sequences of different lengths and a specific adaptation towards a specific training set provided by the user.

5.2. Continuous Mapping from the Raw Data Input

Dynamic actions, such as hand gestures, generally contain three phases that overlaps in times: Preparation, nucleus, and retraction [42,43], of which the nucleus is the most discriminative. The setup proposed is able to capture the discriminative part of a gesture without any explicit instructions about its location within the overall gesture performance. Thanks to this feature, transitions between gestures can be handled autonomously by the ESN. In turn, user input sequences do not need to be artificially and purposefully segmented, allowing for the continuity and natural flowing of the input to be preserved in the ESN output. It is this property of the setup that allows for the user to be placed in the loop of the controlled machine, or robot in the case of this work.

5.3. Geometrical Properties of the Input

Figure 4 shows a detail of three patterns from the condition *Dataset Testing* correctly recognised by the ESN:*CW*, which stands for a clockwise circle pattern performed by the user, *ACW* an anticlockwise circle pattern, and *Up − Down*, an up and down movement of the hand of the user, as recorded by the Leap Motion device. By observing the first segment of the graph and delimited by the first vertical line in the figure around *Time*100, we can see that the network correctly recognises a *CW* gesture in input (the *Value* of *CW* in the figure reaches 1.0). Interestingly, the *ACW* gesture at the same time shows a negative value. This observation can be explained by the fact that the *ACW* pattern is 'opposite' to the *CW* patter. Therefore, it suggests that geometrical properties of the input are retained and the spatial relationship between the two signals is captured and embedded from the network in its output signals.

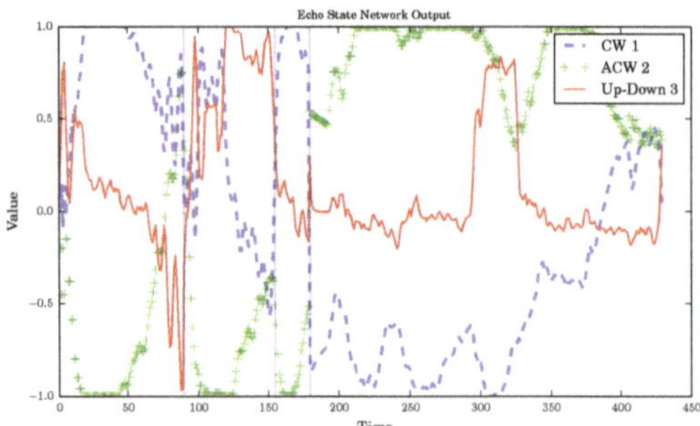

Figure 4. Usage of a trained ESN. The plot highlights how the geometrical properties of the input sequences are retained on the output of the network. The antagonistic behaviour between *Clockwise* (CW) and *Anticlockwise* (ACW) behaviours is shown, while the *Up-Down* motion recognition remains unaffected. In the graph the output of the network is depicted, with each colour and line style representing a pattern recognised in the input. All values are plotted against time.

Similarly, in the next section the input behaviour changes from *CW* to a mixture of both *CW* and *Up − Down*, and ultimately to just *Up − Down* around *Time* 130. This transition is also reflected to the output of the network, but, besides the two behaviours being mixed, *ACW* remains always negative and opposing the values of *CW*. This observation indicates that, although it is possible to mix behaviours, it remains impossible to do so with geometrically opposite ones. This is an interesting feature that, to our knowledge, cannot be found in other models. Similar dynamics can also be observed in the following section of the pattern, where the *Up − Down* recognition settles around 0, the user is performing a *ACW* gesture and, as expected, the opposite *CW* pattern is negative.

Such observation indicates that geometrical properties of the input are propagated to the output. In our example, indeed, the clockwise and an anti-clockwise motion inhibit each other. By assuming that the network has been trained to control a moving robots, it is possible to grasp the importance of this feature. For example, lets assume the CW motion is mapped to the robot moving forward and the ACW backwards. Having opposite behaviours being interpreted as 'opposite' by the system, it provides the network with an 'insight': The user cannot perform two opposite behaviours at the same time, but it can perform the $Up - Down$ gesture in combination with any of the above. At the same time, the fact that the two behaviours are opposite is also maintained in the output. Assuming that the robot behaviours are combined in a linear fashion based on the network outputs, the recognition of 'move forwards' implicitly means for the system that 'move backwards' will hold opposite values (and negative in the specific implementation presented here).

5.4. Recognition Before the End of the Sequence

An important feature of the system presented in this work is that it provides the correct classification before the input sequence is completed. Given the feedback from the output is fed to the reservoir, the network is able to stabilise its dynamics and recognise a given pattern at an early stage of its presentation. This feature allows the system to have a fast response to the sequence in input, making it appealing for real time control cases.

Figure 5 shows an example of the recognition of the *pull* (ID 1) sequence of the dataset. Similar behaviour is also shown by the network for the other sequences as well. That is, the network is able to classify the sequences before their completion. The time steps required for the network to settle to a sequence can vary. This is expected as the sequences do not share the same length.

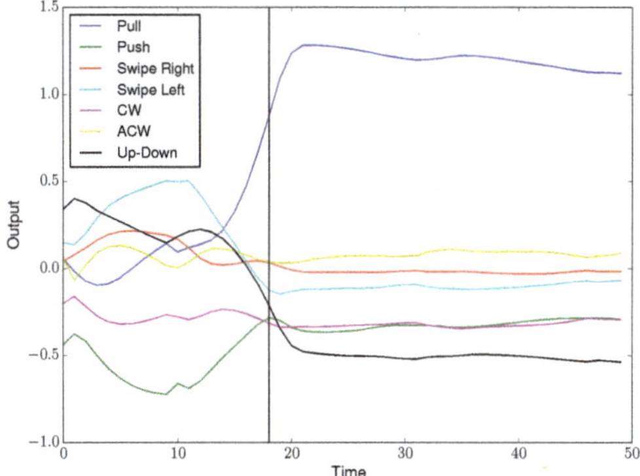

Figure 5. The figure shows the recognition of the *pull* (ID 1) input sequence from the initial presentation until the stabilisation of the network's output. In the graph the output of the network is depicted, with each colour and line style representing a pattern recognised in the input. All values are plotted against time.

For a more comprehensive way of how the proposed architecture captures the dynamics of the input sequence, we tested the recognition with partial input sequences. Each input sequence was used to artificially create four new sequences, each one having 25%, 50%, 75% and 100% of the original sequence. Each sequence resulted from the initial one having the same starting point but a shorter time span, by omitting the remaining elements of the sequence. In this way, a set of 28 sequences were used for testing. The accuracy of the network is measured in the same fashion as before, reporting the percentage of time the network output indicates the correct input. Each sequence was presented to the network independently, resetting the network in between the sequences presented. At the same time, sequences were shuffled so as to eliminate any of their dynamics to be retained in the ESN's reservoir.

Figure 6 shows the results obtained by the test. From the bar chart it is possible to observe that the network produces the correct answer even from the initial 25% of some sequences (i.e., *IDs*1 and 2). At the same time, for most sequences it reaches a good performance with only half (50%) of the sequence being presented. When the 75% of the input sequence is presented the network is able to recognise all input patterns with a high level of accuracy, with the exception of pattern 3, which is the only one that reaches its maximum recognition rate only when the entire 100% of the pattern is presented.

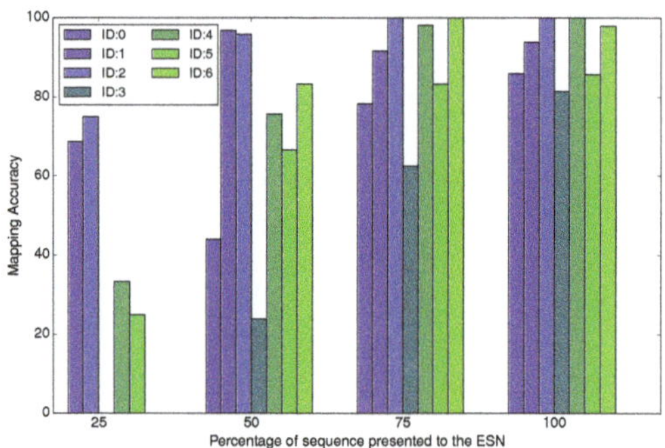

Figure 6. Accuracy of the ESN for partially observed inputs. Each colour represents an input sequence. The bars are grouped in four categories, each one representing the percentage of the signal presented to the network.

Given the unified structure for the detection and recognition provided by the ESN, input patterns are detected before their completion. This allows for fast responses from the system, a feature necessary for real time control. It is shown that humans are very sensitive to the response time of user interfaces, with lags greater than 100ms perceived as annoying [46,47]. Being able to provide feedback within the time span of a given input sequence, that is, during the execution of a gesture, is a challenge that ESN can achieve given the simplicity of the computations performed, which allows for very fast computation in comparison with other methods. Indeed, more complex classification systems perform even more costly computations with similar performances [15].

6. Conclusions

In this paper an echo-state neural architecture for the recognition of continuous time signals is presented, together with a methodology for fast and efficient training. The proposed system is tested under two different paradigms. One to analyse its properties and one to test its real world applications. Through the testing useful properties are highlighted, analysed and their potentials are discussed. Under the scope of human-robot and human-machine interaction the system's applicability is discussed. At the same time the properties of the architecture are discussed independently, in order to allow and encourage usage of the method in other fields.

The findings of this paper show that pattern recognition in continuous time signals is possible without the computational or algorithmic complexity of methods used so far in the field. The particular time signals considered here are coming from the manipulation of input devices within a human-machine interaction framework. The mapping that the proposed architecture provides was tested under a robot navigation task. In the field of robotics such an adaptive mechanism is shown to provide a just-in-time solution for a user centric system, capable of coupling the user's and robot dynamics in real time.

In the field of assistive robotics, ESN can provide a fast and reliable way of adapting the system to the users preferences. This may accommodate cases of increased of decreased mobility and the usage of unorthodox input devices. Being able to capture, train and recognise user behaviours from their preferred input method can be alleviating for use cases that cannot be taken into account in the design procedure.

Author Contributions: C.M.: Conceptualization, data curation, software and writing original draft. D.M.: Conceptualization, writing, review and editing the manuscript.

Conflicts of Interest: The authors declare no conflict of interest.

References

1. Kadous, M.W.; Sheh, R.K.M.; Sammut, C. Effective user interface design for rescue robotics. In Proceedings of the 1st ACM SIGCHI/SIGART Conference on Human-Robot Interaction, Salt Lake City, UT, USA, 2–3 March 2006; pp. 250–257.
2. Shneiderman, B. *Designing the User Interface-Strategies for Effective Human-Computer Interaction*; Pearson Education: Chennai, India, 1986.
3. Melidis, C.; Iizuka, H.; Marocco, D. Intuitive control of mobile robots: an architecture for autonomous adaptive dynamic behaviour integration. *Cognit. Process.* **2018**, *19*, 245–264. [CrossRef]
4. Yin, Y. Real-Time Continuous Gesture Recognition for Natural Multimodal Interaction. Ph.D. Thesis, Massachusetts Institute of Technology, Cambridge, MA, USA, 2014; pp. 1–8.
5. Bodiroža, S.; Stern, H.I.; Edan, Y. Dynamic gesture vocabulary design for intuitive human-robot dialog. In Proceedings of the Seventh Annual ACM/IEEE International Conference on Human-Robot Interaction—HRI '12, Boston, MA, USA, 5–8 March 2012; p. 111. [CrossRef]
6. Mitra, S.; Acharya, T. Gesture recognition: A survey. *IEEE Trans. Syst. Man Cybern. Part C* **2007**, *37*, 311–324. [CrossRef]
7. Neverova, N.; Wolf, C.; Taylor, G.W.; Nebout, F. Multi-scale deep learning for gesture detection and localization. In *Workshop at the European Conference on Computer Vision*; Springer: Cham, Switzerland, 2014; pp. 474–490.
8. Vishwakarma, D.K.; Grover, V. Hand gesture recognition in low-intensity environment using depth images. In Proceedings of the 2017 International Conference on Intelligent Sustainable Systems (ICISS), Palladam, India, 7–8 December 2017; pp. 429–433.
9. Liu, H.; Wang, L. Gesture recognition for human-robot collaboration: A review. *Int. J. Ind. Ergon.* **2018**, *68*, 355–367. [CrossRef]
10. Liarokapis, M.V.; Artemiadis, P.K.; Katsiaris, P.T.; Kyriakopoulos, K.J.; Manolakos, E.S. Learning human reach-to-grasp strategies: Towards EMG-based control of robotic arm-hand systems. In Proceedings of the 2012 IEEE International Conference on Robotics and Automation (ICRA), Saint Paul, MN, USA, 14–18 May 2012; pp. 2287–2292.
11. Bodiroža, S.; Doisy, G.; Hafner, V.V. Position-invariant, real-time gesture recognition based on dynamic time warping. In Proceedings of the 8th ACM/IEEE International Conference on Human-Robot Interaction (HRI), Tokyo, Japan, 3–6 March 2013; pp. 87–88.
12. Ren, Z.; Yuan, J.; Meng, J.; Zhang, Z. Robust Part-Based Hand Gesture Recognition Using Kinect Sensor. *IEEE Trans. Multimedia* **2013**, *15*, 1110–1120, doi:10.1109/TMM.2013.2246148. [CrossRef]
13. Xu, D.; Wu, X.; Chen, Y.L.; Xu, Y. Online dynamic gesture recognition for human robot interaction. *J. Intell. Robot. Syst.* **2015**, *77*, 583–596. [CrossRef]
14. Xu, D.; Chen, Y.L.; Lin, C.; Kong, X.; Wu, X. Real-time dynamic gesture recognition system based on depth perception for robot navigation. In Proceedings of the 2012 IEEE International Conference on Robotics and Biomimetics (ROBIO), Guangzhou, China, 11–14 December 2012; pp. 689–694, doi:10.1109/ROBIO.2012.6491047. [CrossRef]
15. Molchanov, P.; Yang, X.; Gupta, S.; Kim, K.; Tyree, S.; Kautz, J. Online Detection and Classification of Dynamic Hand Gestures With Recurrent 3D Convolutional Neural Network. In Proceedings of the IEEE Conference on Computer Vision and Pattern Recognition, Las Vegas, NV, USA, 27–30 June 2016, pp. 4207–4215.
16. Bailador, G.; Roggen, D.; Tröster, G.; Triviño, G. Real time gesture recognition using continuous time recurrent neural networks. In Proceedings of the ICST 2nd International Conference on Body Area Networks, Florence, Italy, 11–13 June 2007; p. 15.

17. Maung, T.H.H. Real-time hand tracking and gesture recognition system using neural networks. *World Acad. Sci. Eng. Technol.* **2009**, *50*, 466–470.

18. Tsironi, E.; Barros, P.; Wermter, S. Gesture Recognition with a Convolutional Long Short-Term Memory Recurrent Neural Network. In Proceedings of the European Symposium on Artificial Neural Networks, Computational Intelligence and Machine Learning (ESANN), Bruges, Belgium, 27–29 April 2016; pp. 213–218.

19. Jirak, D.; Barros, P.; Wermter, S. *Dynamic Gesture Recognition Using Echo State Networks*; Presses Universitaires de Louvain: Louvain-la-Neuve, Belgium, 2015; p. 475.

20. Wu, D.; Pigou, L.; Kindermans, P.J.; Le, N.D.H.; Shao, L.; Dambre, J.; Odobez, J.M. Deep Dynamic Neural Networks for Multimodal Gesture Segmentation and Recognition. *IEEE Trans. Pattern Anal. Mach. Intell.* **2016**, *38*, 1583–1597. [CrossRef] [PubMed]

21. Čerňanský, M.; Tiňo, P. Comparison of echo state networks with simple recurrent networks and variable-length Markov models on symbolic sequences. In Proceedings of the Artificial Neural Networks—ICANN 2007, Porto, Portugal, 9–13 September 2007; pp. 618–627.

22. Hochreiter, S.; Schmidhuber, J. Long short-term memory. *Neural Comput.* **1997**, *9*, 1735–1780. [CrossRef] [PubMed]

23. Lefebvre, G.; Berlemont, S.; Mamalet, F.; Garcia, C. Inertial gesture recognition with blstm-rnn. In *Artificial Neural Networks*; Springer: Cham, Switzerland, 2015; pp. 393–410.

24. Hu, Y.; Wong, Y.; Wei, W.; Du, Y.; Kankanhalli, M.; Geng, W. A novel attention-based hybrid CNN-RNN architecture for sEMG-based gesture recognition. *PLoS ONE* **2018**, *13*, e0206049. [CrossRef]

25. Sheng, C.; Zhao, J.; Liu, Y.; Wang, W. Prediction for noisy nonlinear time series by echo state network based on dual estimation. *Neurocomputing* **2012**, *82*, 186–195. [CrossRef]

26. Laje, R.; Buonomano, D.V. Robust timing and motor patterns by taming chaos in recurrent neural networks. *Nat. Neurosci.* **2013**, *16*, 925–933. [CrossRef]

27. Sussillo, D.; Abbott, L.F. Generating coherent patterns of activity from chaotic neural networks. *Neuron* **2009**, *63*, 544–557. [CrossRef]

28. Büsing, L.; Schrauwen, B.; Legenstein, R. Connectivity, dynamics, and memory in reservoir computing with binary and analog neurons. *Neural Comput.* **2010**, *22*, 1272–1311. [CrossRef]

29. Nweke, H.F.; Teh, Y.W.; Al-Garadi, M.A.; Alo, U.R. Deep learning algorithms for human activity recognition using mobile and wearable sensor networks: State of the art and research challenges. *Expert Syst. Appl.* **2018**, *105*, 233–261. [CrossRef]

30. Jaeger, H. *The "echo state" Approach to Analysing and Training Recurrent Neural Networks-with an Erratum Note*; Technical Report; German National Research Center for Information Technology (GMD): Bonn, Germany, 2001; Volume 148, p. 34.

31. Jaeger, H.; Haas, H. Harnessing nonlinearity: Predicting chaotic systems and saving energy in wireless communication. *Science* **2004**, *304*, 78–80. [CrossRef]

32. Lukosevicius, M. A Practical Guide to Applying Echo State Networks. In *Neural Networks: Tricks of the Trade*; Springer: Berlin/Heidelberg, Germany, 2012.

33. Jaeger, H. *A Tutorial on Ttraining Recurrent Neural Networks, Covering bppt, rtrl, ekf and the 'Echo State Network' Approach*; Frauenhofer Institue for Autonomous Intelligent: Sankt Augustin, Germany, 2005; pp. 1–46.

34. Manjunath, G.; Jaeger, H. Echo state property linked to an input: Exploring a fundamental characteristic of recurrent neural networks. *Neural Comput.* **2013**, *25*, 671–696. [CrossRef] [PubMed]

35. Reinhart, R.F.; Steil, J.J. Reaching movement generation with a recurrent neural network based on learning inverse kinematics for the humanoid robot iCub. In Proceedings of the 9th IEEE-RAS International Conference on Humanoid Robots, HUMANOIDS09, Paris, France, 7–10 December 2009; pp. 323–330. [CrossRef]

36. Triesch, J. A gradient rule for the plasticity of a neuron's intrinsic excitability. In *Artificial Neural Networks: Biological Inspirations*; Springer: Berlin/Heidelberg, Germany, 2005; pp. 65–70.

37. Steil, J.J. Online reservoir adaptation by intrinsic plasticity for backpropagation–decorrelation and echo state learning. *Neural Netw.* **2007**, *20*, 353–364. [CrossRef] [PubMed]

38. Schrauwen, B.; Wardermann, M.; Verstraeten, D.; Steil, J.J.; Stroobandt, D. Improving reservoirs using intrinsic plasticity. *Neurocomputing* **2008**, *71*, 1159–1171. [CrossRef]

39. Theano Development Team. Theano: A Python framework for fast computation of mathematical expressions. *arXiv* **2016**, arXiv:1605.02688.

40. Koppula, H.S.; Gupta, R.; Saxena, A. Learning human activities and object affordances from rgb-d videos. *Int. J. Robot. Res.* **2013**, *32*, 951–970. [CrossRef]
41. Mici, L.; Hinaut, X.; Wermter, S. Activity recognition with echo state networks using 3D body joints and objects category. In Proceedings of the European Symposium on Artificial Neural Networks, Computational Intelligence and Machine Learning (ESANN), Bruges, Belgium, 27–29 April 2016; pp. 465–470.
42. Gavrila, D.M. The visual analysis of human movement: A survey. *Comput. Vis. Image Underst.* **1999**, *73*, 82–98. [CrossRef]
43. Kendon, A. Current issues in the study of gesture. *Biol. Found. Gestures: Motor Semiot. Asp.* **1986**, *1*, 23–47.
44. Rybok, L.; Schauerte, B.; Al-Halah, Z.; Stiefelhagen, R. "Important stuff, everywhere!" Activity recognition with salient proto-objects as context. In Proceedings of the IEEE Winter Conference on Applications of Computer Vision, Steamboat Springs, CO, USA, 24–26 March 2014; pp. 646–651.
45. Weber, C.; Masui, K.; Mayer, N.M.; Triesch, J.; Asada, M. Reservoir Computing for Sensory Prediction and Classification in Adaptive Agents. In *Machine Learning Research Progress*: Nova publishers: Hauppauge, NY, USA, 2008.
46. Card, S.K.; Robertson, G.G.; Mackinlay, J.D. The information visualizer, an information workspace. In Proceedings of the SIGCHI Conference on Human Factors in Computing Systems, New Orleans, LA, USA, 27 April–2 May 1991; pp. 181–186.
47. Miller, R.B. Response time in man-computer conversational transactions. In Proceedings of the Fall Joint Computer Conference, Part I, San Francisco, CA, USA, 9–11 December 1968; pp. 267–277.

Article

Algorithm for Base Action Set Generation Focusing on Undiscovered Sensor Values

Sho Yamauchi [1,*] **and Keiji Suzuki** [2]

[1] Department of Computer Science, Kitami Institute of Technology, 165 Koencho, Kitami, Hokkaido 090-8507, Japan

[2] Department of Complex and Intelligent Systems, Future university hakodate, 116-2, Kameda-Nakanocho, Hakodate 041-8655, Japan; kjsuzuki@fun.ac.jp

* Correspondence: sho-yama@mail.kitami-it.ac.jp

Received: 26 November 2018; Accepted: 29 December 2018; Published: 4 January 2019

Abstract: Previous machine learning algorithms use a given base action set designed by hand or enable locomotion for a complicated task through trial and error processes with a sophisticated reward function. These generated actions are designed for a specific task, which makes it difficult to apply them to other tasks. This paper proposes an algorithm to obtain a base action set that does not depend on specific tasks and that is usable universally. The proposed algorithm enables as much interoperability among multiple tasks and machine learning methods as possible. A base action set that effectively changes the external environment was chosen as a candidate. The algorithm obtains this base action set on the basis of the hypothesis that an action to effectively change the external environment can be found by observing events to find undiscovered sensor values. The process of obtaining a base action set was validated through a simulation experiment with a differential wheeled robot.

Keywords: action generation; robot motion; undiscovered sensor values; differential wheeled robot

1. Introduction

Previous machine learning algorithms [1,2] such as Q-learning use a given base action set and choose an action from the set repeatedly [3–7]. Pre-defined action sets are commonly used in reinforcement learning, where a robot will choose one action from the set and then use it to execute one action. An action sequence of the robot is generated by repeating this cycle [8,9]. Conventional reinforcement learning methods use a pre-defined action set or acquire actions that depend on the specific task through trial and error processes. Pre-defined actions are designed by hand and there is no clear evidence that these actions are the optimal ones for robots.

Neural networks have been used to obtain actions for achieving specific tasks [10–12]. With this approach, neural networks are given an evaluation function and they decide actions in accordance with this function when a correct teacher signal is unknown. A base action set is also given in such cases.

The base action set and their elements are designed by hand. However, many actuators are used in constructing robots, so its possible actions can become more complicated, which makes it difficult to determine and split ino a suitable number of actions. For these reasons, we propose an algorithm to obtain an action set suitable for the external environment and the robot body. The proposed algorithm obtains a base action set that does not depend on specific tasks and is usable universally. It enables as much interoperability among multiple tasks and machine learning methods as possible and obtains a base action set that effectively changes the external environment.

Many methods to enable locomotion for a complicated task through trial and error processes using reinforcement learning have been proposed [13–19]. Also, a modular self-reconfigurable robot that learns actions for its current configuration and a specific task using multi-agent reinforcement learning method has been reported [20]. The reward function required sophisticated design and acquired actions were only for the specific tasks in these cases, so the aims of this study to get a universal base action set were different.

Although some deep learning methods for complicated tasks have been reported in recent years, the fundamental mechanisms underlying them, such as neural networks and reinforcement learning methods, are still the same. Specifically, these behaviors use a pre-defined action set or acquire actions depending only on specific tasks [21–23].

Emotional behavior generation has also been effective in variation of robot action [24–27]. Several studies have demonstrated richness in variation and mutually independent actions corresponding to human responses. However, they focused on emotional behavior and not on generating effective actions to change the external environment.

The purpose of our research is to develop an action generation algorithm that does not depend on the specific task and to clarify the characteristics of the parameters used in the algorithm. Also, we identify which parameter values are suitable for robot action generation before applying the proposed algorithm to a real robot.

We hypothesize that an action to effectively change the external environment can be found by observing events to find undiscovered sensor values. Thus, we developed an algorithm to obtain a base action set to change the external environment capably.

2. Base Action Set Generation Algorithm Focusing on Undiscovered Sensor Values

2.1. Definition of an Action Fragment

Our intent was to construct an algorithm to acquire a robot's base action set that does not depend on a specific task. Also, we wanted to make this base action set usable in many situations (i.e., universal). We chose a base action set that effectively causes changes in the external environment as a candidate. We hypothesized that an action to effectively change the external environment can be observed indirectly through events to find undiscovered sensor values. Discovering sensor values means situations that have never happened before occurred in the external environment. In other words, a robot can be considered to have caused some changes to it.

On the basis of the above, our intent is to generate effective actions that cause changes in the external environment dynamically by generating actuator output signals without using pre-defined actions. Therefore, we define a unit that combines sensor input signals and actuator output signals for analyzing relationships between them. The "action fragment" is defined as a set of sensor input and actuator output signals for a specific period of time.

The number of sensors and actuators are denoted as m and n, respectively. The action fragment of m sensors and n actuators is denoted as F and the data length of the action fragment F is denoted as l. F contains each of the sensor input values and actuator output signals. The ith sensor input signal is denoted as s_i, and the jth actuator output signal is denoted as a_j. These are expressed as Equations (1)–(3).

$$F = [s_1, \cdots, s_m | a_1, \cdots, a_n]^{\mathrm{T}} \tag{1}$$

$$s_i = [s_i(0), s_i(1), \cdots, s_i(t), \cdots, s_i(l-1)] \tag{2}$$

$$a_j = [a_j(0), a_j(1), \cdots, a_i(t), \cdots, a_j(l-1)] \tag{3}$$

The action fragment is designed as a part of the robot's behavior. Thus, the actuator output signals are generated first, and the robot moves according to the signals when we use the action fragment. Then, the robot's sensor input signals are recorded, and they are combined with the actuator output signals as an action fragment. For example, if we use a differential wheeled robot equipped with two speed-controllable wheels and a forward distance sensor, two actuator output signals are generated and input to the wheels. Then, the sensor input signals are recorded and combined to form the action fragment.

2.2. Action Fragment Operations

Here, we define the union of action fragments. Action fragment F_A is defined as Equations (4), (6) and (7) and action fragment F_B is defined as Equations (5), (8) and (9). Then, action fragment F_C, constructed by combining F_A and F_B, is defined as Equations (10)–(12), where l_A and l_B are the data length of F_A and F_B, respectively. The number of actuators and sensors of F_A and F_B is the same.

$$F_A = [s_{A1}, \cdots , s_{Am} | a_{A1}, \cdots , a_{An}]^{\mathrm{T}} \tag{4}$$

$$F_B = [s_{B1}, \cdots , s_{Bm} | a_{B1}, \cdots , a_{Bn}]^{\mathrm{T}} \tag{5}$$

$$s_{Ai} = [s_{Ai}(0), s_{Ai}(1), \cdots , s_{Ai}(l_A - 1)] \tag{6}$$

$$a_{Aj} = [a_{Aj}(0), a_{Aj}(1), \cdots , a_{Aj}(l_A - 1)] \tag{7}$$

$$s_{Bi} = [s_{Bi}(0), s_{Bi}(1), \cdots , s_{Bi}(l_B - 1)] \tag{8}$$

$$a_{Bj} = [a_{Bj}(0), a_{Bj}(1), \cdots , a_{Bj}(l_B - 1)] \tag{9}$$

$$F_C = F_A + F_B \tag{10}$$

$$s_{Ci} = [s_{Ai}(0), s_{Ai}(1), \cdots , s_{Ai}(l_A - 1), \\ s_{Bi}(0), s_{Bi}(1), \cdots , s_{Bi}(l_B - 1)] \tag{11}$$

$$a_{Cj} = [a_{Aj}(0), a_{Aj}(1), \cdots , a_{Aj}(l_A - 1), \\ a_{Bj}(0), a_{Bj}(1), \cdots , a_{Bj}(l_B - 1)] \tag{12}$$

Also, the extracted part of action fragment F from $t = t_s$ to $t = t_e$ is defined as a sub action fragment and denoted as Equation (13).

$$F[t_s : t_e] \tag{13}$$

2.3. Random Motion Generation Algorithm for Comparison

We define a random motion generation algorithm for comparison before explaining the proposed algorithm. The use of a simple random number as actuator output signals does not work for robot motion in most cases. Therefore, we use a Fourier series to generate random actuator output signals by determining the Fourier coefficients using uniform random numbers. A variety of waves are generated in this way. The jth actuator output signal is determined as Equation (14).

$$a_j(t) = b^{(j)} + \sum_{k=1}^{n_k} \left(c_k^{(j)} cos(kt) + d_k^{(j)} sin(kt) \right) \tag{14}$$

The operation to generate a random action fragment of length l using this method is denoted as $f_{rnd}(l)$. Coefficients $b^{(j)}$, $c_k^{(j)}$, $d_k^{(j)}$ are reset using uniform random numbers each time $f_{rnd}(l)$ is used.

The random motion generation algorithm for comparison is referred to as "Algorithm Random" in this paper, and Algorithm Random uses $f_{rnd}(l)$ multiple times to generate actuator output signals of necessary length (Figure 1).

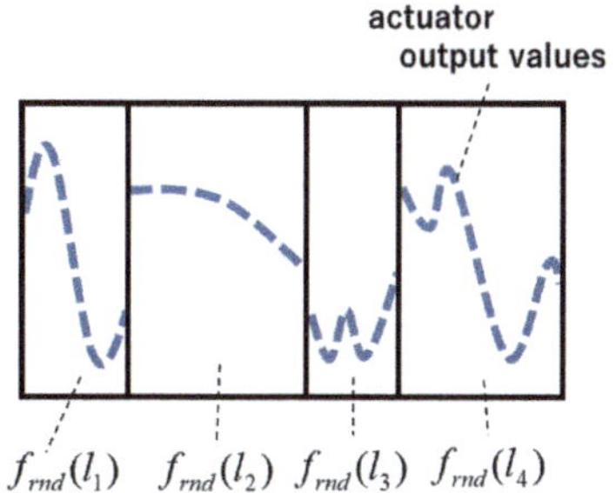

Figure 1. Overview of random motion generation.

2.4. Base Action Set Generation Algorithm to Extract Actions That Cause Changes Effectively in the External Environment and to Combine Those Actions

We developed a base action set generation algorithm focusing on finding processes of undiscovered sensor values.

First, when a robot finds undiscovered sensor values, we assume that some of the actions that effectively change the external environment are generated around that time. Then, the algorithm extracts a part of the actuator output signals before the undiscovered sensor values are found. These extracted parts are used to generate new actuator output signals by combining them. The newly generated signals should effectively change the external environment and enable finding undiscovered sensor values.

A discovery of new sensor values is defined as follows. First, we divide the m-dimensional sensor space of a robot that has m sensors into several parts and assume each part as a bin of a histogram. This histogram is denoted as D_m. Each bin of the m-dimensional histogram D_m is denoted as $b(i_1, i_2, ..., i_m)$, and the number of data in each bin is denoted as $n_b(i_1, i_2, ..., i_m)$. The m-dimensional histogram D_m at time t is denoted as $D_m(t)$. The sensor value at time t $s(t)$ is allocated to the corresponding bin. We assume that the corresponding bin of sensor value $s(t)$ is $b(i_1, i_2, ..., i_m)$. If $n_b(i_1, i_2, ..., i_m) = 0$ in $D_m(t-1)$, the sensor value $s(t)$ is an undiscovered sensor value.

Next, we explain the operations to extract the part of the actuator output signals that have contributed to finding undiscovered sensor values. When a sensor value at time t $s(t)$ is an undiscovered sensor value, this operation extracts a part of the actuator output signals from time $t - l_u$ to time t as a contributed part to find new sensor values. This part is constructed as a sub action fragment of F that records all actuator output signals from time $t = 0$. This sub action fragment I_i is defined as $I_i = F[t - l_u : t]$ (Figure 2a). I_i is added to extracted action fragment set U_I. The part of the actuator output signals that contributed to finding undiscovered sensor values is extracted and maintained using these operations and used to generate new actuator output signals.

This algorithm uses elements of extracted action fragment set U_I when an actuator output signal is newly generated. Now, we describe generating action fragment F_p of length l. F_p is generated according to Equation (15)–(17), where $n(U_I)$ is the number of elements of extracted action fragment set U_I and l_g is a uniform random number in $[l_{min} : l_{max}]$.

$$F_p' = \sum_j g(j) \tag{15}$$

$$F_p = F_p'[0 : l] \tag{16}$$

$$g(j) = \begin{cases} I_k & \left(I_k \in U_I, \text{probability } \frac{(1-r_r)}{n(U_I)}\right) \\ f_{rnd}(l_g) & (\text{probability } r_r) \end{cases} \tag{17}$$

Therefore, this algorithm generates random actuator output signals according to probability r_r, chooses an extracted action fragment according to probability $1 - r_r$, and combines these parts to generate new actuator output signals. The details are shown in Figure 2b.

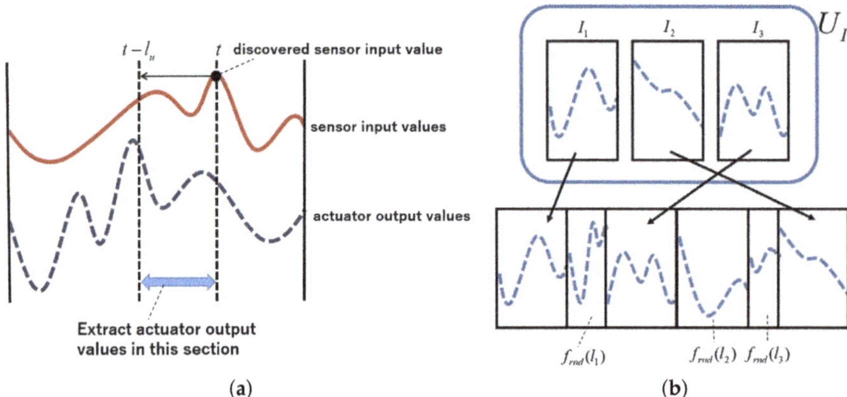

Figure 2. Motion extraction and motion generation. (**a**) Method to extract effective motion for finding undiscovered sensor values; (**b**) Motion generation using extracted effective motion.

2.5. Discard of Extracted Action Fragments

The length of used I_i is denoted as u_i and the length of the contributed parts of I_i to find the undiscovered sensor values is denoted as v_i among all the generated actuator output signals. The parts of I_i that contributed to finding the undiscovered sensor values are the parts of I_i between time $t - l_u$ to time t, where the new sensor value is found at time t. Therefore, v_i is equal to the summation length of the contained parts of action fragment I_i among all the generated action fragments.

Here, we introduce a mechanism that evaluates each action fragment I_i to identify and discard any fragment that has not contributed to finding new sensor values. We define discard criterion w_i for action fragment I_i as

$$w_i = \frac{v_i}{u_i} \tag{18}$$

Action fragment I_i that satisfies $w_i < w$ is discarded from extracted action fragment set U_I, where constant w is a discard criterion threshold.

The meaning of this operation is explained as follows. Discard criterion w_i can be transformed as Equation (19)

$$w_i = \frac{v_i}{u_i} = \frac{\frac{v_i}{l'}}{\frac{u_i}{l'}} = \frac{P(I_i \cap A)}{P(I_i)} = P(A|I_i), \tag{19}$$

where l' is a length of the whole generated actuator output signals. These probabilistic formulations have the following meanings.

- $P(I_i)$: probability to use action fragment I_i in a process of actuator output signal generation.
- $P(I_i \cap A)$: probability that a robot both uses I_i in a process of actuator output signal generation and finds undiscovered sensor values.
- $P(A|I_i)$: probability that a robot finds undiscovered sensor values when it uses action fragment I_i in a process of actuator output signal generation.

Hence, w_i expresses the probability that a robot finds undiscovered sensor values when it uses action fragment I_i in a process of actuator output signal generation. This operation discards actions fragments when the probability $P(A|I_i)$ is below the discard criteria threshold w.

A flowchart of the proposed algorithm is shown in Figure 3.

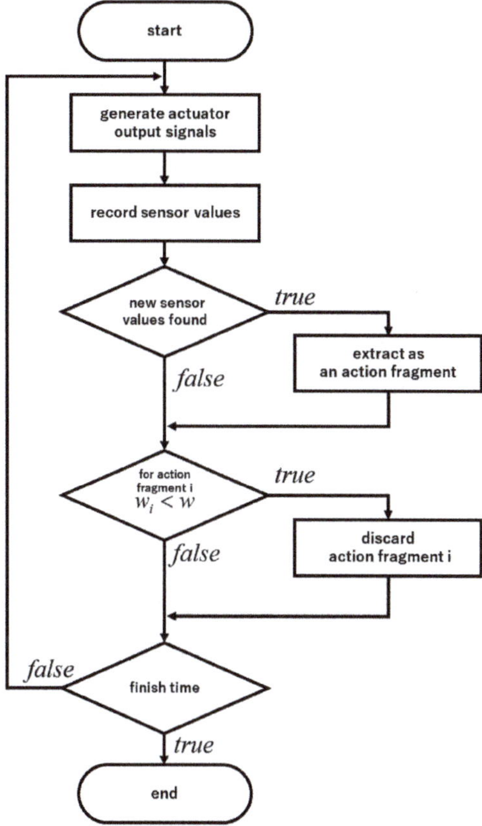

Figure 3. Flowchart of proposed algorithm.

3. Validation of Base Action Set Generation Process Using the Proposed Algorithm through Experiments with a Differential Wheeled Robot

3.1. Experiment Using a Differential Wheeled Robot

We validated the process of base action set generation using the algorithm in a simulation experiment with a differential wheeled robot. The experimental environment is shown in Figure 4a. The experimental field of 10 m × 10 m was surrounded by four walls, and the robot was placed in the center. It had two drive wheels and one caster wheel. The robot weighed 10 kg, and each drive wheel could generate 10 Nm torque at maximum. Boxes of 1 kg were placed around it. The boxes were moved by robot locomotion. The field was divided into the nine spaces shown in Figure 4b, and the regions in which each box was present were calculated. The robot received this information as a sensor input value. For example, it received sensor value $s(t) = \{4,7,2\}$ at time t when Box 1, 2, and 3 were present in regions 4, 7, and 2, respectively. The corresponding bin of $s(t)$ was $b(4,7,2)$ and $s(t)$ was an undiscovered sensor value if $n_b(4,7,2) = 0$ at time $t - 1$. The idea here is to have the algorithm determine which actions will change the external environment and then use them repeatedly to find undiscovered sensor values. A screenshot of the simulation experiment is shown in Figure 5.

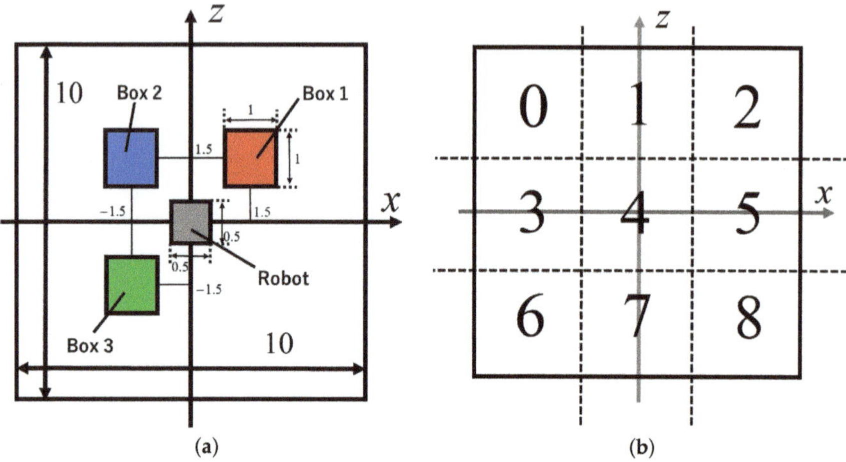

Figure 4. Experiment settings. (**a**) Experimental environment (Unit: meters); (**b**) Sensor division area of boxes.

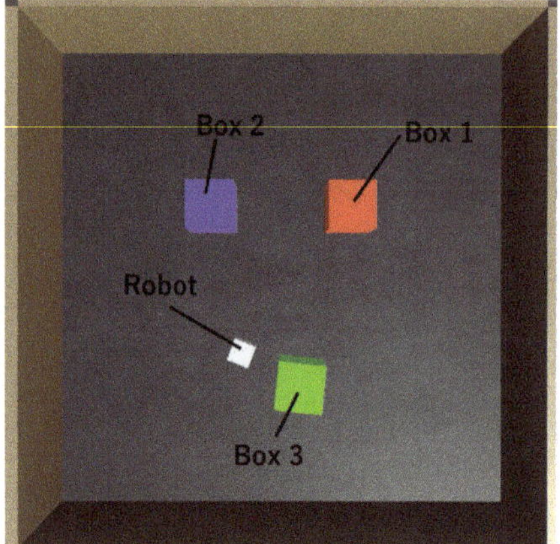

Figure 5. Screenshot of experiment.

1 step of the physics simulation was 0.02 s in this experiment. It is too difficult for the robot to maneuver when boxes are near a wall, so the positions of the boxes and robot were reset to their initial ones every 50,000 step = 50,000 * 0.02 s = 1000 s. Each trial in this experiment was conducted until 200 position resets. We compared three algorithms: A random motion generation algorithm (Algorithm Random), our proposed algorithm(Algorithm A) and Algorithm A without the mechanism for discarding action fragments defined in Section 2.5 (Algorithm A′). Ten trials were executed for each algorithm. In this case, Algorithm Random was the same as Algorithm A without the action fragment extraction and combination mechanism; in other words, Algorithm A maintained its initial state. A three dimensional histogram was prepared for sensor values in Algorithm A and A′. Each dimension of this histogram was divided into nine regions, as shown in Figure 4b, and each sensor value was allocated to a corresponding bin. Thus, this histogram had $9^3 = 729$ bins.

3.2. Viewpoints of the Experiment

We focused on the sensor cover rate and distribution of the robot's motion vectors in this experiment.

3.2.1. Sensor Cover Rate

The sensor cover rate was defined to observe how many variations in sensor value were discovered. The number of bins satisfying $n_b(i_1, i_2, ..., i_m) > 0$ in m-dimensional histogram D_m of sensor values was denoted as n_d, and the number of all bins of D_m was denoted as n_{bin}. Then, sensor cover rate r_s was defined as Equation (20).

$$r_s = \frac{n_d}{n_{bin}} \tag{20}$$

The time shift of sensor cover rate r_s was observed in this experiment.

3.2.2. Distribution of the Robot's Motion Vectors

The motion vectors of the robot at time t $\vec{q}(t)$ were calculated from its position $\vec{p}(t)$, $\vec{p}(t - t')$, $\vec{p}(t - 2t')$ at regular time interval t' as shown in Figure 6.

$$\vec{q}(t) = \vec{p}(t) - \vec{p}(t - t') \tag{21}$$

A magnitude of $\vec{q}(t)$ and an angle $\theta(t)$ between $\vec{q}(t)$ and $\vec{q}(t - t')$ were calculated, and the distribution of these values was expressed as a heat map. We determined the actual motion patterns that the robot generated using these results.

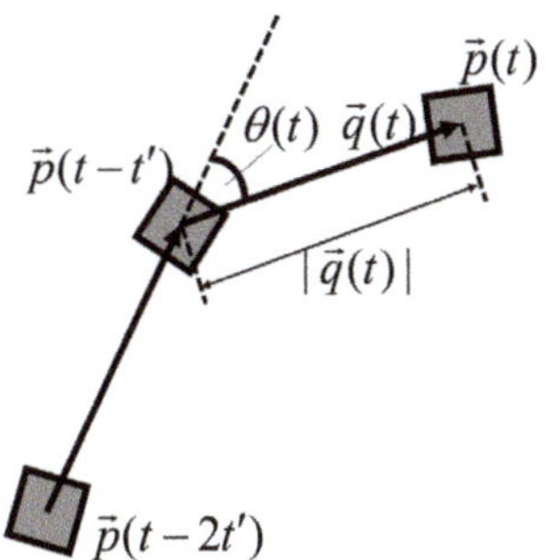

Figure 6. Motion vectors of a robot.

3.3. Parameters Settings

The parameter values used in these experiments are enumerated in Table 1.

Table 1. Experiment parameters.

Action fragment extraction length	$l_u = 50$
Action generation random rate	$r_r = 0.3$
Discard criterion threshold	$w = 0$
Time interval of motion vectors	$t' = 50$
Minimum length of random motion generation	$l_{min} = 10$
Maximum length of random motion generation	$l_{max} = 50$

4. Experimental Results

First, we show our comparison of the time shift of sensor cover rate for each algorithm. The experimental results are shown in Figure 7. Figure 7 shows the average results of ten trials. The sensor cover rate is

$$Algorithm\, A > A' > Random \tag{22}$$

for the entire time and differences among algorithms increased over time. In particular, the difference between Algorithm A and Algorithm Random at the last time (Time = $200(10^3$ s)) was $0.092 = 9.2\%$. The number of all possible sensor value patterns was $9^3 = 729$, so Algorithm A found $0.092 * 729 \simeq 67$ more patterns than Algorithm Random.

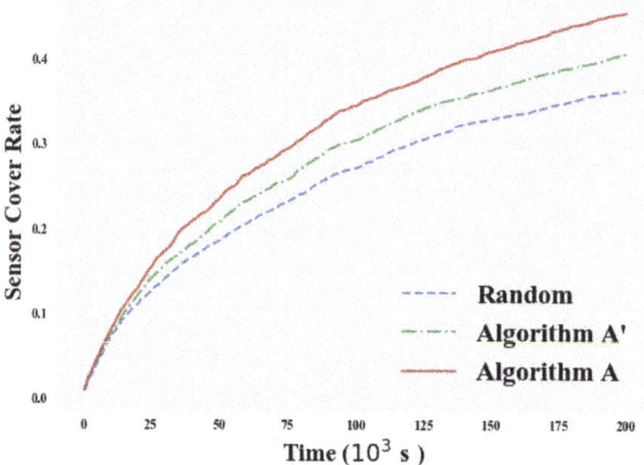

Figure 7. Sensor cover rate of each algorithm

The standard deviations of the final sensor cover rate for each algorithm are listed in Table 2. The maximum standard deviation is seen in the results of Algorithm A', which did not discard the extracted action fragment, and the minimum standard deviation is seen in the results of Algorithm Random.

Table 2. Final standard deviation of each algorithm.

Algorithm	Standard Deviation
Algorithm A	0.0261
Algorithm A'	0.1443
Algorithm Random	0.0102

4.1. Visualization of Robot's Motion Vectors

Next, we enumerated heat maps of the robot's motion vectors at a regular time interval for each algorithm. The motion vectors were calculated once every 50 steps; in other words, 50 * 0.02 s = 1 s. The horizontal axis in these maps denotes the relative angle $\theta(t)$ and the vertical axis denotes the magnitude $|\vec{q}(t)|$. The results of Algorithms A, A', and Random are shown in Figure 8. These results show all the vectors of ten trials for each algorithm.

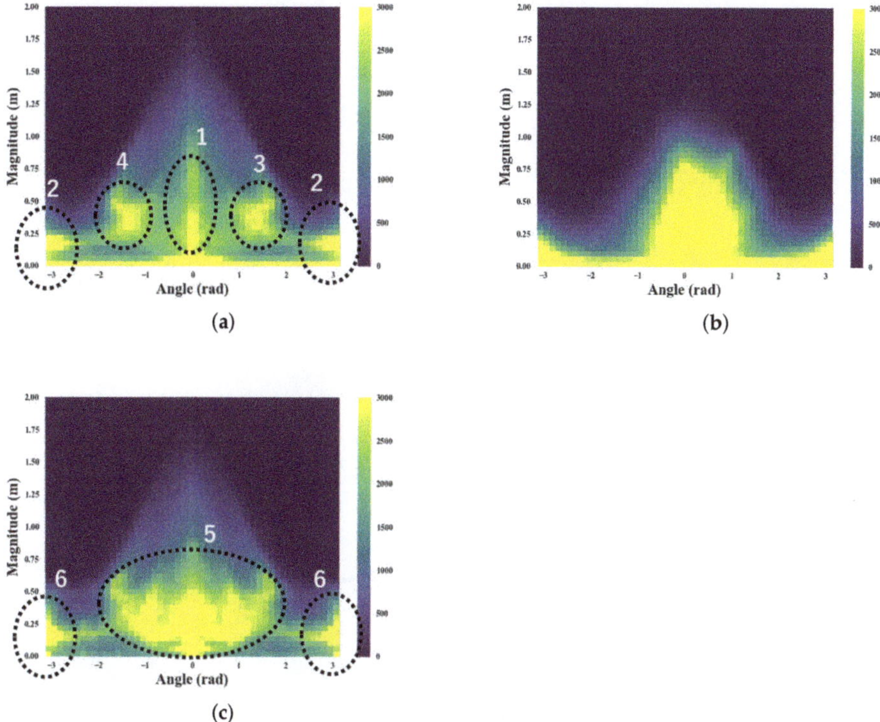

Figure 8. Motion vector heat map of each algorithm. (**a**) Algorithm A; (**b**) Algorithm A'; (**c**) Algorithm Random.

In the results for Algorithm A (Figure 8a), strong responses were shown in four sections: "Go forward" (1 in Figure 8a), "Go backward" (2 in Figure 8a), "Turn 90 degrees right" (3 in Figure 8a), and "Turn 90 degrees left" (4 in Figure 8a). This means the robot used these four locomotions frequently. These locomotion selections were comparable to typical wheeled robot locomotions. However, strong responses were observed evenly from turning 90 degrees right to 90 degrees left except for "Go forward" and "Go backward" (6 in Figure 8c) in the results of Algorithm Random (5 in Figure 8c). This means that the robot could only find about 10% fewer box allocation patterns even though Algorithm Random made more locomotion patterns than Algorithm A. All the heat maps of the ten trials for each algorithm are shown in Figure 9. The distributions of the ten trials were almost the same in Algorithms A and Random. However, the distributions were unstable and variable in Algorithm A'. These results show that discarding action fragments was executed correctly in Algorithm A. Therefore, Algorithm A could determine effective base actions to find undiscovered sensor values. Moreover, Algorithm A discarded actions that did not contribute to finding new sensor values correctly and stabilized its performance.

Next, we divided experiment time $t_{max}(=200(10^3$ s$))$ into four parts and enumerated heat maps of the robot's motion vectors at each part. The four heat maps of Algorithm A with $r_r = 0.3$, which are the results of ten trials, are shown in Figure 10.

As shown in Figure 10, a heat map in the early phase ($[0, t_{max}/4]$) showed strong responses only near the "Go forward" and "Go backward" areas. However, strong responses near "Turn right 90 degrees" and "Turn left 90 degrees" appeared from the second part ($[t_{max}/4, t_{max}/2]$) and remained almost the same from the third part ($[t_{max}/2, 3t_{max}/4]$). These results mean that actions of strong responses in the heat maps were used repeatedly. Therefore, Algorithm A finished extracting the appropriate action fragments until ($[t_{max}/2, 3t_{max}/4]$) and then maintained.

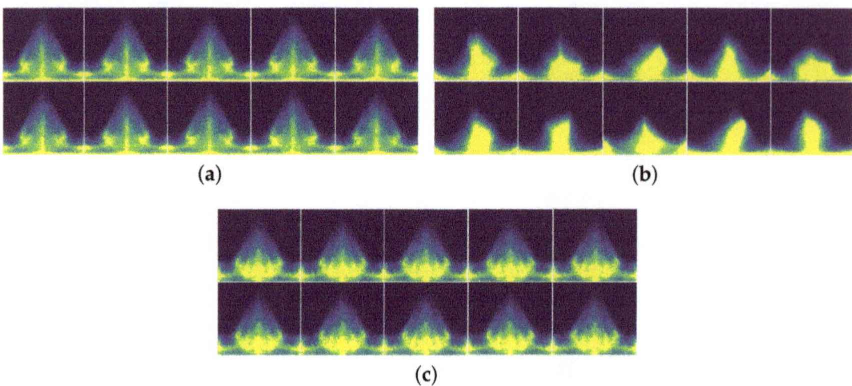

Figure 9. Motion Vector heat map of each trial. (**a**) Algorithm A; (**b**) Algorithm A′; (**c**) Algorithm Random.

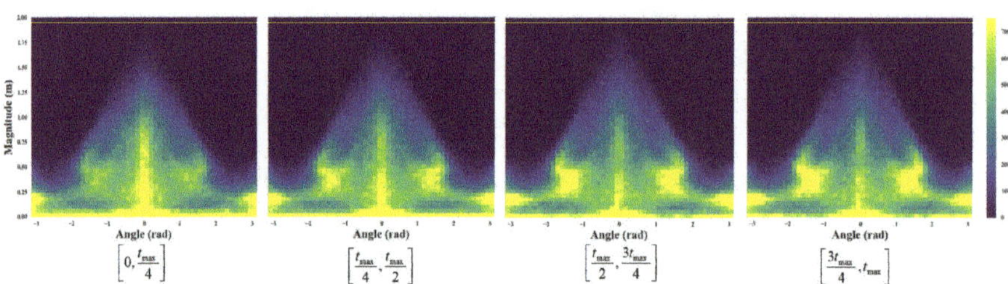

Figure 10. Motion vector heat map of each time section of Algorithm A with $r_r = 0.3$.

Next, we changed the action generation random rate r_r and evaluated its effect on acquiring the appropriate action set in Algorithm A. Average sensor cover rates of Algorithm A with various r_r are given in Figure 11, which includes the average results of ten trials. Average sensor cover rate increased for the entire time in accordance with the decrease of r_r from 0.7 to 0.3. In contrast, they remained almost the same when r_r decreased from 0.3 to 0.1. Heat maps of the robot's motion vectors for each r_r value are shown in Figure 12. Here, as the action generation random rate increases, the results show distributions similar to the result of Algorithm Random ($r_r = 1.0$) that distributed between angle $\theta = -\pi/2$ to $\theta = \pi/2$ uniformly. However, the heat maps of $r_r = 0.1$ and 0.3 are almost the same and showed strong responses in all four areas, as in Figure 8a.

Finally, we changed the interval time t' of the robot's motion vectors and show the heat maps of each case in Figure 13. All motion vectors of ten trials by Algorithm A with $r_r = 0.3$ are shown in Figure 13. The result of $t' = 50$ was the same as the result in Figure 8a and all four areas showed strong responses. However, distributions were located on the specific area and distributed uniformly inside the area when $t' = 20, 100$, and 150. Thus, no specific strong response was observed in those cases. This is because the action fragment extraction length was $l_u = 50$, and generated action sequences have meaning only when $l_u = t'$.

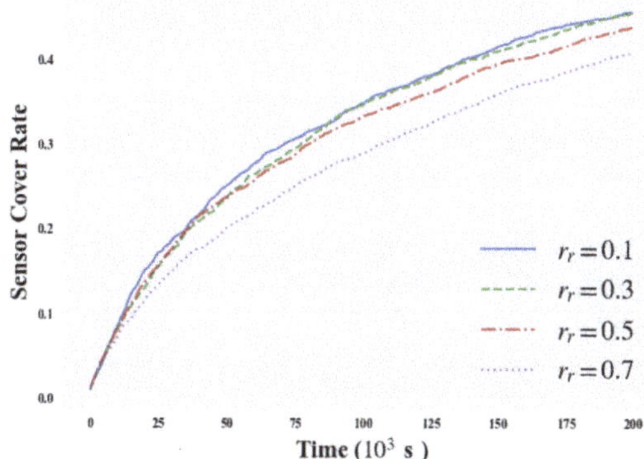

Figure 11. Sensor cover rate of various r_r values.

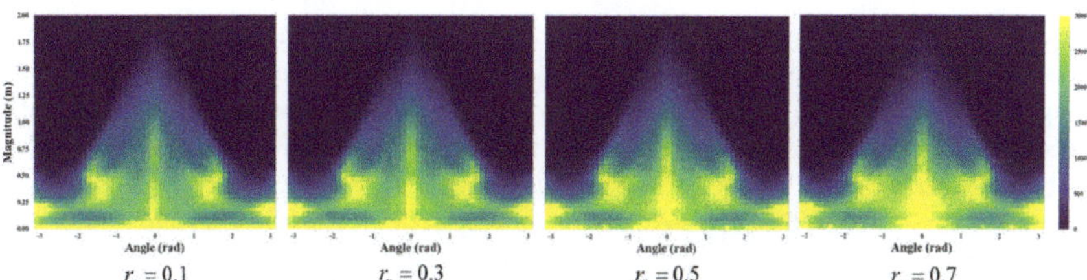

Figure 12. Motion vector heat map of various r_r values.

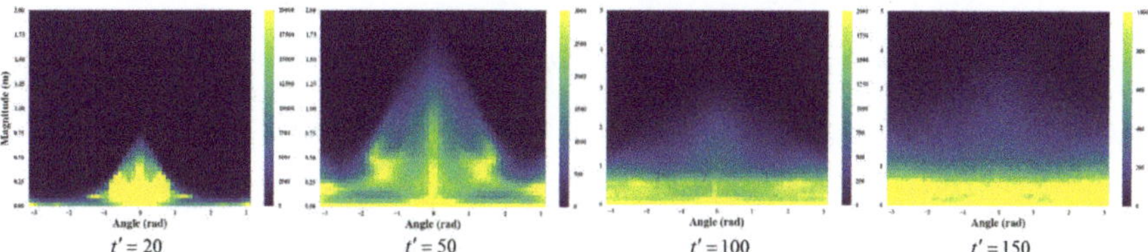

Figure 13. Motion vector heat map of various time intervals.

4.2. Discussion

We found that our proposed algorithm, which features action extraction and combination mechanisms of contributed action fragments, can find more undiscovered sensor values than a random action generation algorithm that is equal to the initial state of the proposed algorithm. Also, appropriate extraction of action fragments in Algorithm A was observed from the results of Figure 10. Four actions showing strong responses were repeatedly used and maintained in Figure 10, and the sensor cover rate of Algorithm A was the highest in Figure 7. Therefore, undiscovered sensor values were found by using those four extracted actions repeatedly, i.e., they were not discarded.

Thus, the action discarding mechanism is effective for removing action fragments that do not contribute to finding undiscovered sensor values and helps stabilize the performance. Algorithm A with the smaller action generation random rate r_r had a better sensor cover rate in Figure 11. This result demonstrates that undiscovered sensor values could be found effectively by using the action generation method that combines extracted action fragments in Algorithm A. However, sensor cover rates were almost the same when $r_r \leq 0.3$ and its effectiveness in finding undiscovered sensor values peaked around $r_r = 0.3$. Thus, r_r should be around 0.3 when we use this algorithm. The robot found undiscovered sensor values effectively, meaning the extracted actions the robot used repeatedly changed the external environment capably and generated new situations. Four types of actions—"Go forward", "Go backward", "Turn right 90 degrees", and "Turn left 90 degrees"—were extracted and used frequently as the base actions of a differential wheeled robot in this experiment. These base actions differ depending on changes in the environment and the robot body. Base actions can easily be set by hand if both the environment and robot body are simple—like they were in this experiment. However, this would not be possible for a complicated robot body and/or a constantly changing environment. In such cases, the algorithm can obtain a base action set to change the external environment capably.

Finally, from the results of Figure 13, generated actions depending on action fragment extraction length l_u and biased distribution of robot action could be observed when $t' = l_u$. This means that l_u should be larger if longer action is needed. Thus, if the robot needs actions of various time scales, l_u should be changed in accordance with the situation. This issue will be addressed in future work.

5. Conclusions

We demonstrated that the proposed algorithm is capable of effectively obtaining a base action set to change the external environment, and that the actions in the base action set contribute to finding undiscovered sensor values. Also, we showed that the algorithm stabilizes its performance by discarding actions that do not contribute to finding undiscovered sensor values. We examined the effects and characteristics of the parameters in the proposed algorithm and clarified the suitable value range of parameters for robot applications. Applying a flexible time scale for extracting action fragments is left for future work. We also intend to investigate the effect of using the action set acquired by the proposed algorithm in conventional learning methods as a base action set. We assume that an action set acquired by the proposed algorithm is both universal and effective. Thus, we will investigate whether the action set acquired by the proposed algorithm can improve the performance of conventional machine learning methods such as reinforcement learning. Also, we will construct a method to share learning data among different tasks by using the universal action set acquired by the proposed algorithm in such cases.

Author Contributions: Investigation, S.Y.; Project administration, K.S.; Writing—original draft, S.Y.; Writing—review and editing, K.S.

Funding: This research received no external funding.

Conflicts of Interest: The authors declare no conflict of interest.

References

1. Lungarella, M.; Metta, G.; Pfeifer, R.; Sandini, G. Developmental robotics: A survey. *Connect. Sci.* **2003**, *15*, 151–190. [CrossRef]

2. Kaelbling, L.P.; Littman, M.L.; Moore, A.W. Reinforcement learning: A survey. *J. Artif. Intell. Res.* **1996**, *4*, 237–285. [CrossRef]

3. Scheier, C.; Pfeifer, R. Classification as sensory-motor coordination. In Proceedings of the Third European Conference on Artificial Life, Granada, Spain, 4–6 June 1995; Springer: Berlin, Germany, 1995; pp. 657–667.

4. Zhu, Y.; Mottaghi, R.; Kolve, E.; Lim, J.J.; Gupta, A.; Li, F.F.; Farhadi, A. Target-driven visual navigation in indoor scenes using deep reinforcement learning. In Proceedings of the 2017 IEEE International Conference on Robotics and Automation (ICRA), Singapore, 29 May–3 June 2017; pp. 3357–3364.

5. Qureshi, A.H.; Nakamura, Y.; Yoshikawa, Y.; Ishiguro, H. Robot gains social intelligence through multimodal deep reinforcement learning. In Proceedings of the 2016 IEEE-RAS 16th International Conference on Humanoid Robots (Humanoids), Cancun, Mexico, 15–17 November 2016; pp. 745–751.

6. Lei, T.; Ming, L. A robot exploration strategy based on q-learning network. In Proceedings of the IEEE International Conference on Real-time Computing and Robotics (RCAR), Angkor Wat, Cambodia, 6–10 June 2016; pp. 57–62.

7. Riedmiller, M.; Gabel, T.; Hafner, R.; Lange, S. Reinforcement learning for robot soccer. *Auton. Robot.* **2009**, *27*, 55–73. [CrossRef]

8.	Matarić, M.J. Reinforcement Learning in the Multi-robot Domain. *Auton. Robot.* **1997**, *4*, 73–83. [CrossRef]

9.	Sutton, R.S.; Barto, A.G.; Barto, A.G.; Bach, F. *Reinforcement Learning: An Introduction*; MIT Press: Cambridge, MA, USA, 1998.

10.	Mitchell, T.M.; Thrun, S.B. Explanation-based neural network learning for robot control. In Proceedings of the 5th International Conference on Neural Information Processing Systems, Denver, CO, USA, 30 November–3 December 1992; pp. 287–294.

11.	Pfeiffer, M.; Nessler, B.; Douglas, R.J.; Maass, W. Reward-modulated hebbian learning of decision making. *Neural Comput.* **2010**, *22*, 1399–1444. [CrossRef] [PubMed]

12.	Hafner, R.; Riedmiller, M. Neural reinforcement learning controllers for a real robot application. In Proceedings of the 2007 IEEE International Conference on Robotics and Automation, Roma, Italy, 10–14 April 2007; pp. 2098–2103.

13.	Kormushev, P.; Calinon, S.; Caldwell, D.G. Reinforcement learning in robotics: Applications and real-world challenges. *Robotics* **2013**, *2*, 122–148. [CrossRef]

14.	Kober, J.; Peters, J.R. Policy search for motor primitives in robotics. In *Advances in Neural Information Processing Systems*; Curran Associates: Vancouver, BC, Canada, 2009; pp. 849–856.

15.	Shen, H.; Yosinski, J.; Kormushev, P.; Caldwell, D.G.; Lipson, H. Learning fast quadruped robot gaits with the RL power spline parameterization. *Cybern. Inf. Technol.* **2012**, *12*, 66–75. [CrossRef]

16.	Ijspeert, A.J.; Nakanishi, J.; Schaal, S. Learning attractor landscapes for learning motor primitives. In *Advances in Neural Information Processing Systems*; Curran Associates: Vancouver, BC, Canada, 2003; pp. 1547–1554.

17.	Kimura, H.; Yamashita, T.; Kobayashi, S. Reinforcement learning of walking behavior for a four-legged robot. *IEEJ Trans. Electron. Inf. Syst.* **2002**, *122*, 330–337.

18.	Shibata, K.; Okabe, Y.; Ito, K. Direct-Vision-Based Reinforcement Learning Using a Layered Neural Network. *Trans. Soc. Instrum. Control Eng.* **2001**, *37*, 168–177. [CrossRef]

19.	Goto, Y.; Shibata, K. Emergence of higher exploration in reinforcement learning using a chaotic neural network. In Proceedings of the International Conference on Neural Information Processing, Siem Reap, Cambodia, 13–16 December 2018; Springer: Berlin, Germany, 2016; pp. 40–48.

20.	Dutta, A.; Dasgupta, P.; Nelson, C. Adaptive locomotion learning in modular self-reconfigurable robots: A game theoretic approach. In Proceedings of the 2017 IEEE/RSJ International Conference on Intelligent Robots and Systems (IROS), Vancouver, BC, Canada, 24–28 September 2017; pp. 3556–3561.[CrossRef]

21.	Lample, G.; Chaplot, D.S. Playing FPS Games with Deep Reinforcement Learning. In Proceedings of the Conference on Artificial Intelligence AAAI, San Francisco, CA, USA, 4–9 February 2017; pp. 2140–2146.

22.	Sallab, A.E.; Abdou, M.; Perot, E.; Yogamani, S. Deep reinforcement learning framework for autonomous driving. *Electron. Imaging* **2017**, *2017*, 70–76. [CrossRef]

23.	Ran, L.; Zhang, Y.; Zhang, Q.; Yang, T. Convolutional neural network-based robot navigation using uncalibrated spherical images. *Sensors* **2017**, *17*, 1341. [CrossRef] [PubMed]

24.	Taki, R.; Maeda, Y.; Takahashi, Y. Generation Method of Mixed Emotional Behavior by Self-Organizing Maps in Interactive Emotion Communication. *J. Jpn. Soc. Fuzzy Theory Intell. Inform.* **2012**, *24*, 933–943. [CrossRef]

25.	Gotoh, M.; Kanoh, M.; Kato, S.; Kunitachi, T.; Itoh, H. Face Generation Using Emotional Regions for Sensibility Robot. *Trans. Jpn. Soc. Artif. Intell.* **2006**, *21*, 55–62. [CrossRef]

26.	Matsui, Y.; Kanoh, M.; Kato, S.; Nakamura, T.; Itoh, H. A Model for Generating Facial Expressions Using Virtual Emotion Based on Simple Recurrent Network. *JACIII* **2010**, *14*, 453–463. [CrossRef]

27.	Yano, Y.; Yamaguchi, A.; Doki, S.; Okuma, S. Emotional Motion Generation Using Emotion Representation Rules Modeled for Human Affection. *J. Jpn. Soc. Fuzzy Theory Intell. Inform.* **2010**, *22*, 39–51. [CrossRef]

Article

Controllers to Chase a High-Speed Evader Using a Pursuer with Variable Speed

Jonghoek Kim

Electrical and Electronic Convergence Department, Hongik University, Sejong 121791, Korea;
jonghoek@hongik.ac.kr

Received: 8 October 2018; Accepted: 17 October 2018; Published: 18 October 2018

Abstract: This paper proposes a chasing controller to enable a pursuer to chase a high-speed evader such that the relative distance between the evader and the pursuer monotonically decreases as time passes. Our controller is designed to assure that the angular rate of Line-of-Sight joining the pair (the pursuer and the evader) is exactly zero at all time indexes. Assuming that the pursuee can readily observe optical flow, but only poorly detect looming, this pursuer's movement is hardly detected by the pursuee. Consider the terminal phase when the pursuer is sufficiently close to the evader. As we slow down the relative speed of the pursuer with respect to the evader, we can reduce the probability of missing the high-speed evader. Thus, our strategy is to make the pursuer decrease its speed in the terminal phase, while ensuring that the distance between the evader and the pursuer monotonically decreases as time passes. The performance of our controller is verified utilizing MATLAB simulations.

Keywords: LOS; motion camouflage control; parallel navigation; missile control system; target tracking; variable speed; high-speed target

---

1. Introduction

This paper proposes a chasing controller so that a pursuer can chase and capture a maneuvering evader which moves at high speed. This problem is related to the challenging missile guidance problem of intercepting a high-speed missile [1–6]. The pursuer must move at high speed to capture a high-speed evader. Consider the terminal phase when the pursuer is sufficiently close to the high-speed evader. The accurate control of the pursuer in the terminal phase is crucial, since it is hard to capture a high-speed evader if the pursuer misses the evader in the terminal phase.

Our strategy in the terminal phase is to slow down the pursuer's speed. As we slow down the relative speed of the pursuer with respect to the evader, we can decrease the probability of missing the high-speed evader. (Consider the case where two spaceships dock each other. It is desirable to slow down the relative speed for safe and accurate docking). Thus, our strategy is to make the pursuer decrease its speed in the terminal phase, while ensuring that the distance between the evader and the pursuer monotonically decreases as time passes. As far as we know, no paper in the literature on chasing targets considered changing the pursuer's speed so as to capture a maneuvering evader in a provably complete manner.

Our controller is designed to assure that the angular rate of Line-of-sight joining the pair (the pursuer and the evader) is exactly zero at all time indexes. This type of movement is called the *motion camouflage with respect to a fixed point at infinity* [7–9].

This motion camouflage is employed by various visual insects and animals to achieve prey capture, mating, or territorial combat [8,10–12]. This movement is a time-optimal solution to capture a pursuee moving with a constant velocity (speed and heading) [10]. In addition [10], argued that this motion minimizes time-to-capture of an unpredictably moving pursuee. Assuming that the pursuee can readily observe optical flow, but only poorly detect looming, this pursuer's movement is hardly detected by the prey [8].

This paper introduces a chasing controller to enable a pursuer to chase a maneuvering evader while not rotating the LOS joining the pair (the evader and the pursuer). We make the pursuer decrease its speed in the terminal phase, while ensuring that the distance between the evader and the pursuer monotonically decreases as time passes.

The literature is abundant with papers on planning path of robots [8,13–27]. Kim [28] developed a path planning algorithm for an underwater robot approaching a static target while not being detected by the target. The path was planned to reduce both the time required to meet the target and the robot's sound measured by the static target.

The authors of [29–31] presented the autonomous tracking and following of a marine vessel by an Unmanned Surface Vehicle (USV) in the presence of dynamic obstacles. In [29,30], the path planning for the USV with International Regulations for Preventing Collisions at Sea (COLREGS) rules was achieved. The authors of [31] presented a trajectory planning and tracking approach for following a differentially constrained target vehicle operating in an obstacle field. Svec [31] predicted the target state several time steps forward in time and generated a collision-free trajectory to allow the USV to safely reach the predicted target state. As far as we know, no paper on chasing targets handled changing the pursuer's speed so as to capture a maneuvering evader in a provably complete manner.

Many controllers have been developed to mimic motion camouflage in nature. To capture an evader [7–9,32] presented a chasing controller based on biologically plausible sensing. Galloway and Raju [9,32] developed a motion camouflage controller in noisy environments. Note that [7–9] only considered a pursuer which moves with a constant speed.

As missile controllers, Proportional Navigation Guidance (PNG) controls and their variations were widely used to let the pursuer capture the evader [1,3–6]. PNG laws enable the pursuer to capture the evader by driving the angular rate of LOS near zero as time passes [4]. But, PNG laws do not make the angular rate of LOS stay at zero at every time index. Note that PNG laws only considered a pursuer which moves with a constant speed.

This paper proves that utilizing our chasing controller, the distance between the evader and the pursuer monotonically decreases, regardless of evader's maneuver or acceleration, if the following assumptions are satisfied: (1) the pursuer speed is bigger than the evader speed; (2) the pursuer can predict the evader's location within two time steps in the future.

Our controller works as follows. The locations of both the pursuer and the evader are accessed at every time index. Considering a robot control system, the location of the pursuer is estimated in real time, since a robot (pursuer) can access the movement of itself utilizing Inertial Navigation Sensor (INS) or Global Positioning System (GPS). The pursuer uses sensor measurements, such as radar, to measure the evader's location in real time.

Based on the accessed evader locations, the pursuer predicts the evader's location two steps forward in time. We acknowledge that sensor measurement noise exists as the pursuer measures the evader's location in real time. Moreover, predicted evader location is related to evader maneuvers and is not easy to conjecture in an accurate manner. The effect of prediction error on the performance of our motion camouflage controller is analyzed in Section 4.2.

Since the maximum acceleration of the evader is bounded, we can assume that the evader's motions are smooth, so the evader's trajectory curves are derivative. Under this assumption, Section 4 presents a fitting method to predict the evader's location two steps forward in time.

After predicting the evader's location two steps forward in time, the pursuer calculates its velocity command, while ensuring that the LOS does not rotate at the next time index. Using deduction, the angular rate of LOS is zero at every time index. In Section 5, the effectiveness of our chasing controller is demonstrated utilizing MATLAB simulations.

This paper is organized as follows. Section 2 presents several definitions and assumptions before presenting our main results. Section 3 presents our chasing controller. Section 4 presents a method to predict the evader location two steps forward in time considering noisy environments. Section 5 introduces MATLAB simulation results to demonstrate the performance of our chasing controller. Section 6 provides Conclusions.

2. The Assumptions and Definitions

2.1. Definitions

Several definitions and assumptions are introduced before presenting our main results. $\angle(\mathbf{v}_1, \mathbf{v}_2)$ is the angle formed by two vectors $\mathbf{v}_1$ and $\mathbf{v}_2$. Mathematically, $\angle(\mathbf{v}_1, \mathbf{v}_2) = \arccos(\frac{\mathbf{v}_1 \cdot \mathbf{v}_2}{\|\mathbf{v}_1\|\|\mathbf{v}_2\|})$. Here, $\angle(\mathbf{v}_1, \mathbf{v}_2)$ exists between 0 and π. T is the sampling interval of our chasing controller in discrete-time systems.

$\mathbf{r}_k^e$ is the evader's location at time index k. $\mathbf{r}_k^p$ is the pursuer's location at time index k. The pursuer is *in the motion camouflage state* at time index $k+1$ in the case where $\angle(\mathbf{r}_{k+1}^p - \mathbf{r}_{k+1}^e, \mathbf{r}_k^p - \mathbf{r}_k^e) = 0$.

v_k^p is the pursuer's speed at time index k. v_k^e is the evader's speed at time index k. The subscript k implies the time index k.

The pursuer's motion model is

$$\mathbf{r}_{k+1}^p = \mathbf{r}_k^p + T v_k^p \mathbf{u}_k. \tag{1}$$

Here, $\mathbf{u}_k$ is a unit vector and indicates the pursuer's heading at time index k. $\mathbf{u}_k$ is determined at every time index k so that the pursuer is in the motion camouflage state at time index k.

In addition, the evader's motion model is

$$\mathbf{r}_{k+1}^e = \mathbf{r}_k^e + T \mathbf{v}_k^e. \tag{2}$$

Here, $\mathbf{v}_k^e$ indicates the evader's velocity vector at time index k.

In order to capture the evader, it is necessary that the evader's speed is slower than the pursuer's speed ([1,2] introduced a variation of PNG controls to capture evaders that are of higher speeds than the pursuer. But, considering a high-speed evader which moves away from a slowly moving pursuer, it is impossible to capture the evader). We control the pursuer's speed so that it is always bigger than that of the evader. This implies that $v_k^e < v_k^p$ at every time index k.

Let a_m denote the pursuer's maximum acceleration. In addition, let $\epsilon > 0$ denote a small constant. The required time interval to decrease the pursuer's speed from v_k^p to $v_k^e + \epsilon$ is

$$T_r = (v_k^p - v_k^e - \epsilon)/a_m. \tag{3}$$

The traversal distance of the pursuer as it decreases its speed from v_k^p to $v_k^e + \epsilon$ is

$$D_k = v_k^p * T_r - 0.5 * a_m * T_r * T_r. \tag{4}$$

We say that the pursuer is in the *terminal phase* in the case where

$$\|\mathbf{r}_k^e - \mathbf{r}_k^p\| < D_k. \tag{5}$$

is met.

L_k is the infinite line (LOS) intersecting both $\mathbf{r}_k^p$ and $\mathbf{r}_k^e$. We draw one infinite line (LOS) $\bar{L}_{k+1}$ intersecting $\mathbf{r}_{k+1}^e$, such that $\bar{L}_{k+1}$ is parallel to L_k. Note that there exists only one infinite line intersecting $\mathbf{r}_{k+1}^e$, such that the line is parallel to L_k.

$\mathbf{c}_k$ is the point on $\bar{L}_{k+1}$, which is the closest to $\mathbf{r}_k^p$. Let $d_k^c = \|\mathbf{r}_k^p - \mathbf{c}_k\|$, and let $\chi_k^e = \angle(\mathbf{r}_k^e - \mathbf{r}_k^p, \mathbf{r}_{k+1}^e - \mathbf{r}_k^e)$. Let $d_k = \|\mathbf{r}_k^p - \mathbf{r}_k^e\|$. See Figure 1 for an illustration of these concepts.

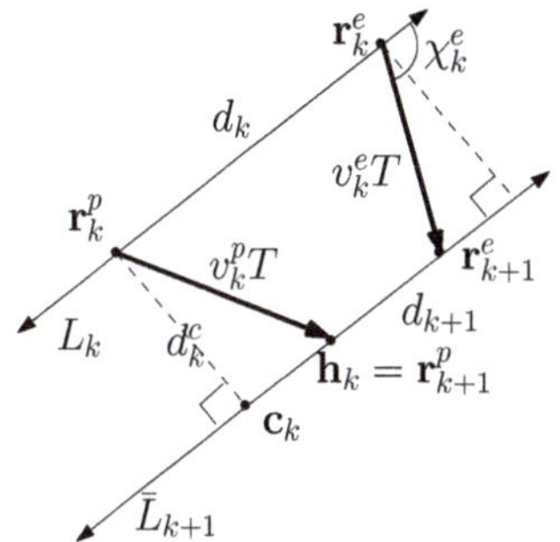

Figure 1. $d_k^c \leq v_k^e T$, and $d_k^c < v_k^p T$.

As depicted in Figure 1, $d_k^c = v_k^e T \sin(\chi_k^e) \leq v_k^e T$. Since $v_k^e < v_k^p$, we have

$$d_k^c < v_k^p T. \tag{6}$$

Utilizing (6), we define $\delta_k > 0$ as follows.

$$\delta_k = \sqrt{(v_k^p T)^2 - (d_k^c)^2}. \tag{7}$$

We also define the *heading point* $\mathbf{h}_k$ as follows:

$$\mathbf{h}_k = \mathbf{c}_k + \frac{\mathbf{r}_k^e - \mathbf{r}_k^p}{\|\mathbf{r}_k^e - \mathbf{r}_k^p\|} \delta_k. \tag{8}$$

Here, δ_k satisfies that $\|\mathbf{h}_k - \mathbf{r}_k^p\|$ is $v_k^p T$. See Figure 1.

At every time index k, the pursuer heads towards the heading point $\mathbf{h}_k$. Consider a circle centered at $\mathbf{r}_k^p$, whose radius is $v_k^p T$. Due to (6), $\bar{L}_{k+1}$ intersects this circle at two points. Between these two points, the heading point $\mathbf{h}_k$ is the point which is the closest to $\mathbf{r}_{k+1}^e$. This way, the pursuer maneuvers to decrease the distance between the pursuer and the evader, while not rotating the LOS.

Let us draw both x-axis and y-axis satisfying the following conditions:

- Both x-axis and y-axis are normal to each other, and they intersect at $\mathbf{c}_k$. $\mathbf{c}_k$ is set as the origin.
- the coordinate of $\mathbf{r}_{k+1}^p = \mathbf{h}_k$ is $(\delta_k, 0)$.
- The coordinate of $\mathbf{r}_k^p$ is $(0, d_k^c)$.

First, we handle the case where $\chi_k^e > \pi/2$. This case, the evader maneuvers to decrease the distance between the pursuer and the evader. Utilizing the geometry in Figure 1, the coordinate of $\mathbf{r}_{k+1}^e$ is

$$co_1(\mathbf{r}_{k+1}^e) = (d_k - \alpha_k, 0). \tag{9}$$

Here, $\alpha_k = \sqrt{(v_k^e T)^2 - (d_k^c)^2}$ is positive.

Next, we handle the case where $\chi_k^e \leq \pi/2$. This case, the evader maneuvers to increase the distance between the pursuer and the evader. Utilizing the geometry in Figure 2, the coordinate of $\mathbf{r}_{k+1}^e$ is

$$co_2(\mathbf{r}_{k+1}^e) = (d_k + \alpha_k, 0). \tag{10}$$

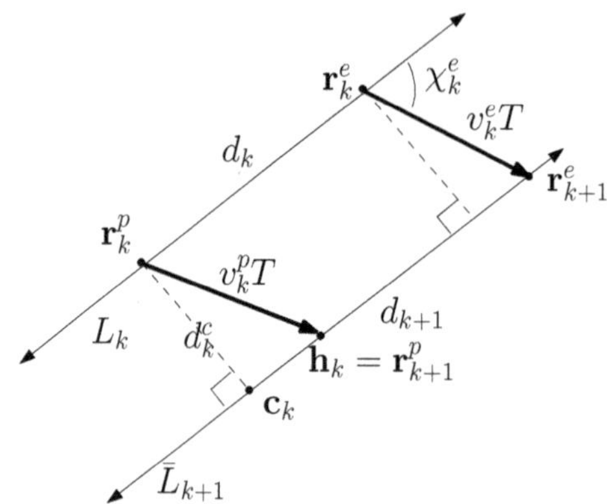

Figure 2. $\chi_k^e \leq \pi/2$.

2.2. Assumptions

In this paper, we assume that the pursuer's location is estimated in real time. In addition, the pursuer can estimate the evader's location at every time index. Therefore, the pursuer can access L_k at every time index k. Since the maximum acceleration of the evader is bounded, we assume that the evader's motions are smooth, so the evader's trajectory curves are derivative. We further assume that the pursuer can predict the evader's location two steps forward in time, which implies that the pursuer at time index k can estimate the evader's velocities $\mathbf{v}_k^e$ and $\mathbf{v}_{k+1}^e$.

3. Control Law

We introduce how to control the pursuer's speed. We assumed that the pursuer at time index k can estimate the evader speed v_k^e. v_k^p can change with respect to k as long as

$$v_k^p > v_k^e \tag{11}$$

is met at each time step k.

Our strategy in the terminal phase is to slow down the pursuer's speed as long as (11) is met. This implies that in the terminal phase, we update the pursuer's speed using

$$v_{k+1}^p = max(v_k^p - a_m T, v_{k+1}^e + \epsilon). \tag{12}$$

Here, $\epsilon > 0$ is a small constant. Note that v_{k+1}^e is available, since the pursuer at time index k can estimate the evader's velocities $\mathbf{v}_k^e$ and $\mathbf{v}_{k+1}^e$. (12) implies that (11) is met at each time step.

We next introduce the heading control $\mathbf{u}_k$ to achieve motion camouflage. $\mathbf{u}_k$ is chosen so that the pursuer moves to $\mathbf{h}_k$. At every time index k, the heading controller is given as follows. *At every time index k, the pursuer selects the new heading command $\mathbf{u}_k$ as* $\frac{\mathbf{h}_k - \mathbf{r}_k^p}{\|\mathbf{h}_k - \mathbf{r}_k^p\|}$.

Consider the situation where the distance between $\mathbf{r}_k^p$ and $\mathbf{r}_{k+1}^e$ is less than $v_k^p T$. In this situation, the pursuer moves towards $\mathbf{r}_{k+1}^e$ directly, while not using $\mathbf{u}_k = \frac{\mathbf{h}_k - \mathbf{r}_k^p}{\|\mathbf{h}_k - \mathbf{r}_k^p\|}$. In this way, the evader is captured at time index $k+1$.

In practice, the pursuer cannot turn with infinite acceleration. Suppose that the maximum turn rate of the pursuer is q radians per second. In the case where the angle formed by $\mathbf{u}_{k-1}$ and the new heading command, $\frac{\mathbf{h}_k - \mathbf{r}_k^p}{\|\mathbf{h}_k - \mathbf{r}_k^p\|}$, is bigger than qT radians, then the heading command $\frac{\mathbf{h}_k - \mathbf{r}_k^p}{\|\mathbf{h}_k - \mathbf{r}_k^p\|}$ cannot be achieved within one sampling interval. This case, we select $\mathbf{u}_k$ using the following method.

Let $\mathbf{u}_k^1 = \mathbf{R}(qT)\mathbf{u}_{k-1}$, where $\mathbf{R}(qT) = \begin{pmatrix} cos(qT) & sin(qT) \\ -sin(qT) & cos(qT) \end{pmatrix}$. In addition, let $\mathbf{u}_k^2 = \mathbf{R}(-qT)\mathbf{u}_{k-1}$. If $\left\| \frac{\mathbf{h}_k - \mathbf{r}_k^p}{\|\mathbf{h}_k - \mathbf{r}_k^p\|} - \mathbf{u}_k^1 \right\| < \left\| \frac{\mathbf{h}_k - \mathbf{r}_k^p}{\|\mathbf{h}_k - \mathbf{r}_k^p\|} - \mathbf{u}_k^2 \right\|$, then we set $\mathbf{u}_k = \mathbf{u}_k^1$. Otherwise, we set $\mathbf{u}_k = \mathbf{u}_k^2$. In this manner, the heading command $\mathbf{u}_k$ can be achieved within one sampling interval.

In the following sections associated with analysis (Sections 3.1, 3.2, and 4), it is assumed that $q = \inf$. This implies that we analyze the performance of our chasing controller, not considering the maximum turn rate of the pursuer. In the Simulation section, we set $q = \pi/6$ radians per second.

3.1. Stability Analysis

Next, the stability of our chasing controller is analyzed. It is derived that $\angle(\mathbf{r}_{k+1}^p - \mathbf{r}_{k+1}^e, \mathbf{r}_k^p - \mathbf{r}_k^e) = 0$ at every time index $k > 1$ under our chasing controller. This implies that the motion camouflage state is achieved at every time index $k > 1$.

Theorem 1. *Under our heading controller u_k, $\angle(\mathbf{r}_{k+1}^p - \mathbf{r}_{k+1}^e, \mathbf{r}_k^p - \mathbf{r}_k^e) = 0$ at every time index $k > 1$.*

Proof. At every time index k, the pursuer sets $\mathbf{u}_k$ as $\frac{\mathbf{h}_k - \mathbf{r}_k^p}{\|\mathbf{h}_k - \mathbf{r}_k^p\|}$. Utilizing the pursuer's motion model in (1), we further get

$$\mathbf{r}_{k+1}^p = \mathbf{r}_k^p + \frac{\mathbf{h}_k - \mathbf{r}_k^p}{\|\mathbf{h}_k - \mathbf{r}_k^p\|} v_k^p T. \tag{13}$$

The heading point $\mathbf{h}_k$ is set utilizing (8). In (8), δ_k satisfies that $\|\mathbf{h}_k - \mathbf{r}_k^p\|$ is $v_k^p T$. By substituting $\|\mathbf{h}_k - \mathbf{r}_k^p\|$ in (13) for $v_k^p T$, we obtain

$$\mathbf{r}_{k+1}^p = \mathbf{h}_k. \tag{14}$$

Here, $\mathbf{h}_k$ is on $\bar{L}_{k+1}$ by its definition. Since $\mathbf{r}_{k+1}^p$ is on $\bar{L}_{k+1}$ which is parallel to L_k, we obtain $\angle(\mathbf{r}_{k+1}^p - \mathbf{r}_{k+1}^e, \mathbf{r}_k^p - \mathbf{r}_k^e) = 0$. $\square$

3.2. Capturability Analysis

Besides achieving motion camouflage, the pursuer must capture the evader in finite time. We prove that the distance between the evader and the pursuer monotonically decreases as time passes.

Theorem 2. *Under our chasing controller, the distance between the evader and the pursuer monotonically decreases until the evader is captured.*

Proof. Before proving the capturability of our chasing controller, the relationship between d_k and d_{k+1} is introduced.

First, we handle the case where $\chi_k^e > \pi/2$. This case, the evader at time index k maneuvers to decrease the distance between the evader and the pursuer. See Figure 1 for an illustration of d_k and d_{k+1}. Since L_k is parallel to $\bar{L}_{k+1}$, the geometry in Figure 1 leads to

$$d_{k+1} = d_k - \alpha_k - \delta_k. \tag{15}$$

This results in

$$d_{k+1} < d_k. \tag{16}$$

Next, we handle the case where $\chi_k^e \leq \pi/2$. This case, the evader at time index k maneuvers to increase the distance between the evader and the pursuer. See Figure 2 for an illustration of d_k and d_{k+1}. Since L_k is parallel to $\bar{L}_{k+1}$, we get

$$d_{k+1} = d_k + \alpha_k - \delta_k. \tag{17}$$

Since $v_k^p > v_k^e$, $\alpha_k - \delta_k$ in (17) is negative. Hence, Ref. (16) is obtained.

Utilizing (16), d_k monotonically decreases as k increases. Therefore, there exists a time index k' such that $d_{k'} > 0$ and that $d_{k'+1} < 0$. We next prove that the distance between $\mathbf{r}_{k'}^p$ and $\mathbf{r}_{k'+1}^e$ is less than $v_k^p T$.

First, handle the case where $\chi_{k'}^e > \pi/2$. Utilizing both $d_{k'+1} < 0$ and (15), the following equation is derived.

$$d_{k'} - \alpha_{k'} < \delta_{k'}. \tag{18}$$

$v_k^p > v_k^e$ leads to

$$\alpha_{k'} - \delta_{k'} < 0. \tag{19}$$

Since $d_{k'}$ is positive, Ref. (19) further results in

$$d_{k'} - \alpha_{k'} > -\delta_{k'}. \tag{20}$$

Utilizing (18) and (20),

$$(d_{k'} - \alpha_{k'})^2 < (\delta_{k'})^2 \tag{21}$$

is derived.

Utilizing (9), the distance between $\mathbf{r}_{k'}^p$ and $\mathbf{r}_{k'+1}^e$ is calculated as

$$\|\mathbf{r}_{k'}^p - \mathbf{r}_{k'+1}^e\| = \|(d_{k'} - \alpha_{k'}, -d_{k'}^c)\|. \tag{22}$$

Utilizing (7), (22) and (21),

$$\|\mathbf{r}_{k'}^p - \mathbf{r}_{k'+1}^e\| < v_k^p T \tag{23}$$

is calculated.

Next, we handle the case where $\chi_{k'}^e \leq \pi/2$. Utilizing both $d_{k'+1} < 0$ and (17), we obtain

$$d_{k'} + \alpha_{k'} < \delta_{k'}. \tag{24}$$

Since $d_{k'} + \alpha_{k'}$ is positive,

$$(d_{k'} + \alpha_{k'})^2 < (\delta_{k'})^2 \tag{25}$$

is calculated.

Utilizing (10), the distance between $\mathbf{r}_{k'}^p$ and $\mathbf{r}_{k'+1}^e$ is calculated as

$$\|\mathbf{r}_{k'}^p - \mathbf{r}_{k'+1}^e\| = \|(d_{k'} + \alpha_{k'}, -d_{k'}^c)\|. \tag{26}$$

Utilizing (7), (26), and (25),

$$\|\mathbf{r}_{k'}^p - \mathbf{r}_{k'+1}^e\| < v_k^p T \tag{27}$$

is derived.

We proved that the distance between $\mathbf{r}_{k'}^p$ and $\mathbf{r}_{k'+1}^e$ is less than $v_k^p T$. At time index k', the pursuer heads towards the evader directly. Thereafter, the pursuer captures the evader at time index $k' + 1$. $\square$

4. Predict the Evader Position Two Steps Forward in Time Considering Noisy Environments

The pursuer utilizes sensor measurements to estimate the evader's location. To implement our chasing controller, the pursuer at time index k must estimate $\mathbf{r}_k^e$, $\mathbf{r}_{k+1}^e$, and $\mathbf{r}_{k+2}^e$. Let $\hat{\mathbf{r}}_k^e$ denote an estimate of $\mathbf{r}_k^e$.

4.1. Estimate $\mathbf{r}_k^e$ and $\mathbf{r}_{k+1}^e$

We discuss how to estimate $\mathbf{r}_k^e$, $\mathbf{r}_{k+1}^e$, and $\mathbf{r}_{k+2}^e$. Let $(x(k), y(k))$ denote the evader's location measured at time index k.

Recall we assumed that the evader's motions are smooth, so the evader's trajectory curves are derivative. Curve fitting methods are used to predict the target's position within two steps in the future. We utilize curve fitting methods for recent measurements: $(x(k - K + 1), y(k - K + 1))$, $(x(k - K + 2), y(k - K + 2))$, ..., $(x(k), y(k))$. Here, $K > 2$. This implies that we require more than two measurements.

Recent x coordinate measurements are as follows. $x(k - K + 1), x(k - K + 2), ..., x(k)$. We fit $x(k - K + 1), x(k - K + 2), ..., x(k)$ using the second order polynomials $x(n) = a_x * n^2 + b_x * n + c_x$. This second order polynomials represent the x coordinate trajectory of the evader within the recent K time indexes. (Similarly, we can use higher order (3 or more) polynomials to represent the x coordinate trajectory of the evader within the recent K time indexes. But, using higher order polynomials does not assure accurate prediction of the evader's position).

To solve this fitting problem, we utilize the following matrix form.

$$A * S = B. \tag{28}$$

Here, $A = \begin{pmatrix} (k - K + 1)^2 & k - K + 1 & 1 \\ (k - K + 2)^2 & k - K + 2 & 1 \\ ... & ... & ... \\ k^2 & k & 1 \end{pmatrix}$, $B = \begin{pmatrix} x_{k-K+1} & x_{k-K+2} & ... & x_k \end{pmatrix}^T$, and $S = \begin{pmatrix} a_x & b_x & c_x \end{pmatrix}^T$. We solve for S using pseudo-inverse methods.

$$S = (A^T * A)^{-1} * A * B. \tag{29}$$

Let $\hat{\mathbf{r}}_k^e[i]$ denote the ith element in $\hat{\mathbf{r}}_k^e$. We estimate the x coordinate of $\mathbf{r}_k^e$.

$$\hat{\mathbf{r}}_k^e[1] = Q_1 * S \tag{30}$$

where $Q_1 = \begin{pmatrix} k^2 & k & 1 \end{pmatrix}$. In addition, we estimate the x coordinate of $\mathbf{r}_{k+w}^e$ where $w \le 2$.

$$\hat{\mathbf{r}}_{k+w}^e[1] = Q_2 * S \tag{31}$$

where $Q_2 = \begin{pmatrix} (k + w)^2 & k + w & 1 \end{pmatrix}$.

Similarly, we estimate the y coordinates of $\mathbf{r}_k^e$, $\mathbf{r}_{k+1}^e$, and $\mathbf{r}_{k+2}^e$ using the second order polynomials $y(n) = a_y * n^2 + b_y * n + c_y$. Recall that this curve fitting method requires that we have more than two measurements, i.e., $K > 2$.

If we have only one measurement, then we set

$$\hat{\mathbf{r}}_1^e = \hat{\mathbf{r}}_2^e = (x(1), y(1)). \tag{32}$$

If we have only two measurements, then we set

$$\hat{\mathbf{r}}_2^e = (x(2), y(2)). \tag{33}$$

In addition, we set

$$\hat{\mathbf{r}}_3^e = (2 * x(2) - x(1), 2 * y(2) - y(1)). \tag{34}$$

Ref. (34) implies that we fit two measurements using the first order polynomials.

4.2. The Relationship Between the Estimate Error and the Controller Performance

We next show the relationship between the estimate error and the performance of our chasing controller. Usually, locations can be measured rather accurately. Hence, we assume that $\mathbf{r}_k^e \approx \hat{\mathbf{r}}_k^e$ in this subsection.

In this subsection, we assume that $q = \inf$. This implies that we analyze the relationship between the estimate error and the performance of our chasing controller, not considering the maximum turn rate of the pursuer.

Due to the estimate error, motion camouflage state ($\angle(\mathbf{r}_k^p - \mathbf{r}_k^e, \mathbf{r}_{k+1}^p - \mathbf{r}_{k+1}^e)$ is zero) cannot be achieved at every time index. We derive the equation for $\angle(\mathbf{r}_k^p - \mathbf{r}_k^e, \mathbf{r}_{k+1}^p - \mathbf{r}_{k+1}^e)$ under our chasing controller.

Theorem 3. *The pursuer moves by applying $\mathbf{u}_k$ at every time index k. Consider the case where the pursuer at time index k cannot access $\mathbf{r}_{k+1}^e$ accurately. Let $\hat{\mathbf{r}}_{k+1}^e$ denote an estimate of $\mathbf{r}_{k+1}^e$. $\mathbf{n}_{k+1} = \mathbf{r}_{k+1}^e - \hat{\mathbf{r}}_{k+1}^e$ is the error in the estimate. Let μ_{k+1} denote $\angle(\hat{\mathbf{r}}_{k+1}^e - \mathbf{r}_{k+1}^e, \mathbf{r}_{k+1}^p - \hat{\mathbf{r}}_{k+1}^e)$ for convenience. $\sin(\angle(\mathbf{r}_k^p - \mathbf{r}_k^e, \mathbf{r}_{k+1}^p - \mathbf{r}_{k+1}^e)) = \frac{\sin(\mu_{k+1})\|\mathbf{n}_{k+1}\|}{d_{k+1}}$. Here, d_{k+1} is $\|\mathbf{r}_{k+1}^p - \mathbf{r}_{k+1}^e\|$.*

Proof. See Figure 3 for an illustration of μ_{k+1}. Utilizing the geometry in this figure, we derive

$$\sin(\angle(\mathbf{r}_{k+1}^p - \hat{\mathbf{r}}_{k+1}^e, \mathbf{r}_{k+1}^p - \mathbf{r}_{k+1}^e)) = \frac{\sin(\mu_{k+1})\|\mathbf{n}_{k+1}\|}{d_{k+1}}. \tag{35}$$

Since the pursuer at time index k can only access $\hat{\mathbf{r}}_{k+1}^e$ instead of $\mathbf{r}_{k+1}^e$, we draw $\bar{L}_{k+1}$ intersecting $\hat{\mathbf{r}}_{k+1}^e$ such that $\bar{L}_{k+1}$ is parallel to L_k. See Figure 3 for an illustration. Utilizing our chasing controller $\mathbf{u}_k$, $\mathbf{r}_{k+1}^p$ is the heading point on $\bar{L}_{k+1}$. Utilizing the same argument as in Theorem 1, we obtain

$$\angle(\mathbf{r}_{k+1}^p - \hat{\mathbf{r}}_{k+1}^e, \mathbf{r}_k^p - \mathbf{r}_k^e) = 0. \tag{36}$$

Utilizing (35) and (36), we derive

$$\sin(\angle(\mathbf{r}_k^p - \mathbf{r}_k^e, \mathbf{r}_{k+1}^p - \mathbf{r}_{k+1}^e)) = \frac{\sin(\mu_{k+1})\|\mathbf{n}_{k+1}\|}{d_{k+1}}. \tag{37}$$

$\square$

Theorem 3 depicts that in the case where d_{k+1} is too small compared to $\sin(\mu_{k+1})\|\mathbf{n}_{k+1}\|$, $\angle(\mathbf{r}_{k+1}^p - \mathbf{r}_{k+1}^e, \mathbf{r}_k^p - \mathbf{r}_k^e)$ is large. In other words, the rotation rate of LOS is large. This implies that the pursuer may leave the motion camouflage state as the distance between the evader and the pursuer is too small.

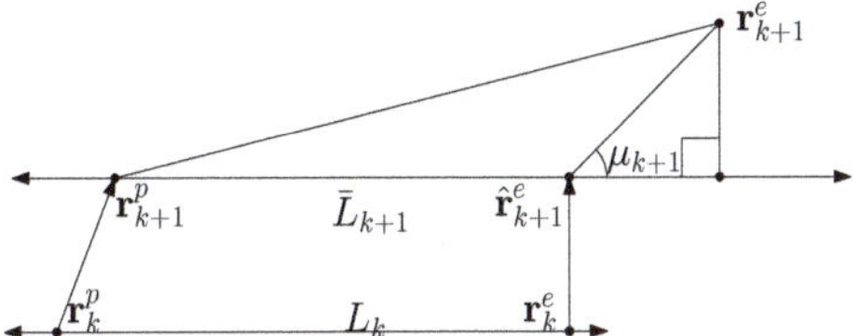

Figure 3. An illustration of μ_{k+1}.

5. MATLAB Simulation Results

In this section, we demonstrate the effectiveness of our chasing controller utilizing simulations. $(x(k), y(k)) - \mathbf{r}_k^e$ represents sensor measurement noise and is a vector with two elements. Each element in this vector has a Gaussian distribution with mean 0 and covariance 0.01. To predict the evader's location two steps forward in time, we utilize recent $K = 5$ measurements.

Initially, the evader is at $(0, 5000)$, and the pursuer is at the origin. $T = 2$ s. In addition, the initial speed of the pursuer is 150 m/s. In the terminal phase, v_k^p decreases with respect to k until v_k^p converges to 110 m/s. The pursuer's maximum acceleration is $a_m = 10$ m/s^2.

Initially, the pursuer's heading is $\mathbf{u}_0 = (1, 0)$. In the simulations, we set the maximum turn rate of the pursuer as $q = \pi/6$ radians per second.

MATLAB simulation runs for finite time, at the end of which the distance between the pursuer and the evader is less than 100 m. Recall that the evader's motion model is presented in (2).

The authors of [8] introduced Frenet-Serret frames [33] to model the movement of an evader.

$$\begin{aligned}
\mathbf{r}_{k+1}^e &= \mathbf{r}_k^e + T * v_k^e * \mathbf{x}_k^e \\
\mathbf{x}_{k+1}^e &= \mathbf{x}_k^e + T * v_k^e * u_k^e * \mathbf{y}_k^e \\
\mathbf{y}_{k+1}^t &= \mathbf{y}_k^e - T * v_k^e * u_k^e * \mathbf{x}_k^e,
\end{aligned} \tag{38}$$

where u_k^e is the steering (i.e., curvature) control of the evader at time index k, and v_k^e is the speed of the evader at time index k. Recall that $\mathbf{r}_k^e$ is the location of the evader at time index k. Moreover, $\mathbf{x}_k^e$ is the unit tangent vector to the trajectory of the evader at time index k, and $\mathbf{y}_k^e$ is the corresponding unit normal vector at time index k. We utilize (38) to simulate the motion of the evader.

As the first scenario, we set the evader's speed as $v_k^e = 100$ in m/s. In addition, the evader does not maneuver, i.e., $u_k^e = 0$. Figure 4 shows pursuer and evader trajectories. In this figure, the pursuer is depicted with red points, and the evader is depicted with blue points. Initially, the pursuer turns towards the evader with the maximum turn rate q. Thereafter, the pursuer converges to the motion camouflage state.

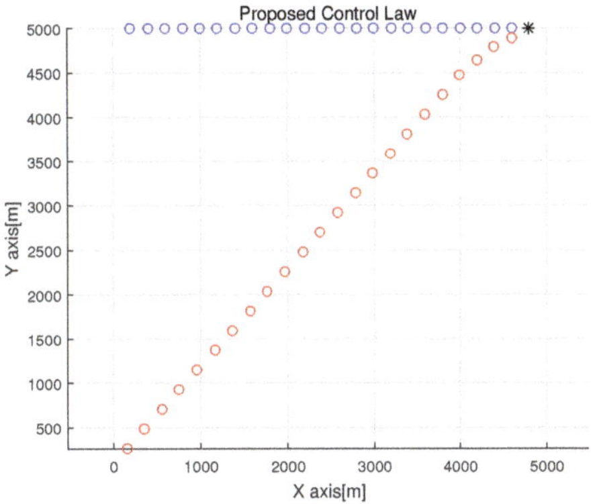

Figure 4. The system behavior for a constant velocity evader with $u_k^e = 0$ and $v_k^e = 100$.

In Figure 5, $l = \frac{r^e - r^p}{\|r^e - r^p\|}$ is plotted. $UnitX - proj$ and $UnitY - proj$ indicate the projection of l on the x-axis and y-axis respectively. The change of these values with respect to time indicates the rotation rate of LOS at every time index. For a pursuit-evasion system in the motion camouflage state, l converges. Hence, $UnitX - proj$ and $UnitY - proj$ also converge.

Figure 5 depicts that a pursuit-evasion system converges to the motion camouflage state. The chasing controller is designed to converge to the motion camouflage state at time index 0. But, we set $q = \pi/6$ radians per second in the simulation. Due to the constraint on the maximum turn rate, the pursuer cannot converge to the motion camouflage state at time index 0. The pursuer turns with the maximum turn rate until reaching the motion camouflage state. As the pursuer reaches the motion camouflage state, it stays in the state.

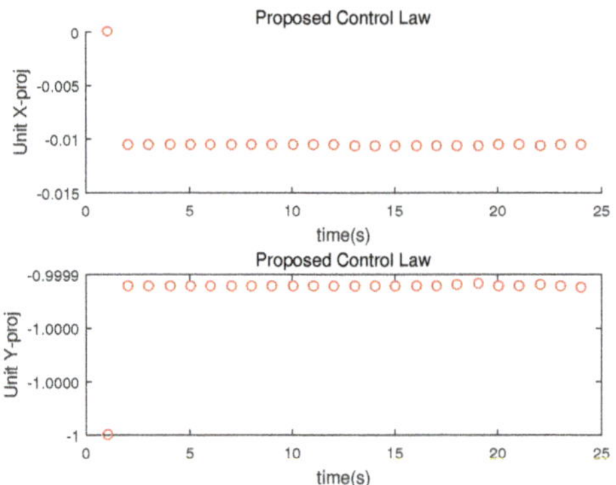

Figure 5. Plot of l for a constant velocity evader.

The top figure in Figure 6 depicts the pursuer's speed with respect to time. See that in the terminal phase, the pursuer adjusts its speed so that it can capture the evader with low speed. Since the relative speed of the evader with respect to the pursuer is low, the probability to the evader is also low. The bottom figure in Figure 6 depicts the distance between the evader and the pursuer with respect to time. Even though the pursuer slows down in the terminal phase, the distance monotonically decreases as time passes.

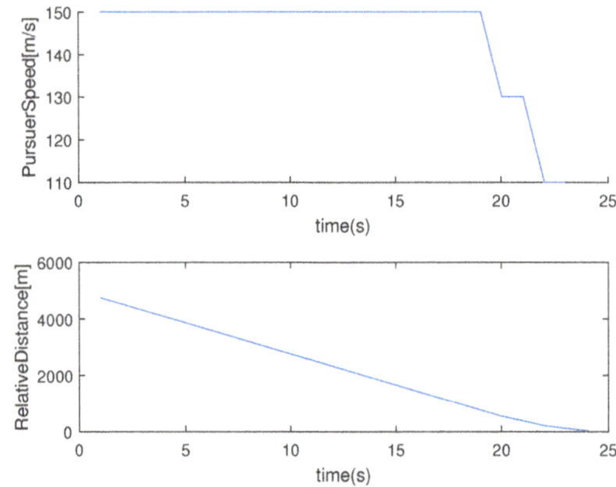

Figure 6. The (**top**) figure depicts the pursuer's speed with respect to time. The (**bottom**) figure depicts the distance between the evader and the pursuer with respect to time (constant velocity evader).

As the second scenario, we set the evader's speed as $v_k^e = 70 + 30 \times \sin(T \times k \times 0.01)$ in m/s. This implies that the evader changes its speed. In addition, the evader maneuvers using $u_k^e = 0.005 \times \sin(T \times k/100)$. Figure 7 shows pursuer and evader trajectories.

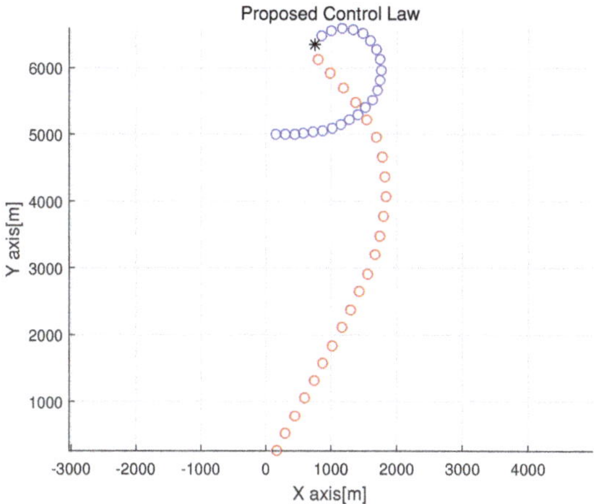

Figure 7. We set the evader's speed as $v_k^e = 70 + 30 \times \sin(T \times k \times 0.01)$ in m/s. In addition, the evader maneuvers using $u_k^e = 0.005 \times \sin(T \times k/100)$.

In Figure 8, $\mathbf{l} = \frac{\mathbf{r}^e - \mathbf{r}^p}{\|\mathbf{r}^e - \mathbf{r}^p\|}$ is plotted. $UnitX - proj$ and $UnitY - proj$ indicate the projection of $\mathbf{l}$ on the x-axis and y-axis respectively. In the motion camouflage state, $\mathbf{l}$ converges. Hence, $UnitX - proj$ and $UnitY - proj$ also converge. Figure 8 depicts that a pursuit-evasion system converges to the motion camouflage state. But, the pursuer leaves the motion camouflage state as the distance between the evader and the pursuer is too small. This phenomenon can be explained using Theorem 3.

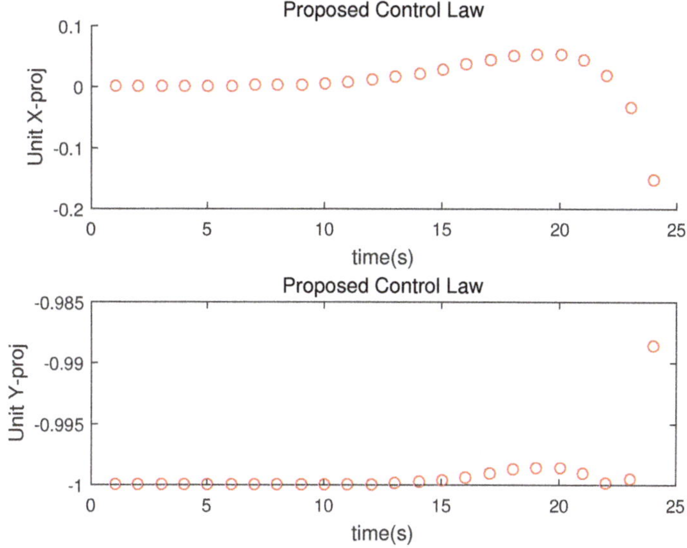

Figure 8. Plot of $\mathbf{l}$ for a maneuvering evader.

The top figure in Figure 9 depicts the pursuer's speed with respect to time. See that in the terminal phase, the pursuer adjusts its speed so that it can capture the evader with low speed. The bottom figure in Figure 9 depicts the distance between the evader and the pursuer with respect to time. Even though the pursuer slows down in the terminal phase, the distance monotonically decreases as time passes.

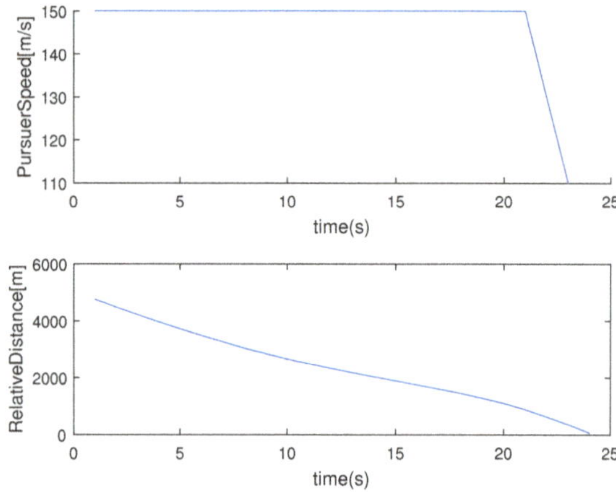

Figure 9. The (**top**) figure depicts the pursuer's speed with respect to time. The (**bottom**) figure depicts the distance between the evader and the pursuer with respect to time (maneuvering evader).

Analysis

We analyze the computational load of our control law. To compute the pursuer's heading vector $\mathbf{u}_k$ at each time index k, it takes 0.03 s. This is negligible compared to the sampling time interval $T = 2$ s. Hence, we can argue that latency due to the computational load is negligible.

Next, we run simulations while varying the initial heading angle of the pursuer. We set the evader's speed as $v_k^e = 100$ in m/s. In addition, the evader does not maneuver, i.e., $u_k^e = 0$.

We introduce two metrics, T_f and E_f. $T_f = k_f T$ is the spent time(seconds) to capture the evader. Here, k_f is the time index when the pursuer captures the evader. $E_f = \sum_{k=1}^{k_f} \|v_k^p \angle(\mathbf{u}_k, \mathbf{u}_{k-1})/T\|$. Here, $\angle(\mathbf{u}_k, \mathbf{u}_{k-1})/T$ is the pursuer's angular acceleration at time index k. E_f can be regarded as the energy (acceleration sum in m/s^2) required to capture the evader.

The pursuer's initial location is the origin. We calculate both T_f and E_f while changing the initial heading angle, say $H = atan2([0,1] \times \mathbf{u}_0, [1,0] \times \mathbf{u}_0)$, of the pursuer from 0 to 180 degrees. Here, the initial heading angle is measured counter-clockwise from the x-axis. Table 1 shows T_f and E_f while changing H. See that the evader is captured regardless of the initial heading angle of the pursuer.

Table 1. Results.

H (degrees)	T_f (s)	E_f (m/s^2)
0	47	238
30	49	321
60	49	242
90	49	164
120	49	242
150	49	321
180	51	405

6. Conclusions

This paper introduces a chasing controller to capture a high-speed evader with variable velocity. We handle the case where the pursuer moves with a variable speed and the angular acceleration of the pursuer is controllable. In the terminal phase, we slow down the pursuer's speed, while assuring that the distance between the evader and the pursuer monotonically decreases as time passes. By slowing down the relative speed of the pursuer with respect to the evader, we can reduce the probability of missing the high-speed evader. We demonstrate the performance of our chasing controller utilizing simulations. As our future works, we will verify the performance of our controller using experiments with mobile robots.

In this paper, we handled motion camouflage with respect to a fixed point at infinity. As our future works, we will develop control laws to handle motion camouflage with respect to a fixed point which is not at infinity. Assuming that the evader can readily observe optical flow, but only poorly detect looming, this pursuer's movement is hardly detected by the evader.

Conflicts of Interest: The author declares no conflict of interest.

References

1. Prasanna, H.M.; Ghose, D. Retro-Proportional-Navigation: A new guidance law for interception of high-speed targets. *J. Guid. Control Dyn.* **2012**, *35*, 377–386. [CrossRef]
2. Ghosh, S.; Ghose, D.; Raha, S. Capturability analysis of a 3D Retro-PN guidance law for higher speed nonmaneuvering targets. *IEEE Trans. Control Syst. Technol.* **2014**, *22*, 1864–1874. [CrossRef]
3. Shneydor, N.A. *Missile Guidance and Pursuit- Kinematics, Dynamics and Control*; Woodhead Publishing: Cambridge, UK, 1998.
4. Oh, J.; Ha, I. Capturability of the 3-dimensional pure PNG law. *IEEE Trans. Aerosp. Electron. Syst.* **1999**, *35*, 491–503.
5. Song, S.; Ha, I. A lyapunov-like approach to performance analysis of 3-dimensional pure PNG laws. *IEEE Trans. Aerosp. Electron. Syst.* **1994**, *30*, 238–248. [CrossRef]
6. Urakubo, T.; Kanade, T. Capturability analysis of a three dimensional guidance law with angular acceraltion input. In Proceedings of the American Control Conference(ACC), Washington, DC, USA, 27–29 June 2012; pp. 2551–2556.
7. Reddy, P.V.; Justh, E.W.; Krishnaprasad, P.S. Motion camouflage in three dimensions. In Proceedings of the IEEE Conference on Decision and Control, San Diego, CA, USA, 13–15 December 2006; pp. 3327–3332.
8. Justh, E.; Krishnaprasad, P. Steering laws for motion camouflage. *Proc. R. Soc. A* **2006**, *462*, 3629–3643. [CrossRef]
9. Galloway, K.S.; Justh, E.W.; Krishnaprasad, P.S. Motion camouflage in a stochastic setting. In Proceedings of the IEEE International Conference on Decision and Control (CDC), New Orleans, LA, USA, 10–11 December 2007; pp. 1652–1659.
10. Ghose, K.; Horiuchi, T.K.; Krishnaparasad, P.S.; Moss, C.F. Ecolocating bats use a nearly time-optimal strategy to intercept prey. *PLos Biol.* **2006**, *4*, 865–873. [CrossRef] [PubMed]
11. Mizutani, A.; Chahl, J.; Srinivasan, M. Insect behaviour: Motion camouflage in dragonflies. *Nature* **2003**, *423*, 604. [CrossRef] [PubMed]
12. Anderson, A.J.; McOwan, P.W. Model of a predatory stealth behaviour camouflaging motion. *Proc. R. Soc. B* **2003**, *270*, 489–495. [CrossRef] [PubMed]
13. Hernandez, J.D.; Moll, M.; Vidal, E.; Carreras, M.; Kavraki, L.E. Planning Feasible and Safe Paths Online for Autonomous Underwater Vehicles in Unknown Environments. In Proceedings of the IEEE/RSJ International Conference on Intelligent Robots and Systems (IROS), Daejeon, Korea, 9–14 October 2016; pp. 1313–1320.
14. Yilmaz, N.K.; Evangelinos, C.; Lermusiaux, P.F.J.; Patrikalakis, N.M. Path Planning of Autonomous Underwater Vehicles for Adaptive Sampling Using Mixed Integer Linear Programming. *IEEE J. Ocean Eng.* **2008**, *33*, 522–537. [CrossRef]

15. Maki, T.; Mizushima, H.; Kondo, H.; Ura, T.; Sakamaki, T.; Yanagisawa, M. Real time path-planning of an AUV based on characteristics of passive acoustic landmarks for visual mapping of shallow vent fields. In Proceedings of the OCEANS, Aberdeen, Scotland, 18–21 June 2007; pp. 1–8.

16. Lekkas, A.; Fossen, T. Integral LOS Path Following for Curved Paths Based on a Monotone Cubic Hermite Spline Parametrization. *IEEE Trans. Control Syst. Technol.* **2014**, *22*, 2287–2301. [CrossRef]

17. Breivik, M. Topics in Guided Motion Control of Marine Vehicles. Ph.D. Dissertation, Department Engneering Cybernetics, Norweigian University Science Technollogy, Trondheim, Norweigian, 2010.

18. Oh, S.; Sun, J. Path following of underactuated marine surface vessels using line-of-sight based model predictive control. *Ocean Eng.* **2010**, *37*, 289–295. [CrossRef]

19. Snider, J.M. *Automatic Steering Methods for Autonomous Automobile Path Tracking*; CMU-RI-TR-09-08; Robotics Institute Carnegie Mellon University: Pittsburgh, PA, USA, 2009.

20. Szwaykowska, K.; Zhang, F. Trend and Bounds for Error Growth in Controlled Lagrangian Particle Tracking. *IEEE J. Ocean. Eng.* **2014**, *39*, 10–25. [CrossRef]

21. Hoy, M.; Matveev, A.S.; Savkin, A.V. Algorithms for collision-free navigation of mobile robots in complex cluttered environments: A survey. *Robotica* **2015**, *33*, 463–497. [CrossRef]

22. Kim, J.; Zhang, F.; Egerstedt, M. Curve Tracking Control for Autonomous Vehicles with Rigidly Mounted Range Sensors. *J. Intell. Robot. Syst.* **2009**, *56*, 177–197. [CrossRef]

23. Petres, C.; Pailhas, Y.; Patron, P.; Petillot, Y.; Evans, J.; Lane, D. Path planning for autonomous underwater vehicles. *IEEE Trans. Robot.* **2007**, *23*, 331–341. [CrossRef]

24. Choset, H.; Lynch, K.; Hutchinson, S.; Kantor, G.; Burgard, W.; Kavraki, L.; Thrun, S. *Principles of Robot Motion*; MIT Press: Cambridge, MA, USA, 1988.

25. Lavalle, S.M. *Planning Algorithms*; Cambridge University Press: Cambridge, UK, 2006.

26. Lolla, T.; Ueckermann, M.P.; Yigit, K.; Haley, P.J.H., Jr.; Lermusiaux, P.F.J. Path planning in time dependent flow fields using level set methods. In Proceedings of the IEEE Conference on Robotics and Automation, Guangzhou, China, 11–14 December 2012; pp. 166–173.

27. Feit, A.; Toval, L.; Hovagimian, R.; Greenstadt, R. A travel-time optimization edge weighting scheme for dynamic re-planning. In Proceedings of the AAAI Workshop: Bridging the Gap Between Task and Motion Planning, Atlanta, GA, USA, 11 July 2010; pp. 26–32.

28. Kim, J.; Kim, S.; Choo, Y. Stealth Path Planning for a High Speed Torpedo-Shaped Autonomous Underwater Vehicle to Approach a Target Ship. *Cyber Phys. Syst.* **2018**, *4*, 1–16. [CrossRef]

29. Agrawal, P.; Dolan, J.M. Colregs-compliant target following for an unmanned surface vehicle in dynamic environments. In Proceedings of the IEEE/RSJ International Conference on Intelligent Robots and Systems (IROS), Hamburg, Germany, 28 September–2 October 2015; pp. 1065–1070.

30. Svec, P.; Shah, B.C.; Bertaska, I.R.; Alvarez, J.; Sinisterra, A.J.; von Ellenrieder, K.; Dhanak, M.; Gupta, S.K. Dynamics-aware target following for an automomous surface vehicle pperating under Colregs in civilian traffic. In Proceedings of the IEEE/RSJ International Conference on Intelligent Robots and Systems (IROS), Tokyo, Japan, 3–8 November 2013; pp. 3871–3878.

31. Svec, P.; Thakur, A.; Raboin, E.; Shah, B.C.; Gupta, S.K. Target following with motion prediction for unmanned surface vehicle operating in cluttered environments. *Auton. Robot.* **2014**, *36*, 383–405. [CrossRef]

32. Raju, V.; Krishnaprasad, P.S. Motion camouflage in the presence of sensory noise and delay. In Proceedings of the 2016 IEEE 55th Conference on Decision and Control (CDC), Las Vegas, NV, USA, 12–14 December 2016, pp. 2846–2852.

33. Carmo, M.D. *Differential Geometry of Curves and Surfaces*; Prentice Hall: Upper Saddle River, NJ, USA, 1976.

Article

A Real-Time Hydrodynamic-Based Obstacle Avoidance System for Non-holonomic Mobile Robots with Curvature Constraints

Pei-Li Kuo, Chung-Hsun Wang, Han-Jung Chou and Jing-Sin Liu

Institute of Information Science, Academia Sinica, Nangang, Taipei 115, Taiwan; r01525004@ntu.edu.tw (P.-L.K.); frankwang50302@gmail.com (C.-H.W.); r02221012@gmail.com (H.-J.C.)
* Correspondence: liu@iis.sinica.edu.tw

Received: 11 September 2018; Accepted: 18 October 2018; Published: 2 November 2018

Abstract: The harmonic potential field of an incompressible nonviscous fluid governed by the Laplace's Equation has shown its potential for being beneficial to autonomous unmanned vehicles to generate smooth, natural-looking, and predictable paths for obstacle avoidance. The streamlines generated by the boundary value problem of the Laplace's Equation have explicit, easily computable, or analytic vector fields as the path tangent or robot heading specification without the waypoints and higher order path characteristics. We implemented an obstacle avoidance approach with a focus on curvature constraint for a non-holonomic mobile robot regarded as a particle using curvature-constrained streamlines and streamline changing via pure pursuit. First, we use the potential flow field around a circle to derive three primitive curvature-constrained paths to avoid single obstacles. Furthermore, the pure pursuit controller is implemented to achieve a smooth transition between the streamline paths in the environment with multiple obstacles. In addition to comparative simulations, a proof of concept experiment implemented on a two-wheel driving mobile robot with range sensors validates the practical usefulness of the integrated system that is able to navigate smoothly and safely among multiple cylinder obstacles. The computational requirement of the obstacle avoidance system takes advantage of an a priori selection of fast computing primitive streamline paths, thus, making the system able to generate online a feasible path with a lower maximum curvature that does not violate the curvature constraint.

Keywords: obstacle avoidance system; harmonic potential field; curvature constraint; non-holonomic mobile robot

---

1. Introduction

Due to advances in sensing, actuation, communication, computing, storage, and AI technologies with affordable costs, increasing deployment and applications of intelligent autonomous mobile robots or ground vehicles for long-term operation are prevalent. The integrated platforms at our disposal are intended for missions such as SAR (search and rescue), autonomous driver-less driving, manufacturing, and surveillance in cluttered environments [1–40]. Along with an increasingly heavy interaction between human and robots, update-to-date but cost-effective implementation or prototyping of intelligent mobile robot systems to accomplish missions autonomously employing recent advances in localization, mapping, and navigation have been the focus of some endeavors put into the robotic systems, as shown, for instance, in References [39,41–43].

Obstacle avoidance and motion planning [2–4,31] are essential for the completion of missions in a smooth and optimal manner. A variety of obstacle avoidance algorithms are designed and implemented for navigating a mobile robot to avoid static and dynamic obstacles in open space or in narrow passages. The artificial potential field (APF) approach is one of the most well-known reactive online obstacle avoidance methods applicable in known or unknown environments [5]. The goal position of the robot is assigned as an artificial attractive potential and obstacles are applied as artificial repulsive forces. Then, the collision avoidance path is derived by using the gradient of the linear superposition of each potential. However, by following the gradient path of APF, navigation routes generated by APFs suffer from the local minima due to the presence of obstacles, i.e., the set of trap positions where the gradient of APF vanishes so that the robot gets stuck there, thus preventing the robot from reaching the target or lower the navigation efficiency. The number of obstacles affects the number of local minima significantly [44]. In general, it is desired that no other local minima except the goal exist in APF so that the navigation is efficient. For this purpose, hydrodynamic or harmonic (or velocity) potential functions (HPFs) (see References [7,9,15,17]) derived from the velocity potential of the solution to the boundary value problem of Laplace's Equation with appropriate boundary conditions in a computational domain are proposed as an appealing class of APFs for navigation. The properties of HPFs such as min-max principle and superposition in the context of path planning were proved in the foundation work [7,15,17]. To ensure the repulsion of the flow from the obstacles, one effective way is to impose two types of boundary conditions on the obstacle and domain boundaries for the solution to the Laplace's Equation in a computational domain: Dirichlet type (the potentials on the boundaries of obstacles and domain are assigned a constant high value, i.e., the fluid motion on the obstacle boundary is along the normal direction of the obstacle boundary/wall) and Neumann type (i.e., the flow cannot pass through the boundary, or the fluid motion on the obstacle boundary and wall is parallel to the tangential direction of the obstacle boundary since the normal component is null). For path planning, the HPF has the property of the min-max principle, so that no local minima other than the goal in the interior of cluttered or bounded environments with state constraints of a point robot are defined by Dirichlet boundary conditions or Neumann boundary conditions on the borders of the obstacles and computational domain.

Motion planning based on hydrodynamic potential applies (1) different fundamental elements such as a point sink (representing the goal), a point source (representing the robot location), or a uniform flow (defined as a flow with constant speed in a prespecified direction) plus a doublet (representing the obstacle), and their superposition, or (2) velocity potential solution to the Laplace's Equation with appropriate boundary conditions, to create a new HPF. The gradient of the HPF or the streamline that defines the vector field of the path at every point can be computed efficiently; analytically for a simple obstacle shape such as a circle or numerically. There are a few simulations showing that a vehicle modeled as a point (fluid) particle could smoothly navigate without collision with the (circular, elliptical, rectangle, or arbitrary-shaped) obstacles [6,8,11–18,26,30] by following streamlines from a variety of start points in an environment composed of multiple obstacles. To build an APF via hydrodynamics potential, Reference [15] proposed a panel method by first approximating an arbitrarily shaped obstacle by an enclosing polygon (set of panels). Each obstacle panel is treated as a source/sink with the strength adjusted to make the obstacle a repelling potential function by summing the HPF of each panel, i.e., with nonzero outer normal velocity at the obstacle panel representative point. Wang et al. [16] introduced a reactivity parameter to adjust the amplitude of the path's deflection around an obstacle and an optimal 3D path is obtained by a genetic algorithm. For path planning on an unstructured terrain consisting of meshes of different size and geometry of computational domain discretized by finite element, Reference [25] proposed to use a graph search to generate an initial path from the start to the goal, then used streamlines to smooth the initial path.

Planning and optimization of state and input trajectories for stabilizing non-holonomic mobile robots studied in this paper or more general non-holonomic systems such as a car with trailers [33] in multiple obstacles environment subject to bounded state constraints (such as the environment constraint and path and its derivative constraints) and input constraints is challenging. The motion planners have to plan in the full state space (the space of position, heading, speed, curvature, and/or curvature derivative) and deal simultaneously with constraints of collision avoidance and non-holonomic constraint and admissible input (such as the computation of the invariant set or reachable set of the system [31]). Approaches that can either achieve the optimal or near-optimal position and heading and velocity or higher order derivatives simultaneously or sequentially are proposed for non-holonomic mobile robots. A dipolar potential field is combined with discontinuous state feedback for navigating a non-holonomic mobile robot in the presence of obstacles in Reference [32]. Pontryagin Maximum Principle in variational form was employed in Reference [33] to obtain a convergent sequence of controls for optimal motions of general non-holonomic systems with state and input constraints. New methods to non-holonomic path planning are given in Reference [34], which proposed to steer the non-holonomic mobile robot without additional constraints via rotation and linear translation using trigonometric switch inputs, a generic APF-based method via deforming a feasible path of non-holonomic systems (e.g., a mobile robot with trailers) without the violating non-holonomic constraint [35], and the ones based on partial differential equations other than the Laplace's Equation coined by References [7,24], which applied a parabolic partial differential equation, and Reference [29], which used the Navier–Stokes equation of viscous flow for path planning. It is noted that HPF is interpreted in heat conduction, instead of hydrodynamics, in a recent work [28] with a demonstration of a real-time mobile robot navigation experiment.

To produce a feasible path that avoids obstacles in a cluttered environment, waypoint navigation method has been a viable approach used to generate a sequence of waypoints connected via a path primitive [36,37] for reactive trajectory generation, while the entire path satisfies the smoothness requirement or some other criterion and respects the kinodynamic constraints imposed on the vehicle motion such as the constraints on the instantaneous velocities that can be achieved [2,22]. However, the path primitives have to be recomputed whenever some of the waypoints built between the start and the goal are changed. Furthermore, the entire path could be lengthy because of many detours. HPF-based non-holonomic path planning for a mobile robot subject to kinodynamic constraints [17,22,26] takes advantages of features of streamlines that are very appropriate for building a directional navigation system:

(i)   Streamlines are rich, thus, enabling the selection of appropriate paths for the planner. The desired state trajectory to be followed by a robot is determined only by the input defined by HPFs, i.e., along a streamline-based trajectory compatible with the kinodynamic constraints of the motion, even in high-speed motion [12].

(ii) In many applications, the smoothness of trajectories is essential. Trajectories generated by HPF approaches are the integral curves of the gradient vector field of HPF. The trajectories that are smooth are readily executable.

(iii) Streamlines can be computed offline systematically based on prior obstacle information (distribution, i.e., shape, size, location, and number) without the waypoints and path primitives, thus, being more predictable.

(iv) Notably, the HPF-based path planner is a complete [7] and anytime algorithm [10], in which the streamlines generated cover the free regions of the workspace.

Lau et al. [18] provide a streamline-based kinodynamic motion planning approach to avoid elliptical obstacles, guaranteeing that both velocity and curvature are within limits by adjusting the strength of a source and a sink if a portion of the 3D trajectory violates the kinematic constraints. Recent work [26] used the gradient of HPF as an additional input to alter the motion pattern of a two-wheeled drive mobile robot based on the stabilizing controller design using an invariant manifold to avoid the obstacles, thus, extending the guidance method of Reference [22] based on the HPF only. Along this line of HPF-based non-holonomic path planning work, this work starts with the point that the curvature constraint, viewed as a constrained input to (1), significantly restricts the selection or generation of allowable paths to be followed in a cluttered environment. We elaborate on demonstrating the potential of the hydrodynamics-based motion planning approach of (1) subject to the curvature constraint. An obstacle avoidance system using three primitive streamline-based paths and a path selection strategy that makes the avoidance of small obstacle easier is proposed.

For avoiding multiple obstacles, a pure pursuit algorithm [20,21] is implemented in this paper for the purpose of enabling a smooth and safe transition using streamline changing between obstacles without violating the curvature constraint. The experiment is conducted for a constant speed circular non-holonomic mobile robot to navigate smoothly in real-world partially unknown environments cluttered with cylinder-shaped obstacles. The Dr. Robot X80 robot (Dr Robot Inc. in Markham, ON, Canada), which was equipped with a low-cost sonar and infrared sensors to detect obstacles during motion, is used as the navigation platform.

The paper is organized as follows. Section 2 gives a brief introduction of a mobile robot with two independently driven wheels for the experiment. Section 3 mentions the harmonic potential field approach for avoiding cylindrical obstacles. Then, we propose three primitive paths based on streamlines and a distance-based path selection strategy of a primitive path for obstacle avoidance of curvature-constrained non-holonomic mobile robots. In Section 4, we propose a new real-time obstacle avoidance system using primitive paths and streamline-changing via the pure pursuit algorithm. Comparisons and the proof of concept experimental result in a simple cluttered environment are presented in Section 5. Section 6 ends with the conclusions of the paper.

2. The Mobile Robot with Range Sensors

2.1. Wheeled Mobile Robot System

We implemented our real-time hydrodynamics-based obstacle avoidance algorithm on Dr. robot X80, a wireless two-wheeled drive mobile robot platform. Figure 1 shows the configuration and the front view of the mobile robot. The X80 mobile robot is an integrated electronic and software robotic system. It can be designed through a set of ActiveX control components (SDK) developed for C/C++. A DUR5200 Ultrasonic Sensor and GP2Y0A21YK Sharp Infrared Sensor are equipped on the mobile robot. The robot platform is further modified to be equipped with a laser scanner, Kinect, and a laptop. The navigation algorithm runs directly on the remote PC through wireless communication.

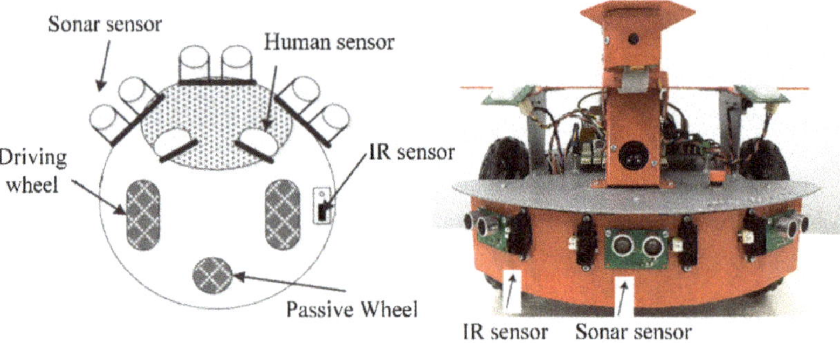

Figure 1. The outline of the circular non-holonomic mobile robot Dr. Robot X80 used for the experiment.

2.2. Kinematic Model

Figure 2 illustrates the kinematic model of the differential-drive circular mobile robot with radius r_{Robot}. The mobile robot is modeled as a representative point of the circular mobile robot so that its non-holonomic constraint of rolling without the slipping of wheels is described by the kinematics of the non-holonomic unicycle (1).

$$
\begin{aligned}
\dot{x} &= U\cos\theta \\
\dot{y} &= U\sin\theta \\
\dot{\theta} &= \omega
\end{aligned}
\tag{1}
$$

where (x, y) denotes the coordinates and θ denotes the orientation (heading), U and ω denote the velocity and angular velocity, respectively. As long as the velocity vector is tangent to the path the unicycle follows, the non-holonomic constraint (1) is automatically satisfied. The non-holonomic constraint (1) limits the maneuverability of the vehicle motion so that no instantaneous lateral motion is allowed. The mobile robot is controlled by the low level velocity control of two wheels driven by DC motors independently. The velocities of the mobile robot are determined by the actuated wheel velocities via the one-to-one correspondence given by

$$
\begin{bmatrix} U \\ \omega \end{bmatrix} = \begin{bmatrix} \frac{r}{2} & \frac{r}{2} \\ -\frac{r}{d} & \frac{r}{d} \end{bmatrix} \begin{bmatrix} w_L \\ w_R \end{bmatrix}
$$

where w_L and w_R denote the left and right wheel velocity, respectively; r is the wheel radius; d is the distance between the two wheels.

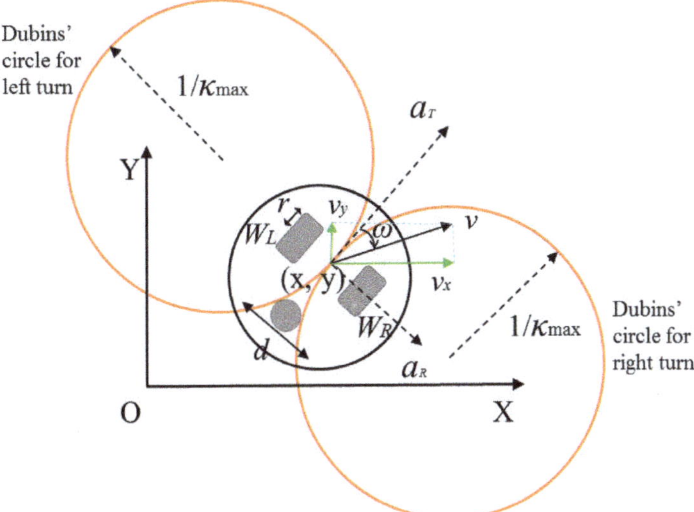

Figure 2. The mobile robot's kinematic model with the curvature constraint, where a positive curvature denotes a right (clockwise) turn. The robot is assumed as a circle. The robot velocity U and its x-component v_x, y-component v_y, the radial acceleration a_R, and tangential acceleration a_T, are shown. The O-XY coordinate system is the global coordinate frame and a local frame is attached to a representative point (the center) of the circular mobile robot.

2.3. Obstacle Detector

Obstacle avoidance relies on the detection of obstacles for the real-time operation of mobile robots. The sensor configuration is shown in Figure 3. A DUR5200 Ultrasonic Sensor and GP2Y0A21YK Sharp Infrared Sensor are equipped on the mobile robot. There are three sonars (Sonar 1, Sonar 2, and Sonar 3) and four infrared sensors (IR 1, IR 2, IR 3, and IR 4) on the mobile robot, where θ_1 equals $12°$, θ_2 equals $18°$, and θ_3 equals $15°$. The detecting range of an ultrasonic sensor is from 4 to 255 cm, while the detecting distance range of the IR sensor is between 10 and 80 cm. The sensors' update rate are both 10 Hz. We assume the obstacle is located in the direction of the sensor if only a single sensor detects an obstacle. Otherwise, when two sensors detect an obstacle at the same time, we assume that the obstacle lies on the bisector of these two sensors' directions. Figure 4 shows the detection of an obstacle. For instance, if Sonar 2 detects an obstacle, the obstacle is located in front of the robot with an azimuth angle $0°$. Likewise, Sonar 1, Sonar 3, IR 1, IR 2, IR 3, and IR 4 detect the obstacle with azimuths of $45°$, $-45°$, $-30°$, $-12°$, $12°$, and $30°$, respectively. If both Sonar 2 and IR 3 detect an obstacle, then the obstacle's direction will be 6°, which is between Sonar 2 and IR 3. The estimated obstacle location is transformed into the global frame for the motion planner. The estimation and localization errors are accommodated by a safety distance r_{Safe} in the practical implementation of the navigation system, as in our experiment presented in Section 5.

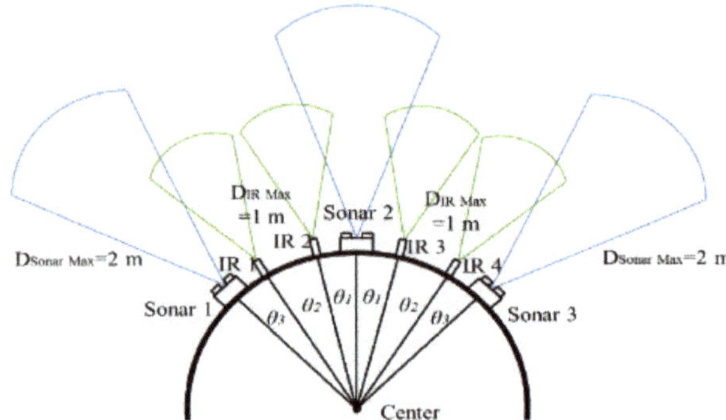

Figure 3. The configuration of the sonar and infrared sensors and their sensing ranges in Dr. Robot X80.

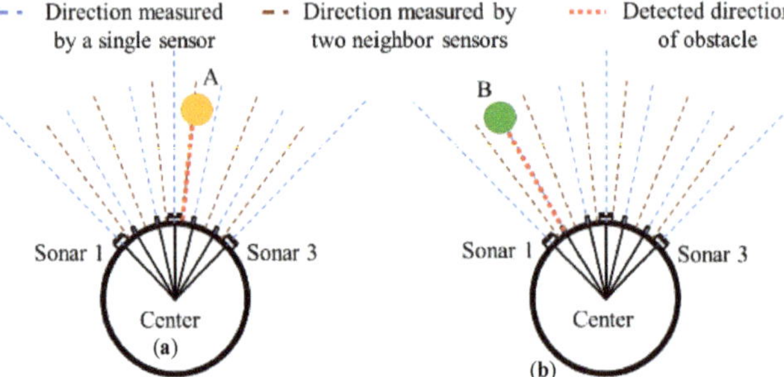

Figure 4. The scenarios of obstacle detection. The ray of each sensor is presented as a dashed line. As the rays intersect with an obstacle, the nearest intersection point is retained. (**a**) An obstacle is in front of the robot. Sonar 2 detects an obstacle with distance less than 150 cm, the robot knows the obstacle is located in front of the robot with azimuth angle 0°. (**b**) Both Sonar 2 and IR 3 detect the obstacle, and the obstacle's direction will be 6 °, which is between Sonar 2 and IR 3. The distance of the obstacle is the average of the data received from the two sensors.

3. Obstacle Avoidance Model by Harmonic Potential Field with Curvature Constraint

Extracting the topologically different candidate trajectories from a set of trajectories based on the equivalence relations and optimization schemes is a means of efficient online local trajectory optimization [44]. To provide a collision-free path rapidly to the planner, it is beneficial to extract a priori the streamlines that specify the essential motion pattern details of desired navigation trajectories such as curvature constraint and free of multiple local minima. In this section, we briefly summarize the C^2 smooth path produced by the streamlines of the harmonic potential field. Then we present a mobile robot navigation system based on the streamlines of HPF that satisfy the curvature constraint. The system is based on three primitive paths extracted from the streamlines and pure pursuit algorithm for streamline-changing.

3.1. Harmonic Potential Function and Streamlines

Harmonic potential functions are solutions to Laplace's Equation, so functions generated by Laplace's Equation do not exhibit local minima [7]. In a two-dimensional computational domain **D** of

Euclidean space, the velocity potential $\phi \in C^2(\mathbf{D})$ is a solution of Laplace's Equation $\nabla^2 \phi = 0$ with the given boundary conditions that govern the flow of the non-viscous, incompressible, irrotational fluid particle motion at every point of the domain. Laplace's Equation can be solved analytically in simple cases or by numerical methods in a general situation. For numerical methods, Laplace's Equation in a computational domain $\mathbf{D}$ could be discretized using both the finite difference method or the finite element method, resulting in a system of linear equations for the solution of the potential value in a grid or mesh environment. The Jacobi iteration, Gauss-Seidel iteration, and SOR (successive over-relaxation) iteration methods in a grid environment can be employed to solve the linear equations with an efficient accuracy. A log-space algorithm with GPU acceleration was proposed to fix the numerical precision problem of the numerical solution of a linear system of linear equations at the grid points that have nearly vanishing gradients resulting from discretized Laplace's Equation via the finite difference methods [10]. A streamline indicates the local flow direction: its tangent at every point (vector field) is in the direction of the local fluid velocity associated with the flow defined in Equation (2).

$$\mathbf{u} = \nabla \phi(x, y) \tag{2}$$

That is, the gradient of the obtained potential values gives the streamline or the direction of velocity at each grid [7], offering an explicit specification of the heading of a smooth, natural-looking path for navigation. Higher order path characteristics such as curvature can also be obtained for streamlines, thus, being more predictable.

Consider a mobile robot at $\mathbf{x} = [x\ y]^{\mathrm{T}}$ modeled as a fluid particle moving with velocity $\left[v_x, v_y\right]^{\mathrm{T}}$ in the Cartesian space. It moves in the $+x$-axis direction with a forward/longitudinal speed U to avoid a circular obstacle of radius $r_{Obstacle}$ (or an enlarged r_{Obs} for safety) located at the origin. The velocity potential field $\phi(x, y)$ can be represented as the superposition of a uniform rectilinear flow and a doublet [19] as

$$\phi(x, y) = U + \frac{A}{x^2 + y^2} x \tag{3}$$

where $A = Ur^2$. According to [19], the robot's velocity $\left[v_x, v_y\right]^{\mathrm{T}} := [u, v]^{\mathrm{T}}$ in the Cartesian coordinate is determined by the gradient $\nabla \phi(x, y)$

$$u = \frac{\partial \phi(x, y)}{\partial x} = U - \frac{2Ax^2}{\left(x^2 + y^2\right)^2} + \frac{A}{x^2 + y^2}, \quad v = \frac{\partial \phi(x, y)}{\partial y} = -\frac{2Axy}{\left(x^2 + y^2\right)^2} \tag{4}$$

In practice (and in our experiment in Section 5), we assume that the linear speed $U = \sqrt{u^2 + v^2}$ is normalized to unity while its direction of motion is preserved. Then, the normalized velocity and the corresponding acceleration of each point on the streamline in the uniform flow are given by (5) and (6).

$$u_N = \frac{u}{U}, \quad v_N = \frac{v}{U}, \tag{5}$$

$$a_x = \frac{\delta u_N}{\delta x} u_N + \frac{\delta u_N}{\delta y} v_N, \quad a_y = \frac{\delta v_N}{\delta x} u_N + \frac{\delta v_N}{\delta y} v_N \tag{6}$$

Furthermore, curvature and deviation of curvature can be derived by velocity and acceleration as

$$\begin{aligned}
\kappa &= \frac{u_N a_y - v_N a_x}{\left(u_N^2 + v_N^2\right)^{3/2}} \\
&= -2Ay \sqrt{A^2 + 2AU(-x^2 + y^2) + U^2(x^2 + y^2)^2} \\
&\quad \times \frac{\left[A^2 + 2AU(x^2 + y^2) + U^2(-3x^4 - 2x^2 y^2 + y^4)\right]}{(x^2 + y^2)\left[A^2 + 2AU(-x^2 + y^2) + U^2(x^2 + y^2)^2\right]^2}
\end{aligned} \tag{7}$$

$$\dot{\kappa} = \frac{\delta\kappa}{\delta x}\frac{dx}{dt} + \frac{\delta\kappa}{\delta y}\frac{dy}{dt} = \frac{\delta\kappa}{\delta x}u_N + \frac{\delta\kappa}{\delta y}v_N$$

$$= \frac{24AUxy\left[\begin{array}{c} A^4\left(x^2-y^2\right)^4 - U^4\left(x^2-y^2\right)\left(x^2+y^2\right) \\ -2A^3U\left(x^2+y^2\right)^2 + 2AU^3\left(x^2+y^2\right)^4 \end{array}\right]}{\left(x^2+y^2\right)^2\left[A^2+2AU\left(-x^2+y^2\right)+U^2\left(x^2+y^2\right)^2\right]^3}$$

From the above equations, given a start position and a circle with a known radius at the origin, a streamline (integral curve) can be derived by numerical integration of the velocity vector field that specifies the tangent of the path or the robot heading at each point. In addition, the higher order path characteristics such as curvature are easily computable as well, making the path to be followed more predictable. That is, the velocity vector field of streamlines serves as a vector field for the guidance of the robot's motion everywhere in the obstacle-free region. Therefore, in path planning applications, streamlines provide a pool of systematic paths that have an explicit or analytic vector field for the tangent to the path as the path specification, and their higher order path properties such as curvature could be easily computed as well. However, there are several drawbacks for robots to purely follow streamline paths. First of all, we assume that the curvature of unicycle kinematics (1) is constrained by its upper bound. The curvature of streamline it follows has a larger curvature as the distance to the obstacle gets closer. For example, Figure 5 depicts that streamline starting from $(-5, 0.2)$ exhibits the maximum curvature. Second, the paths with a smaller curvature are longer and keep an unnecessary distance with the obstacle. Moreover, for paths with an initial position further than the radius of the obstacle in the direction of the x-axis, it is not necessary to follow a streamline path because a robot can pass the obstacle straightly, such as the paths with a start position further than the radius of obstacle with the x-axis in Figure 5. Therefore, we provide an improved streamline-based approach in the following section.

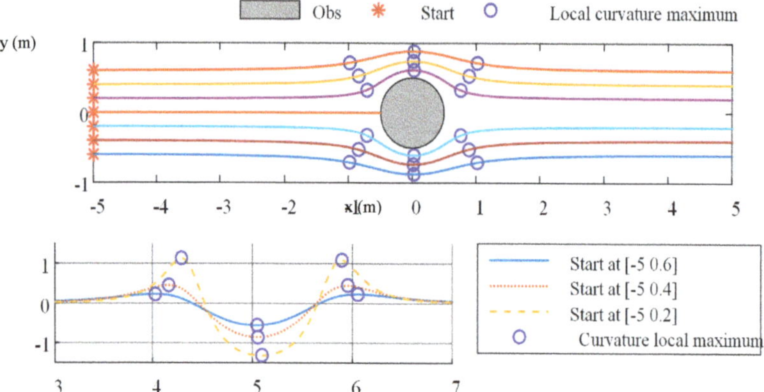

Figure 5. The streamline paths (upper plot) with different start positions and their curvatures (lower plot) for a uniform flow around a circular obstacle. The path to avoid an obstacle is similar to the streamline of the fluid flow around a cylinder. The very large turning radii of some of the paths closer to the obstacle may become infeasible when the curvature constraint imposed on the vehicle motion is accounted for.

3.2. Three Primitive Paths with Curvature Constraint

In Figure 6 we show a scenario of three alternative options in which another obstacle is encountered immediately after circumventing one obstacle. Therefore, the robot has to turn sharply to prevent an imminent collision, as shown in Figure 6a. On the other hand, passing the first obstacle with a low curvature streamline allows the robot to pass these two obstacles from the same side while there is not enough space to pass through between the two obstacles using a streamline and complying with the curvature constraint, as shown in Figure 6b. The third way shown in Figure 6c is passing the obstacle from the farther side by a sharp turn. The example demonstrates that the different timings for applying these two strategies are related to the upcoming obstacle's position after passing the first obstacle.

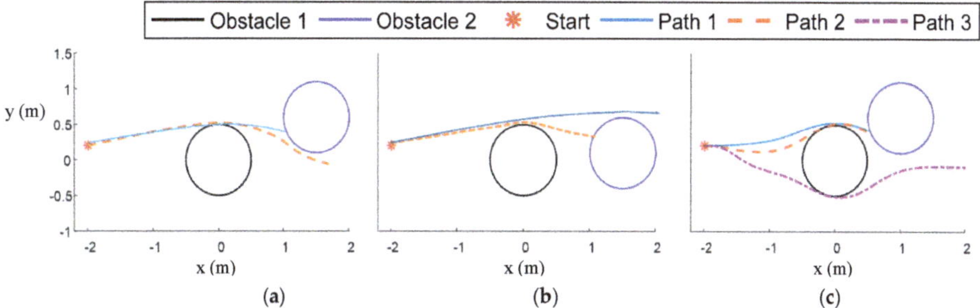

Figure 6. Sequential obstacle avoidance via sharp turn and low-curvature turn in the situation of different obstacle configurations (**a**) From the space between two obstacles; (**b**) From the same side of both obstacles; (**c**) From farther side.

There are two strategies for local obstacle avoidance based on streamlines. For local obstacle avoidance, a point robot could circumvent an obstacle from its left or its right side. Alternatively, it could also pass an obstacle with maximum curvature via a sharp turn. We first describe the three primitive paths for the avoidance of a single circular obstacle. For multiple obstacles, the same strategy is employed sequentially for every detected obstacle that is to be avoided. The following are the details and procedures of a sharp turn and low curvature turn for avoiding the obstacle by using streamline paths. Three collision-free primitive paths most aligned to the current robot's heading that achieve compliance with the curvature constraint of the mobile robot are proposed by exploiting the richness of the streamlines that cover the computational domain of Laplace's Equation and the rather intuitive property that the deflection and curvature of a streamline become smaller as it is farther from the obstacle. For simplicity of illustration, we refer to Figure 7. It is assumed that the robot is moving in the $+x$-direction and a circular obstacle of radius r_{obs} is located at the origin so that the maximum curvature of streamline occurs at the y-axis.

(A) Continuous-curvature sharp left or right turn

Consider the avoidance of the nearest obstacle within the sensing range in front of the robot. Two curvature-constrained streamline-based left or right turn paths could be used as two primitive paths to ensure the safe navigation from the left or right side of the obstacle based on the obstacle's radius. In particular, we look for the two streamlines corresponding to the left turn and right turn with a curvature maximum equal to the maximum curvature κ_{max}. The desired streamline is obtained by shifting the selected streamline in a parallel way until its curvature maximum point grazes the obstacle boundary.

Figure 7 illustrates an example of a hydrodynamic streamline path with different curvature constraints. From the curvatures of flow depicted in Figure 7a (similar to the scenario of Figure 5), if the initial y-position is further away from the center of the obstacle, a collision-free path is a nearly straight streamline with a smaller curvature. In addition, we identify that the points with local maximum curvature on a streamline are located at the y-axis, assuming that for simplicity the robot is moving forward in the x-direction. In order to verify the curvature constraint for a streamline path, the points of maximum curvature of a streamline have to be found so that it is sufficient to search over the streamlines that satisfy the curvature constraint. In the scenario depicted in Figure 7, the maximum curvature of a streamline path is smaller when further from the obstacle, and there are two points with a local maximum curvature in a single path. They lie on the y-axis and are identified first by a binary search presented in Algorithm 1. Algorithm 1 starts with a point $(0, yLow\_max)$ with curvature (7) not larger than the maximum

$$\kappa(0, yLow\_max) \leq \kappa_{max} \tag{8}$$

This point *yLow_max* with the largest magnitude |*yLow*| is identified by increasing the y coordinate from the upper-most border point $(0, a)$ of the obstacle centered at $(0, 0)$ with radius a. Then, Algorithm 1 of binary search is used to locate the point $(0, yLow\_min)$ between $(0, yLow\_max - a)$ and $(0, yLow\_max)$ with a curvature κ equal to the maximum allowed curvature κ_{max}. Then, the streamline path is generated with a velocity U in (5) by numerical integration initialized with $(0, yLow\_min)$. In this way, we can find two streamlines S_1, S_2 with a given maximum curvature with starts at a different location from the current robot position. The streamlines in the region between S_1, S_2 do not comply with the curvature constraints. The modified streamlines obtained by pulling the streamlines S_1, S_2 back to the current robot position are the paths, as shown in Figure 7b.

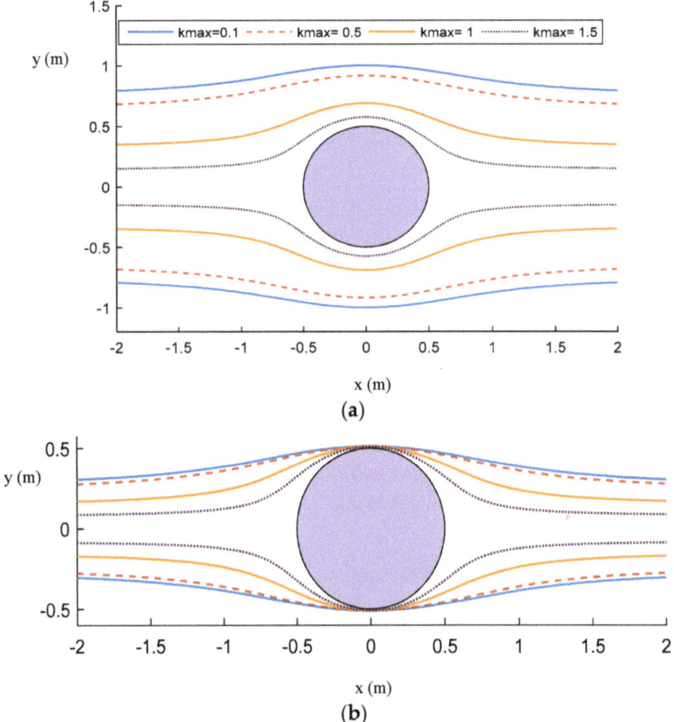

Figure 7. The concept of sharp turn streamline paths to circumvent a circular obstacle with the center placed at the origin in the *x*-*y* plane. (**a**) Streamline paths with different maximum curvatures. The curvature of the streamline is larger as it is closer to the circle. (**b**) Feasible paths compatible with different curvature constraints are pulled back to graze the obstacle border.

Algorithm 1. Bisection for searching *yLow_max, yLow_min*

Input: Maximum allowed curvature κ_{max}, a circular obstacle with center (0,0) and radius *a*
Output: Maximum curvature point *y* in the *y*-axis and its curvature κ

//First find (i) *yLow_max* whose curvature is not larger than κ_{max}, then find (ii) *yLow_min* whose curvature
is κ_{max}.
While $\kappa(0,\ yLow) > \kappa_{max}$//$\kappa$(x, y)
yLow = *yLow* + *a*
endwhile
yHigh = *yLow* − *a*
   //Curvature maximum point is found by binary search on the interval [*yHigh,yLow*]
While $\kappa(0,\ yHigh) - \kappa(0,\ yLow) < \varepsilon$ //ε: tolerance
y = (*yLow* + *yHigh*)/2
   If $\kappa(0,\ y) > \kappa_{max}$, **then** *yLow* = *y*
   Else *yHigh* = *y*
   end if
end while
return *y*

(B)   Continuous-curvature low curvature turn

Among the possible streamlines that can pass a given obstacle in front of the mobile robot, a feasible path with a low curvature using Algorithm 1can be found. This is done by using Algorithm 1 for searching the unique streamline passing (0, *yLow*_max) with the largest |*yLow*| satisfying Equation (9). This streamline is called the low curvature turn path. A low-curvature collision-free path is found by Algorithm 1, a lateral displacement vertical to the moving direction (as Figure 8 shows, along ±*y* direction) is used to pull the low curvature streamline path back to the current robot position. Figure 8 shows the concept of the low curvature path.

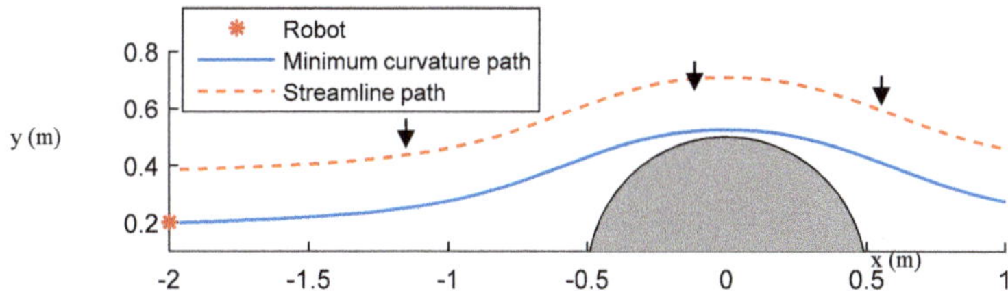

Figure 8. The concept of the low curvature path.

Given an obstacle's radius and the robot's maximum allowed curvature, we can derive three primitive paths which satisfy the curvature constraints as described above. For all three primitive paths, the robot needs to keep a sufficient longitudinal distance with the obstacle to achieve the pursuit of the primitive paths with a stricter curvature constraint, i.e., a lower maximum curvature. In addition, while the maximum curvature constraint decreases, it is more difficult to achieve primitive paths. Excessively restrictive curvature constraint causes no feasible path is found. Moreover, the lateral distance to the obstacle will influence the ability to find a feasible path. Figure 9 demonstrates primitive paths correspond to various initial positions (or relative distance to the obstacle) and curvature constraints.

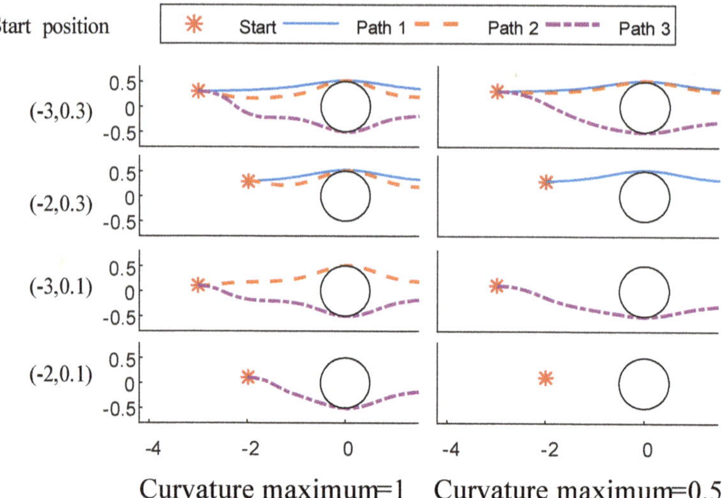

Figure 9. The single obstacle avoidance path for different start positions and the curvature maximum constraint in the *x-y* plane. The radius of the obstacle is 0.5 m. The robot forward moving direction is the positive *x*-axis (toward the right).

3.3. Distance-Based Obstacle-Avoiding Path Selecting Strategy

The selecting strategy of quickly computing three primitive streamlines has to identify the situation the robot encounters, depending on the reaction distance to the upcoming obstacle position or lateral displacement relative to the size of the obstacle. The strategy is illustrated in the scenario of Figure 10. The robot is initially located at $x = -2$ with a different lateral distance y related to a cylindrical obstacle at the origin. Let $b+$ and $b-$ be the points which two sharp turn paths intersect with the line $x = -2$. The interval $[-r, r]$, denoted by $[d-, d+]$ in Figure 10 at the vertical line $x = -2$ is partitioned into intervals B+~B− by the labeled points $d-$ to $d+$ according to the start points of the primitive paths, symmetrically with respect to the current robot position. We define L_{sharp} as the distance between points $c+$ (the start point with a right sharp turn path) and $c-$ (the start point with a low curvature path).

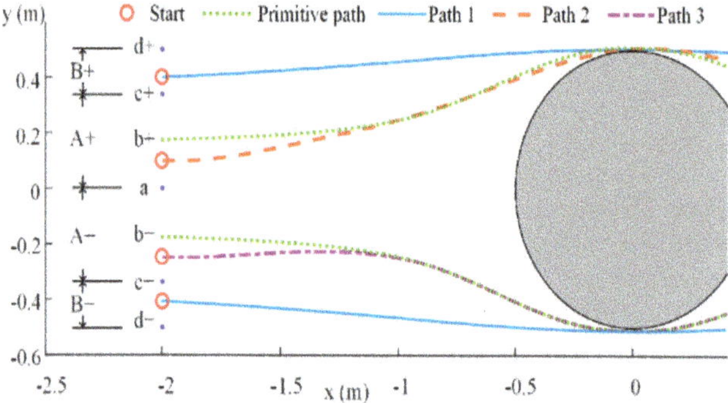

Figure 10. The illustration of the primitive path selecting strategy. The range $[-r_{obs}, r_{obs}]$ according to the obstacle size is partitioned manually into three intervals from far to near to reflect the reaction distance. Robots with different lateral displacements related to an obstacle will pursue different primitive streamline paths aligned with the current robot heading. Path 1 and Path 2 tangentially traverse the enlarged circular obstacle boundary.

Specifically, given a current robot configuration and an obstacle of known size and location in front of the robot forward route, we propose the following rules

$$\begin{cases} d_{lat} \in A + A- \rightarrow \text{Path 1} \\ d_{lat} \in B + B- \rightarrow \text{Path 2} \\ d_{lat} \in \text{special cases} \rightarrow \text{Path 3} \end{cases} \tag{9}$$

The strategy is according to the relative lateral distance $d_{lat} \leq r_{obs}$ measured from the center of the obstacle (x_{obs}, y_{obs}) in front of the robot, where the current robot location is at a fixed longitudinal distance. The range $[-r_{obs}, r_{obs}]$ according to the obstacle size is partitioned manually into three intervals from far to near to reflect the reaction distance as Figure 10 shows. Once an obstacle is sensed by the obstacle detector, the motion planner determines the proper primitive path via Equation (9) for the mobile robot to follow to circumvent the obstacle. This strategy is designed to use a low curvature turn in Intervals A+ and A−, while it pursues a sharp left turn in Interval B+ and a sharp right turn in Interval B− as the obstacle is closer. In addition, this strategy makes the avoidance of a small obstacle easier (with smaller $r_{Obstacle}$). Note that for obstacles with the same radius, the paths generated by streamlines of Laplace's Equation according to the same position are identical. Hence, a set of streamline paths can be a priori computed for circular obstacles with different radii and stored and maintained in the dataset. The path obtained by transforming a sample path computed for an obstacle located at the origin to the estimated or true obstacle position could then be re-used to online plan or re-plan the collision-free movement with a reduced time-complexity. In practice, the localization error requires the motion planner to online update the primitive path according to a proposed lateral distance-based path selection strategy.

4. Real-Time Streamline-Based Obstacle Avoidance Strategy

4.1. Overview of the Obstacle Avoidance System

Figure 11 depicts the building blocks of the obstacle avoidance system. The real-time obstacle avoidance system is built by three subsystems, which are the obstacle detector, the motion planner that incorporates the curvature constraint, and the pure pursuit controller used to control the robot to follow the specific primitive path with an allowable angular velocity satisfying the curvature constraint. For the part of the robot's hardware, we used the robot's kinematic model to estimate the robot's own kinematics and the sensors to detect the surrounding environment. The obstacle's global location is estimated by the range data received from sonar and infrared sensors. Our motion planner initially selects a streamline starting from the robot's start position, which is generated based on an a priori known obstacle distribution. The obstacle's location is updated based on new sensor data during the robot's forward motion, and the motion planner will decide whether to enable local re-planning based on a path selection strategy or to retain the original path for the robot to follow. Local re-planning is performed by generating and updating a local subgoal that is on a new primitive collision-free path and the smooth transition between streamlines is enabled via pure pursuit.

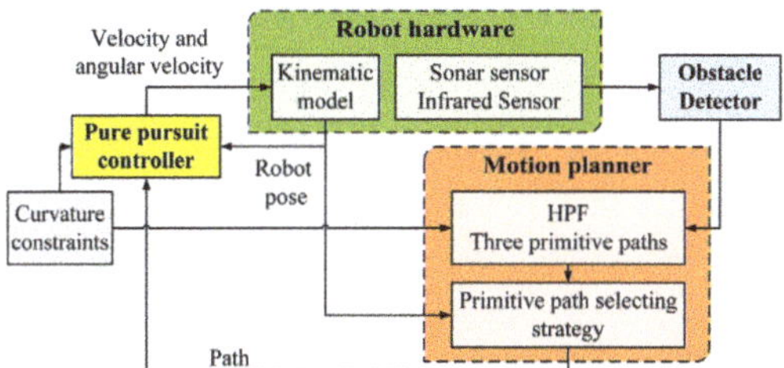

Figure 11. The flowchart of the online obstacle avoidance system for a curvature constrained non-holonomic mobile robot based on the primitive streamline paths, a path selection strategy, and a pure pursuit algorithm for streamline changing.

4.2. Pure Pursuit Controller for Mobile Robots

We assume that an initial streamline is chosen based on the initial robot configuration and an a priori known obstacle distribution, taking into consideration the curvature constraint. This initial path may collide with obstacles. To ensure the safe and smooth navigation, one subsystem of our obstacle avoidance system in Figure 11 is to make the robot redirect from following one streamline to an alternative streamline at a look-ahead distance via a local, online pure pursuit algorithm without violating curvature constraint. Figure 12 shows the plots of a streamline-changing circular path for redirecting the mobile robot from its current pose to a subgoal on another streamline via pure pursuit. The details are as follows.

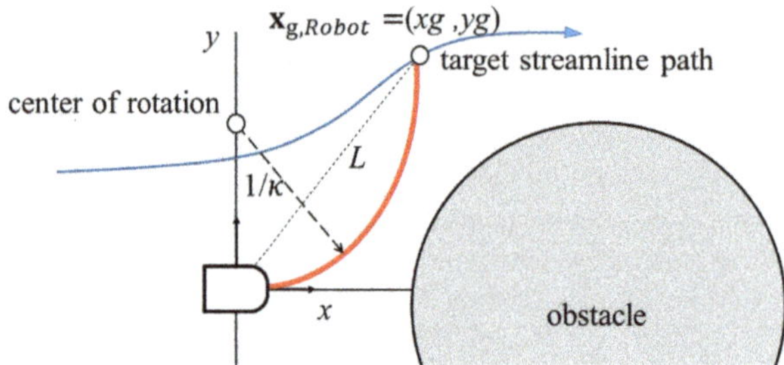

Figure 12. The smooth transition for redirecting the mobile robot from its current pose on current the streamline to a subgoal on the target streamline via pure pursuit.

Pure pursuit is a path tracking method by calculating the curvature of a new circular path for a vehicle to pursuit a subgoal position ahead of the vehicle by leaving the initially planned path from its current position [20,21], where the orientation of the subgoal is not concerned. Due to the fact that non-holonomic mobile robots cannot directly move in the lateral direction, a robot pursues a subgoal position ahead of the robot to redirect from its current position along an arc of curvature κ via pure pursuit. Since the obstacle avoidance path can be computed analytically according to the relative position of the robot and obstacle and shape and size of the obstacle, this method has an implementation advantage.

Once a streamline is selected for streamline-changing, the next step is to find a subgoal point $\mathbf{X}_g$ which is located after the closest point in the global coordinate. Let a local coordinate system be attached to the robot with its origin set as the rotation center of the robot and the $+x$-axis of the local frame aligned with the forward motion direction. Then transform a subgoal point $\mathbf{X}_g$ in the global coordinate to $\mathbf{x}_{g,Robot} = [x_g\ y_g]^T$ in the local frame with x_g and y_g denoting the longitudinal and the lateral displacements, respectively:

$$\mathbf{x}_{g,Robot} = R(\theta)(\mathbf{X}_g - \mathbf{X}_{Robot}), \ R(\theta) = \begin{bmatrix} \cos(\theta) & \sin(\theta) \\ -\sin(\theta) & \cos(\theta) \end{bmatrix} \tag{10}$$

where $\mathbf{X}_{Robot}$ is the current robot position in the global frame, and θ is the heading angle of the robot in the global frame. The subgoal point $\mathbf{x}_{g,Robot}$ in the vehicle coordinates can be represented with curvature κ and α by geometry:

$$\begin{aligned} x_g &= r_c(\cos(\alpha) - 1) = \frac{\cos(\alpha) - 1}{\kappa} \\ y_g &= r_c \sin(\alpha) = \frac{\sin(\alpha)}{\kappa} \end{aligned} \tag{11}$$

where α is the angle of the arc between the vehicle and the subgoal (see Figure 12). The subgoal point to pursue keeps a specific look-ahead distance $L = \sqrt{x_{g,Robot}^2 + y_{g,Robot}^2}$, since non-holonomic robots cannot correct errors directly with respect to the nearest point on the path.

Now the equations of the pure-pursuit curvature control law are derived. The curvature κ of the vehicle is defined as the inverse of the distance r_c, also called the radius of curvature, between the vehicle's frame origin and its instantaneous CoR (center of rotation). Second, the curvature also represents the instantaneous change of the vehicle heading angle $d\theta$ with respect to the traveled distance ds. Hence, curvature is formally defined as follows:

$$\kappa = \frac{1}{r_c} = \frac{d\theta}{ds} \tag{12}$$

In implementation, curvature can be defined as the instantaneous change of the heading angle $\Delta\theta$ with respect to the travel distance $U \cdot \Delta t$ in one sampling time Δt. Curvature then could be further related to the robot's velocity and angular velocity. Therefore, the angular velocity of a robot moves along a path at a constant forward speed U could be computed from the path curvature via the following relation (Equation (12)):

$$\kappa = \frac{1}{r_c} = \frac{\omega}{U} = \frac{\Delta\theta}{U \cdot \Delta t} \tag{13}$$

where $\kappa = y_{g,Robot}/L^2$ [20]. To satisfy the maximum allowable curvature, we regularize the signed curvature as

$$\kappa_{constraints} = Sign(\kappa) \cdot \kappa_{max} \quad if \quad \kappa > \kappa_{max} \tag{14}$$

Thus, the motion in the local frame of the robot can be applied to command the motion controller via the inverse kinematics of Equation (1). The displacement ΔX_g in the global frame can be derived by exploiting the displacement Δx_{Robot} in the local frame

$$\Delta X_g = R(\theta + \Delta\theta)\Delta x_{Robot}$$
$$\Delta X_g = \begin{bmatrix} -\sin(\Delta\theta) \\ \cos(\Delta\theta) \end{bmatrix} U \cdot \Delta t, \tag{15}$$

A pure pursuit algorithm summarizing this subsection is shown in Algorithm 2.

Algorithm 2. The pure pursuit streamline path.

Input: robot initial pose, target streamline path, look-ahead distance, maximum allowable curvature
Output: status, pursuit path

While (not *timeout* **or** *status*) **do**
let $R(\theta)$ represent the transformation to robot coordinate
pClosest$\leftarrow$ {(x, y) | min{| (x, y)–poseRobot |} **and** (x, y) in
          pursuitpath}
$X_g \leftarrow$ {(x, y) | min{| (x, y) − pClosest |} **and** (x, y) in pursuit path
        after pClosest}
$x_{g,Robot} \leftarrow R(\theta)(X_g − X_{Robot})$ (9)
Calculate the curvature $\kappa \leftarrow y_{g,Robot}/L^2$
Regularize the curvature constraints (Equation (14))
Set the steering angle of the robot (Equation (13))
Update robot's heading direction, poseRobot
  if poseRobot and curvature == streamline path **do**
    status$\leftarrow$**TRUE**
    store path into pathArray
  if collide with obstacle **do**
    status$\leftarrow$**FALSE**
end while

4.3. Setting Lookahead Distance

Look-ahead distance is the only parameter in the pure pursuit algorithm, and the reason for the look-ahead distance is that non-holonomic robots cannot correct errors directly with respect to the nearest point on the path [20]. The pure pursuit procedure is summarized in Algorithm 2. The common practice for setting look-ahead distance takes into consideration its effect on path geometry and tracking performance. A longer look-ahead distance results in smoother paths but a worse tracking accuracy. In contrast, a shorter look-ahead can reduce tracking errors more quickly. Yet, due to the curvature constraint, the pure pursuit controller may not be able to follow steering commands, and the robot motion becomes unstable. Therefore, a suitable look-ahead is needed for both stability and tracking performance.

Figure 13 illustrates that the pure pursuit path is sensitive to the setting of look-ahead distance. In this example, there is an angle between the *x*-axis and the line connecting the point robot and the obstacle center. The path with a look-ahead of 0.5 m has an effective and efficient tracking ability. However, the path with too large of a look-ahead distance, 1 m, for example, is smoother but unable to track the path accurately. On the contrary, the path with a shorter look-ahead 0.1 m responds to tracking errors quickly, but the robot's motion is unstable and overdamping because of the curvature constraint. Both paths obtained with a look-ahead of 1 m and 0.1 m lead to a collision with the obstacle.

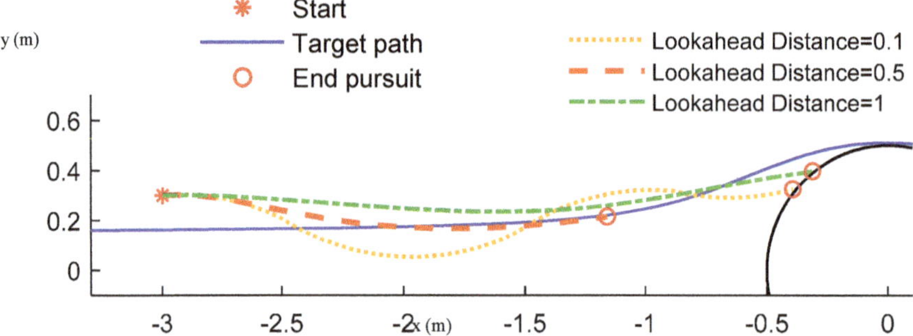

Figure 13. Replanning via pure pursuit paths starting at a fixed position with a set of subgoals determined by different look-ahead distances. The pursuit paths with look-ahead distance 0.1 and 1 intersect with the obstacle, while the pure pursuit path with a look-ahead distance of 0.5 is collision-free.

4.4. Multiple Obstacles Avoidance Strategies

In an environment composed of multiple obstacles, previous researchers provided several different methods to create a guidance vector field. The weighted superposition of a single obstacle is the most commonly used method for multiple obstacles (e.g., Reference [6,8,11,12,15,18]). Though the sum of HPFs is also HPF (hence, free of local minima), the superposition has no guarantee to satisfy the curvature constraint. Hence, we propose a new avoidance strategy for multiple obstacles.

(1) Superposition of multiple obstacles

A weighted velocity field of each obstacle vector field not only guarantees no local equilibria in the workspace, but also satisfies the zero Neumann boundary condition on every boundary of an obstacle. In multiple obstacles, the path tangent or vector field at each point corresponds to the streamline of the flow with velocity defined by the weighted superposition of velocity which is induced by an individual obstacle. The total influence of all obstacles in an environment with N obstacles on the velocity field $\mathbf{V}_{total}$ can be expressed as the weighted sum of N velocity fields of $\mathbf{V}_1, \ldots, \mathbf{V}_N$ for each obstacle

$$\mathbf{V}_{total}\mathbf{V}_{total} = \sum_{i=1}^{N} w_i \mathbf{V}_i \tag{16}$$

where w_i is the position-dependent weighting function for obstacle i. One can design $w_i = \prod_{j \neq i}^{N} \frac{d_i}{d_i + d_j}$ with d_i denoting the shortest distance between the robot and the obstacle i. This design makes the closest obstacle have the largest weight. In real-time applications, Equation (16) is calculated for only all of the obstacles detected within the sensing range or a user-defined safety zone.

(2)   The proposed strategy

We identify three primitive paths once the robot's initial pose, maximum allowable curvature, obstacle's position, and radius are given. Figure 14 summarized the discussions so far as a flowchart of hydrodynamic path planning in combination with pure pursuit in a multiple obstacles situation. In the multiple obstacles situation, we initialize a queue called *poseRobotArray* to store the robot's initial pose. While the queue is not empty, we assign the first element of *poseRobotArray* to *poseRobot* and pop the first element of the queue. Then, we move the robot forward to the target to check whether the path is collision-free. If no obstacles are detected on the path, we store the path into *pathArray*. Otherwise, we generate three primitive paths to avoid the detected obstacle. For each generated path, we check if it collides with any other obstacle. If it is still collision-free, we store the robot's final pose into *poseRobotArray*. On the other hand, we check if the collision happened before or after passing the first obstacle. If the collision happened before the first obstacle, we generate three primitive paths to avoid the new obstacle from the robot's initial pose. In contrast, we set the location of the closest point on the path to the original obstacle as *poseRobot* and then avoid the new obstacle. Finally, we select the optimized path from *pathArray*.

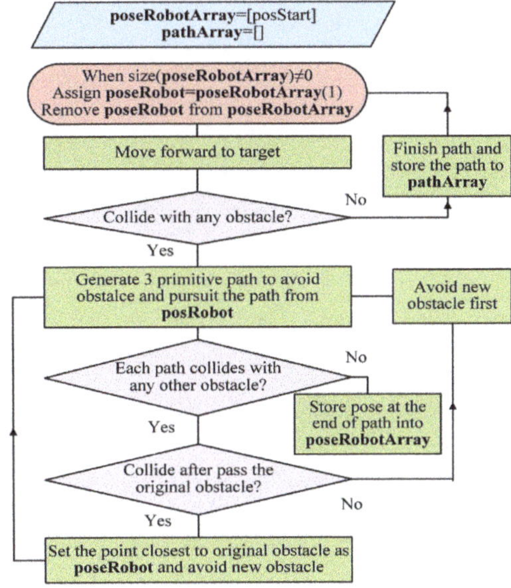

Figure 14. The flowchart of sequential obstacle avoidance by the hydrodynamic path planning with pure pursuit in the multiple obstacles situation.

5. Comparisons and Experiment

In this section, we demonstrate the proposed algorithm for navigation within multiple circular obstacles via comparisons with other methods to show the planner's performance in a cluttered environment and a proof of concept experiment to show the feasibility. The speed of the robots was set as 1 m/s for all scenarios and the initial heading direction is aligned in the positive x-axis defined as the forward direction. Two different cases are discussed.

- Pure pursuit method vs. lane hopping method
- Multiple obstacles environment

5.1. Comparison of the Pure Pursuit Method and Lane Hopping Method

In order to leave an initially planned streamline and change to another one, lane hopping (streamline changing) [14] is enabled in case the mobile robot (1) is too close to the obstacle (risk of imminent collision) or the current streamline the robot follows violates the curvature constraint. Lane hopping requires that the x coordinate in the two streamline paths before and after hopping is almost the same. In the lane-hopping method, a 2×2 filter matrix K_{filter} is used to generate the lane-hopping paths. In contrast to the filter matrix, the pure pursuit strategy is easier in application, for only a single value of the look-ahead distance is needed to be tuned. Thus, it is easier to find feasible paths by pure pursuit. Furthermore, the pure pursuit method is also designed to satisfy the curvature constraint for the part of the streamline-changing path, in addition to smoothness. Figure 15 presents that the pursuit of a target path both by pure pursuit and lane hopping. Both the filter matrix $K_{filter} = 0.1I$ with I the 2×2 identity matrix in lane-hopping, and look-ahead distance $L = 0.5$ m in pure pursuit are selected manually so that the pursuit paths can achieve their respective subgoal on the selected new streamline to be followed. The maximum curvature of lane hopping (about 10 (1/m)) is remarkably larger than that of pure pursuit (1(1/m)) at the beginning of streamline changing, and the lane-hopping path can achieve the target streamline earlier.

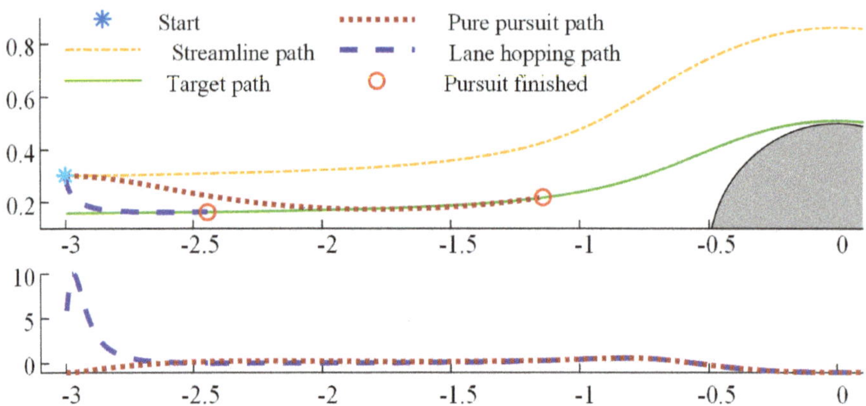

Figure 15. The comparison of pure pursuit and lane hopping: path (upper plot) and curvature (lower plot). Horizontal axis is the forward +x direction.

5.2. Multi-Obstacles Environment

5.2.1. Comparison with Streamline Path by Weighting Velocity of Each Single Obstacle

In Figure 16, the proposed method is compared with the streamline path obtained via the weighting method. The maximum curvature of the pure pursuit path and streamline path are 0.50 (1/m) and 1.44 (1/m), respectively. The maximum curvature of the pure pursuit path is smaller than the streamline path.

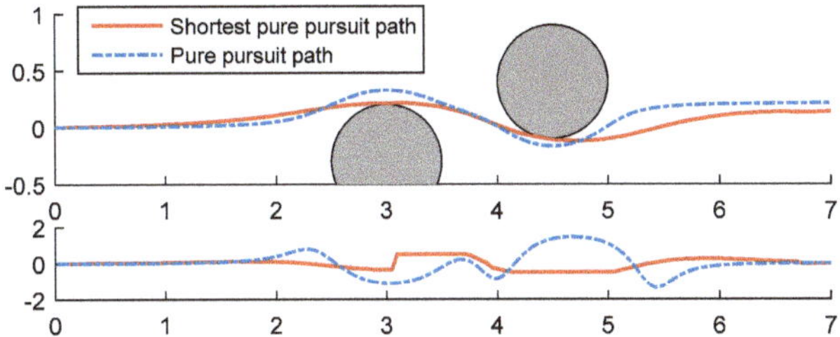

Figure 16. The two obstacle environment. Comparison of the pure pursuit path and weighting streamline path.

5.2.2. Clutter Case: Comparison with Lau's Approach

In this example, we compare our methods with a fluid motion planner provided by Lau et al. [18] in Figure 17. In the clutter case, there are two feasible paths with curvature constraints of 0.3 (1/m) and 0.8 (1/m), respectively, by our proposed method. On the other hand, the streamline path with curvature constraints 1 (1/m) collides with an obstacle and fails due to a late response. In sum, our proposed method performed well and can have a higher probability to get feasible paths with a small maximum curvature.

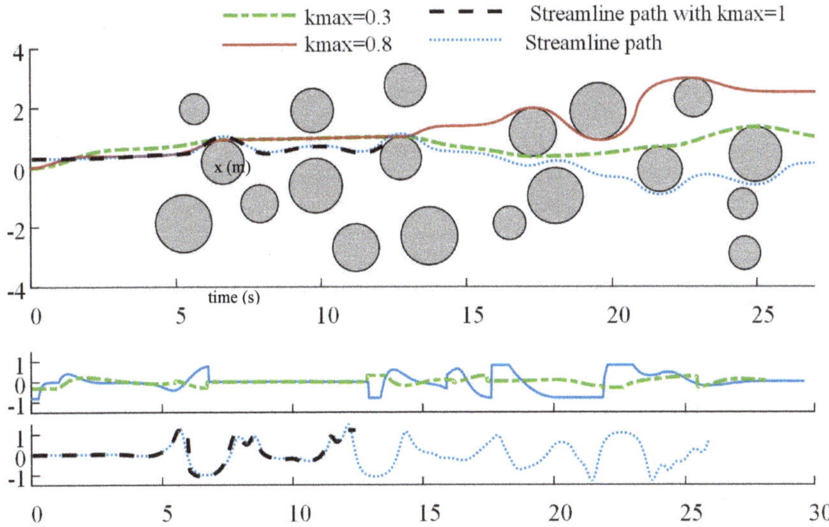

Figure 17. A cluttered environment: upper plot = the paths, lower plots = the curvature. Black dashed path denotes the path generated by the method of superposition that collides with the obstacle.

5.3. Proof of Concept Experiment and Discussion

In the performed indoor experiment, three aspects related to the feasibility of trajectory generation are presented, which are (1) curvature constraint, (2) arrangement of obstacles, and (3) sensor detection error. First, while the maximum curvature constraint decreases, it is more difficult for robots to achieve primitive paths. Second, a robot needs to keep enough clearance from the obstacle to ensure the pursuit of all the three primitive paths possible. Third, we rely on low-cost sonar and infrared sensors to estimate the obstacle location. Furthermore, in order to guarantee obstacle avoidance, the obstacles are arranged so that only one obstacle is detected within a pre-specified look-ahead distance of the mobile robot's current location at a time. Table 1 lists the parameters of the robot platform related to the experiment.

Table 1. The parameters of the mobile robot Dr. Robot X80 (Figure 1) used for the experiment.

Wheel's radius r	12.5 (cm)
Distance between two wheels d	25 (cm)
Robot radius r_{Robot}	20 (cm)
Height	25.5 (cm)
Weight	3.5 (kg)
Operating (U)/Max.speed (V_{max})	0.5/1 (m/s)
Cycle time	200 (ms)
Safety distance r_{Safe}	0.1 (m)
Curvature constraint κ_{max}	1.5 (1/m)

5.3.1. Experimental Setting

The robot and the obstacle are modeled as a circle defined by its radius r_{Robot} = 0.2 m (Table 1), $r_{Obstacle}$ = 0.1 m, respectively. The mobile robot is initially located at the origin of the global coordinate system and its initial forward moving direction is the + x-axis. The mobile robot is modeled as a kinematic unicycle (1) with curvature as a constrained input bounded by the maximum curvature (Figure 2) while the tangential speed is held constant with v_x = U = 0.5 m/s. The robot pose is obtained by numerically integrating Equation (1), which is directly controlled by the path curvature or angular velocity with the maximum allowable curvature for the mobile robot set as 1.5 (1/m), as shown in Table 1. Thus, the angular speed is upper bounded by $\omega \leq \kappa_{max}$U = 1.5 * 0.5 = 0.75 rad/s. The pure pursuit command rate is 10 Hz, and the safety distance is r_{Safe} = 0.1 m, which is the robot displacement in a period to allow a sharp turn in case of an imminent collision. Given a safety distance r_{Safe} between a robot and a detected obstacle, the radius of the obstacle is enlarged to account for the robot radius as r_{Obs} = $r_{Obstacle}$ + r_{Robot} + r_{Safe} = 0.4 m for guaranteed local obstacle avoidance. In our experiment, the look-ahead distance is equal to the robot radius L = r_{Robot} = 0.2 m. Note that if different subgoals between the obstacles are selected online for streamline changing from the current robot pose, we obtain topologically different paths with different curvatures for sequential obstacle avoidance in an environment with multiple obstacles. This is seen in the experiment in the following.

The distance between the robot's center and the obstacle's center is D = D_{Sensor} + $r_{Obstacle}$ + r_{Robot}, where D_{Sensor} is the range value measured by the sensor. In the case of a circular robot and a circular obstacle, which is the case studied in this paper, the collision-free criterion for safe navigation is that the distance D between the point robot (robot center) and the obstacle center is larger than r_{Obs}, i.e., D > 0.4. In the experiment reported here, a left/right turn is the default action and must maintain at least 0.4 m to a detected obstacle to be avoided. A narrow passage is not considered in this proof of concept experiment, since this default action may cause navigation difficulty, if not an impossibility in such a situation.

5.3.2. Online static cylinder obstacles avoidance

The experimental setup for the task of avoidance of multiple obstacles is operating a mobile robot Dr. Robot X80 in a cluttered indoor environment. Similar to Reference [11], there are four cylinder obstacles, all assumed to be identical cylinders with radius $r_{Obstacle}$ = 0.1 m, placed at (1, 0), (1.8, −0.6), (2.6, 0) and (2.6, −1.2) in meters, which are distributed around the forward motion route. The velocity field generated by the gradient of the HPF solution to Laplace's Equation for this map is depicted in Figure 18. We assume the following:

(i)   We do not consider the navigation between very tight spaces, thus, the obstacles are arranged far apart to avoid the difficulty of APF-based navigation within a narrow passage. We arrange the clearance between any two adjacent obstacles smaller than the sensing range of the sensors but large enough to allow the pure pursuit algorithm to generate a local collision-free path for navigation. Specifically, the minimum distance between two obstacles' centers is $2r_{Obs}$ = 0.8 m, wide enough for the robot to pass between two obstacles.

(ii)   The projection of all obstacles onto the ground plane is an identical circle, but the number of static obstacles and their locations are unknown.

The navigation trajectories generated in the experiment are depicted in Figure 19, along with Figure 20 showing the corresponding velocities and the path curvature profiles of two navigation paths with nearly the same starting point at the origin. The velocity field covers the free space and does not pass through the obstacles as desired. For the experiment shown here in Figure 19, the robot starts at the origin and initially follows a straight streamline trajectory with a speed of 0.5 m/s along the $+x$-axis. Then it confronts the detected obstacles in the front, and two feasible smooth navigation paths are generated for online safe and smooth navigation via the proposed HPF-based planner during the experiment. Different paths are obtained due to the slight variation in the initial position and heading of the mobile robot. It is noted that the portion of the two smooth paths is between two obstacles, thus no collision is guaranteed, and the maximum curvature of either path does not violate the maximum curvature. As remarked in Reference [28], the smoothness of the trajectory and its gradient is generic due to the physical characteristics of the continuously differentiable velocity potential solution to Laplace's Equation mentioned in Section 3. The map of the environment is created by Hector SLAM, an open source SLAM algorithm available in ROS [23] based on the laser range data processed after the experiment. The robot can autonomously localize itself once new sensor readings are available (within 1 ms). The location of the detected obstacle fluctuates because of sensor noise and detection error. In this scenario, the primitive paths can be computed within 0.2 ms in our implementation. The obstacle avoidance system based on a priori selection of fast computing primitive streamline paths is computationally efficient, and the reaction time is able to handle the situation as obstacles in the front are detected, thus, making the system real-time. We remark that other options of a wide range of more involved trajectories such as B-spline [45] are applicable for the proposed obstacle avoidance system depicted in Figure 11, not only pertinent to streamlines. The main computational requirement is that the trajectories employed as primitive paths compatible with curvature constraint are a priori computed and could be online selected based on some rules, thus, preventing the generation of infeasible paths.

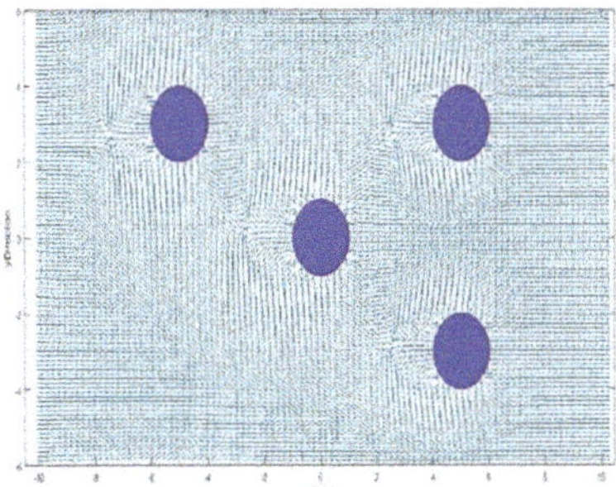

Figure 18. The smooth velocity field generated by Laplace's Equation with Dirichlet boundary conditions in the experiment scenario depicted in Figure 19, assuming a bounded rectangular domain with circular obstacles sufficiently apart.

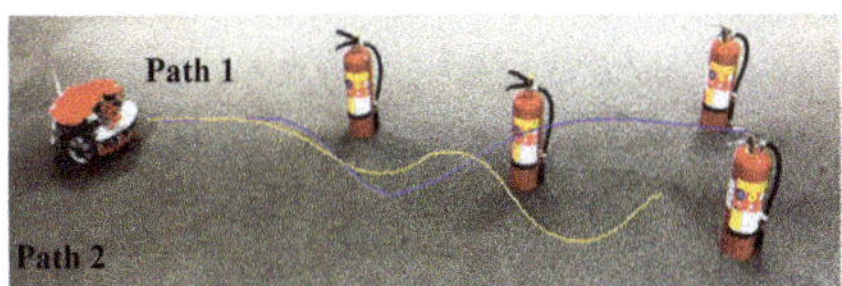

Figure 19. *Cont.*

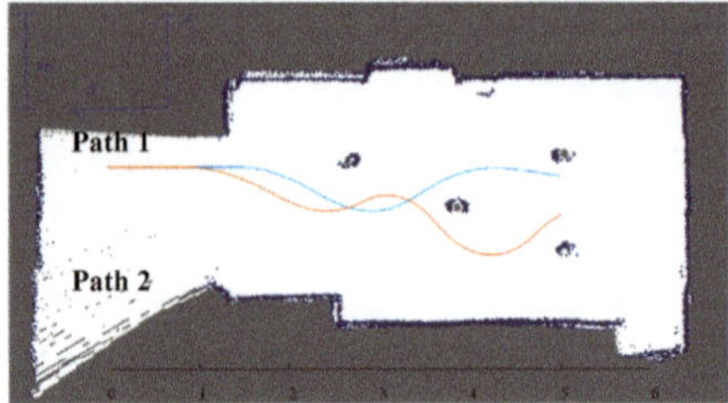

Figure 19. The experimental setup for operating a mobile robot in a cluttered environment (**upper**: the photo, **lower**: the mapping) and two resulting obstacle-avoiding navigation paths generated online by the streamline-based approach.

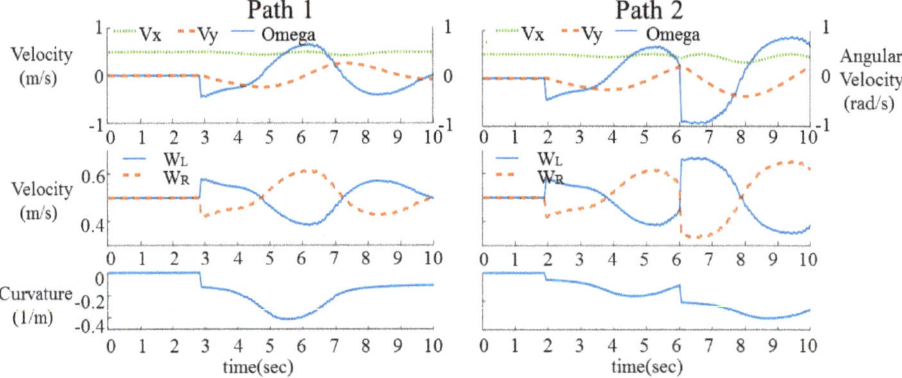

Figure 20. The linear and angular velocities, speeds of the left and right wheels and curvature profiles of the two directional navigation paths depicted in Figure 19, where $v_x = 0.5$ is the desired constant reference longitudinal/forward/tangential speed and v_y is the lateral speed of the robot.

6. Conclusions

Recent developments on navigation methodologies for autonomous unmanned vehicles show that the HPF-based navigation algorithm is an example of the unifying view or framework of a vector field or dynamical systems based vehicle navigation pattern generators inspired from the scalar function of APFs (e.g., Reference [38]). The streamlines provide a pool of systematic, predictable smooth primitive paths that are integral curves of explicit and easily computable vector fields useful for the specification of the path tangent for directional navigation guidance and higher order path characteristics such as the curvature at each point of the path followed by autonomous vehicles. We demonstrate the practical usefulness of an HPF-based method to generate smooth paths for non-holonomic mobile robots to avoid obstacles via an obstacle avoidance system based on streamlines with a curvature constraint in this paper. First, streamlines extracted from the harmonic potential field are used to design three primitive smooth paths satisfying the curvature constraint for a single obstacle avoidance along with their application situations according to a lateral distance relative to obstacle size. Second, the pure pursuit algorithm is implemented to pursuit streamline-changing paths satisfying the curvature constraint for local multiple-obstacle avoidance situations. The simulation results show that pure pursuit paths in combination with initially planned streamlines can find feasible paths with smaller curvature constraints compared to previous hydrodynamics-based approaches. Furthermore, a proof of concept experiment was conducted to validate that the obstacle avoidance system based on the a priori selection of fast computing primitive streamline paths is computationally efficient and that the reaction time is able to handle the situation of an obstacle being detected in front of it online. Future work is planned to focus on the extension to 3D space, and toward safe navigation in environments of increasing size and complexity such as moving obstacles and multiple vehicles scenarios with the complete real-time HPF-based path planner. Additionally, it is also interesting to see the navigation performance under different sensor qualities such as the RGB-D sensor.

Author Contributions: P.-L.K. wrote the initial draft, P.-L.K. and C.-H.W. conducted the experiment, H.-J.C. worked on HPF simulation, J.-S.L. worked as PI of the whole project.

Funding: This research was funded by IIS, Academia Sinica.

Conflicts of Interest: The authors declare that there is no conflict of interest.

References

1. Dong, J.F.; Sabastian, S.E.; Lim, T.M.; Li, Y.P. Autonomous In-door Vehicles. In *Handbook of Manufacturing Engineering and Technology*; Springer: London, UK, 2015.
2. Adouane, L. *Autonomous Vehicle Navigation: From Behavioral to Hybrid Multi-Controller Architectures*; CRC Press: Boca Raton, FL, USA, 2016.
3. Kunchev, V.; Jain, L.; Ivancevic, V.; Finn, A. Path planning and obstacle avoidance for autonomous mobile robots: A review. In *Knowledge-Based Intelligent Information and Engineering Systems, Proceedings of the 10th International Conference, KES 2006, Bournemouth, UK, 9–11 October 2006*; Springer: Berlin, Germany, 2006; pp. 537–544.
4. Minguez, J.; Lamiraux, F.; Laumond, J.-P. Motion planning and obstacle avoidance. In *Handbook of Robotics*; Springer: Basel, Switzerland, 2016; pp. 1177–1202.
5. Khatib, O. Real-time obstacle avoidance for manipulators and mobile robots. In Proceedings of the IEEE International Conference on Robotics and Automation, St. Louis, MO, USA, 25–28 March 1985; pp. 500–505.
6. Fahimi, F. Obstacle avoidance using harmonic potential functions. In *Autonomous Robots*; Springer: Boston, MA, USA, 2009; pp. 1–49.
7. Connolly, C.I.; Grupen, R.A. On the applications of harmonicfunctionsto robotics. *J. Robot. Syst.* **1993**, *10*, 931–946. [CrossRef]
8. Kaľavský, M.; Ferková, Ž. Harmonic potential field method for path planning of mobile robot. *ICTIC* **2012**, *1*, 41–46.
9. Akishita, S.; Kawamura, S.; Hayashi, K. New navigation function utilizing hydrodynamic potential for mobile robot. In Proceedings of the IEEE International Workshop on Intelligent Motion Control, Istanbul, Turkey, 20–22 August 1990; pp. 413–417.
10. Wray, K.H.; Ruiken, D.; Grupen, R.A.; Zilberstein, S. Log-space harmonic function path planning. In Proceedings of the IEEE/RSJ International Conference onIntelligent Robots and Systems, Daejeon, Korea, 9–14 October 2016; pp. 1511–1516.
11. Waydo, S.; Murray, R.M. Vehicle motion planning using stream functions. In Proceedings of the IEEE International Conference on Robotics and Automation, Taipei, Taiwan, 14–19 September 2003; pp. 2484–2491.
12. Daily, R.; Bevly, D.M. Harmonic potential field path planning for high speed vehicles. In Proceedings of the American Control Conference, Seattle, WA, USA, 11–13 June 2008; pp. 4609–4614.
13. Owen, T.; Hillier, R.; Lau, D. Smooth path planning around elliptical obstacles using potential flow for non-holonomic robots. In *Robot Soccer World Cup*; Springer: Berlin/Heidelberg, Germnay, 2011; pp. 329–340.
14. Palm, R.; Driankov, D. Fluid mechanics for path planning and obstacle avoidance of mobile robots. In Proceedings of the 11th International Conference on Informatics in Control, Automation and Robotics (ICINCO), Vienna, Austria, 1–3 September 2014; pp. 231–238.
15. Kim, J.O.; Khosla, P. Real-time obstacle avoidance using harmonic potential functions. In Proceedings of the IEEE International Conference on Robotics and Automation, Sacramento, CA, USA, 9–11 April 1991; pp. 790–796.
16. Wang, H.; Lyu, W.; Yao, P.; Liang, X.; Liu, C. Three-dimensional path planning for unmanned aerial vehicle based on interfered fluid dynamical system. *Chin. J. Aeronaut.* **2015**, *28*, 229–239. [CrossRef]
17. Connolly, C.I.; Grupen, R.A. *Nonholonomic Path Planning Using Harmonic Functions*; National Science Foundation: Alexandria, VA, USA, 1994.
18. Lau, D.; Eden, J.; Oetomo, D. Fluid Motion Planner for Nonholonomic 3-D Mobile Robots with Kinematic Constraints. *IEEE Trans. Robot.* **2015**, *31*, 1537–1547. [CrossRef]

19. Munson, B.R.; Young, D.F.; Okiishi, T.H. *Fundamentals of Fluid Mechanics*; Fowley, D., Ed.; John Wiley & Sons, Inc.: New York, NY, USA, 1990.

20. Coulter, R.C. *Implementation of the Pure Pursuit Path Tracking Algorithm*; DTIC Document; Defense Technical Information Center: Fort Belvoir, WV, USA, 1992.

21. Morales, J.; Martínez, J.L.; Martínez, M.A.; Mandow, A. Pure-pursuit reactive path tracking for nonholonomic mobile robots with a 2D laser scanner. *EURASIP J. Adv. Signal Process.* **2009**, *2009*, 935237. [CrossRef]

22. Masoud, A.A. Kinodynamic motion planning. *IEEE Robot. Autom. Mag.* **2010**, *17*, 85–99. [CrossRef]

23. Kohlbrecher, S.; von Stryk, O.; Meyer, J.; Klingauf, U. A flexible and scalable slam system with full 3d motion estimation. In Proceedings of the IEEE International Symposium on Safety, Security, and Rescue Robotics, Kyoto, Japan, 1–5 November 2011.

24. Belabbas, M.A.; Liu, S. New method for motion planning for non-holonomic systems using partial differential equations. In Proceedings of the American Control Conference, Seattle, WA, USA, 24–26 May 2017; pp. 4189–4194.

25. Gingras, D.; Dupuis, E.; Payre, G.; de Lafontaine, J. Path planning based on fluid mechanics for mobile robots using unstructured terrain models. In Proceedings of the IEEE International Conference on Robotics and Automation, Anchorage, AK, USA, 3–7 May 2010; pp. 1978–1984.

26. Motonaka, K. Kinodynamic Motion Planning for a Two-Wheel-Drive Mobile Robot. In *Handbook of Research on Biomimetics and Biomedical Robotics*; IGI Global: Hershey, PA, USA, 2018; pp. 332–346.

27. Liu, C.A.; Wei, Z.; Liu, C. A new algorithm for mobile robot obstacle avoidance based on hydrodynamics. In Proceedings of the IEEE International Conference on Automation and Logistics, Jinan, China, 18–21 August 2007; pp. 2310–2313.

28. Golan, Y.; Edelman, S.; Shapiro, A.; Rimon, E. Online Robot Navigation Using Continuously Updated Artificial Temperature Gradients. *IEEE Robot. Autom. Lett.* **2017**, *2*, 1280–1287. [CrossRef]

29. Louste, C.; Liégeois, A. Path planning for non-holonomic vehicles: a potential viscous fluid field method. *Robotica* **2002**, *20*, 291–298. [CrossRef]

30. Pedersen, M.D.; Fossen, T.I. Marine vessel path planning & guidance using potential flow. *IFAC Proc. Vol.* **2012**, *45*, 188–193.

31. LaValle, S.M. *Planning Algorithms*; Cambridge University Press: New York, NY, USA, 2006.

32. Tanner, H.G.; Loizou, S.; Kyriakopoulos, K.J. Nonholonomic stabilization with collision avoidance for mobile robots. In Proceedings of the IEEE/RSJ International Conference on Intelligent Robots and Systems, Maui, HI, USA, 29 October–3 November 2001; pp. 1220–1225.

33. Galick, M. The planning of optimal motions of non-holonomic systems. *Nonlinear Dyn.* **2017**, *90*, 2163–2184. [CrossRef]

34. Li, L. Nonholonomic motion planning using trigonometric switch inputs. *Int. J. Simul. Model. (IJSIMM)* **2017**, *16*, 176–186. [CrossRef]

35. Lamiraux, F.; Bonnafous, D. Reactive trajectory deformation for nonholonomic systems: Application to mobile robots. In Proceedings of the IEEE International Conference on Robotics and Automation, Washington, DC, USA, 11–15 May 2002; pp. 3099–3104.

36. Kelly, A.; Nagy, B. Reactive nonholonomic trajectory generation via parametric optimal control. *Int. J. Robot. Res.* **2003**, *22*, 583–601. [CrossRef]

37. Ho, Y.-J.; Liu, J.-S. Collision-free curvature-bounded smooth path planning using composite Bezier curve based on Voronoi diagram. In Proceedings of the IEEE International Symposium on Computational Intelligence in Robotics and Automation, Daejeon, Korea, 15–18 December 2009; pp. 463–468.

38. Panagou, D. Motion planning and collision avoidance using navigation vector fields. In Proceedings of the IEEE International Conference on Robotics and Automation, Hong Kong, China, 31 May–7 June 2014; pp. 2513–2518.

39. Papoutsidakis, M.; Piromalis, D.; Neri, F.; Camilleri, M. Intelligent algorithms based on data processing for modular robotic vehicles control. *WSEAS Trans. Syst.* **2014**, *13*, 242–251.

40. Bekey, G.A. *Autonomous Robots: From Biological Inspiration to Implementation and Control*; MIT Press: Cambridge, MA, USA 2005.

41. Lo, C.-W.; Wu, K.-L.; Lin, Y.-C.; Liu, J.-S. An intelligent control system for mobile robot navigation tasks in surveillance. In *Robot Intelligence Technology and Applications 2*; Springer: Berlin, Germany, 2014; pp. 449–462.

42. Schoof, E.; Manzie, C.; Shames, I.; Chapman, A.; Oetomo, D. An experimental platform for heterogeneous multi-vehicle missions. In Proceedings of the International Conference on Science and Innovation for Land Power, Adelaide, Australia, 5–6 September 2018.
43. Giftthaler, M.; Sandy, T.; Dörfler, K.; Brooks, I.; Buckingham, M.; Rey, G.; Kohler, M.; Gramazio, F.; Buchli, J. Mobile robotic fabrication at 1:1 scale: The in situ fabricator. *Constr. Robot.* **2017**, *1*, 3–14. [CrossRef]
44. Rösmann, C.; Hoffmann, F.; Bertram, T. Integrated online trajectory planning and optimization in distinctive topologies. *Robot. Auton. Syst.* **2017**, *88*, 142–153. [CrossRef]
45. Kano, H.; Fujioka, H. B-Spline Trajectory Planning with Curvature Constraint. In Proceedings of the Annual American Control Conference, Milwaukee, WI, USA, 27–29 June 2018; pp. 1963–1968.

 *applied sciences*

Article

Alpine Skiing Robot Using a Passive Turn with Variable Mechanism

Takuma Saga [1] and Norihiko Saga [2],*

[1]   Graduate School of Frontier Biosciences, Osaka University, 1-3 Yamadaoka, Suita, Osaka 565-0871, Japan; takuma.saga@gmail.com

[2]   Department of Human System Interaction, School of Science and Technology, Kwansei Gakuin University, 2-1 Gakuen, Sanda 669-1337, Japan

*   Correspondence: saga@kwansei.ac.jp; Tel.: +81-79-565-7042

Received: 5 November 2018; Accepted: 11 December 2018; Published: 17 December 2018

Abstract: Recently, the number of alpine ski junior players in Japan has drastically decreased. The causes include a decrease in ski areas and instructors, along with difficulty of early childhood alpine ski guidance. The alpine ski competition is not simply a glide on a slope. It requires understanding of ski deflection and skier posture mechanics. Therefore, a passive ski robot without an actuator was developed for junior racers of the alpine ski competition to facilitate understanding of the turn mechanism. Using this robot can elucidate factors affecting ski turns, such as the position of the center of gravity (COG) and the ski shape. Furthermore, a mechanism for changing the COG height, the edge angle and the ski deflection is added to the passive turn type ski robot. The developed ski robot can freely control the turn by changing those parameters during sliding.

Keywords: actuatorless; alpine ski; human–robot interaction; mechanism; passive skiing turn; skiing robot

---

1. Introduction

Since the establishment of sports engineering as a research field, the role of engineering in sports has grown. For example, by considering physical properties such as mechanical properties, sports equipment might be designed and developed to, to improve the performance of all athletes or a single athlete [1,2]. In recent years, it has become possible to analyze motion three-dimensionally from an image captured using a high-speed video camera and to pursue an ideal form in a competition. Engineering has become indispensable for the development of sports. The study of all sports has been pursued vigorously. Regarding skiing, the turn mechanism can be elucidated from various viewpoints such as skier motion analysis, dynamic model simulation and experimentation using robots [3]. During skier motion analysis, difficulties of coupling with skis and difficulties of biological loads are addressed using multiple CCD cameras and force sensors. Moreover, the effects of various elements such as mechanical properties of the ski and snow surface on the turn can be examined [4–7]. Simplified simulation of the skier turn position has also been performed; highly reliable simulation methods and skier models have been proposed [8–12]. Furthermore, in studies using ski robots [13–17], a mechanical turn is clarified by having the robot reproduce a turn using the human turn position as code. Nevertheless, because many approximations are used in the dynamic models of turns derived in these studies and because nonlinear differential equations are solved, the actual turns of skiers have not been elucidated sufficiently. It is difficult to apply the study results directly to skiers.

Therefore, for this study, to examine the competition form of alpine skiing, a simple passive-turn type ski robot was fabricated particularly emphasizing the turn position. This ski robot is suitable for examining basic operations required for the turn because the ski robot can repeat the turn continuously, simply by the gravitational movement of its center of gravity (COG) without using the power of a motor or other device. Because the COG movement timing during turning can be assessed visually, it is easy for junior athletes to understand its results. Future development and strengthening of the athlete can be expected. As described in this paper, the effects of different ski shapes and the COG position of the ski during the turn are examined experimentally using a passive turn type ski robot. Their efficient operation in alpine ski turns is clarified. Also, mechanisms for changing the COG height, the ski edge angle and the ski deflection are added to the passive turn type ski robot to develop a robot that can freely control the turn by changing them during sliding.

2. Alpine Ski Turning

Alpine ski turns are used by carving turns: turning is done while tilting the upper body to make an angle between the snow surface and the ski. For this reason, the ski has a concave shape called a side curve [18], the ski edge is made to stand on the snow surface to make use of this side curve (the angle formed by the snow surface and the ski at this time is called the edge angle). While maintaining the speed using the centrifugal force generated from the sliding speed, the ski is turned using the arc track generated by the ski. Heretofore, skitting turns have been performed using lateral deviation by turning a leg portion of a straight-shaped ski plate with the outer leg of the turn facing inward. Whereas carving turns performed by alpine skiers, who compete for velocity, can make a sharp turn greater than the arc of the side curve, the skitting turn using the turning motion of the legs has no left or right leg pointing in the same direction. Therefore, the turn is accompanied by lateral deviation, thereby suppressing the speed.

Figure 1 shows a carving turn in an actual competition. Figure 1a shows the first-ranked junior high school girl competitor in Akita Prefecture. Figure 1b shows a fourth-grade elementary school student during the first year of the competition. Assuming that the starting point (yellow mark) is the point at which the lumbar center position changes from the left foot side to the right foot side, the top point is the point at which the ski faces almost rightward below. The finishing point is the point at which the lumbar center position changes from the right foot side to the left foot side. The player depicted in Figure 1a reaches the top point beside the flag gate, whereas the player in Figure 1b reaches the top point after passage through the flag gate. The finish point is delayed because of the effect. It is also readily apparent that the player in Figure 1a turns at an angle between the snow surface and the board while flexing the ski.

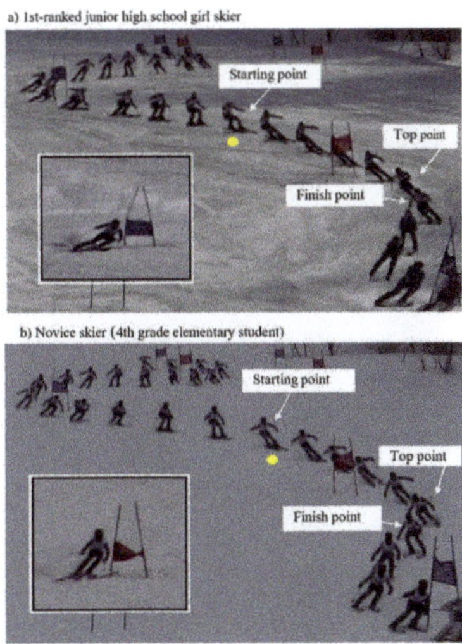

Figure 1. Turn used in alpine skiing.

3. Passive Turn Type of Ski Robot

3.1. Composition of Ski Robot

Next, we examine a carving turn. To do so, we produced a passive-turn type of ski robot [19,20] that can use some angle between the snow surface and the skis by shifting its COG. Using it, we can verify experimentally how much of an effect is attributable to ski turning by the ski shape and by the COG position. The passive-turn type of ski robot presented in Figure 2 is made entirely of ABS resin. It weighs 50 g with skis attached. The robot is structured such that its legs and body are connected by hinges, with its left and right legs separated stably using a support bar attached at the body part.

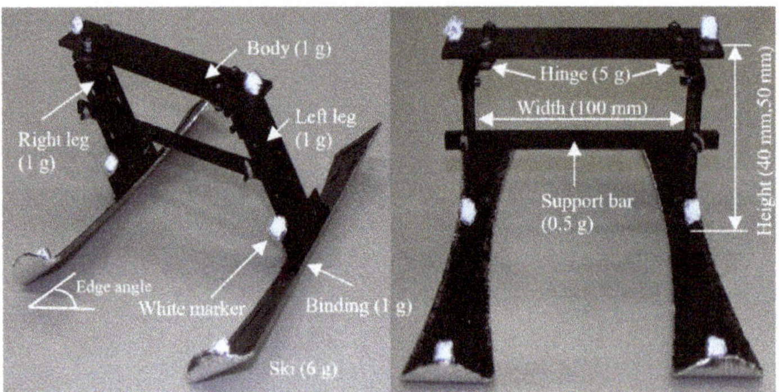

Figure 2. Structure of passive turn type ski robot (50 g).

The connections are made using hinges. Therefore, the body part can move freely only in a sideways direction. We set the gap separating the left and right skis using a support bar such that the edge angle can be as great as ±35 deg. A ski is connected firmly to each robot leg so that each leg is always perpendicular to each associated ski surface in its width direction, thereby enabling determination of a ski edge angle. Stainless steel tape was adhered to the lower surfaces of the skis to facilitate sliding. Six white markers are attached to the robot, each at a different location. They are used for motion analyses.

3.2. Profiles of Skis

Figure 3 shows two kinds of skis, each of which is made of ABS resin. They are 1 mm thick, 170 mm long and 35 mm wide at the front end and back end, with respective side curve radii R of 400 mm and 800 mm. Because the ski robot we use for this experiment cannot bend the skis similarly to a real skier, we created and used two pairs of skis, each with a different side curve radius R, by which we were able to assume that the skis were bent.

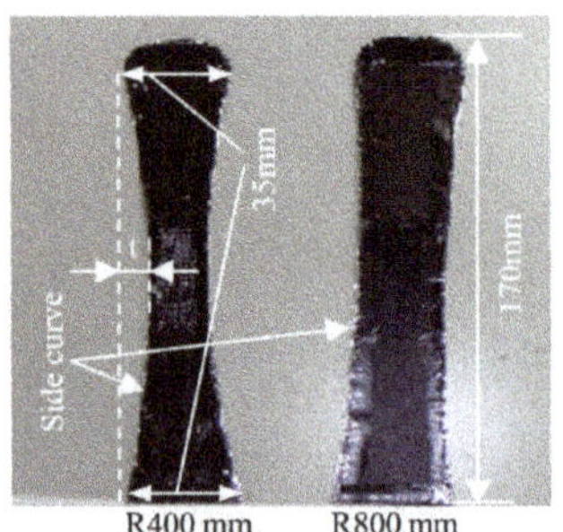

Figure 3. Ski design.

3.3. Passive-Turn Mechanism

To conceptualize the turn mechanism of the ski robot, a ski slope was assumed as portrayed in Figure 4. As viewed from directly above, the locus of a skiing turn locus was assumed as a circle.

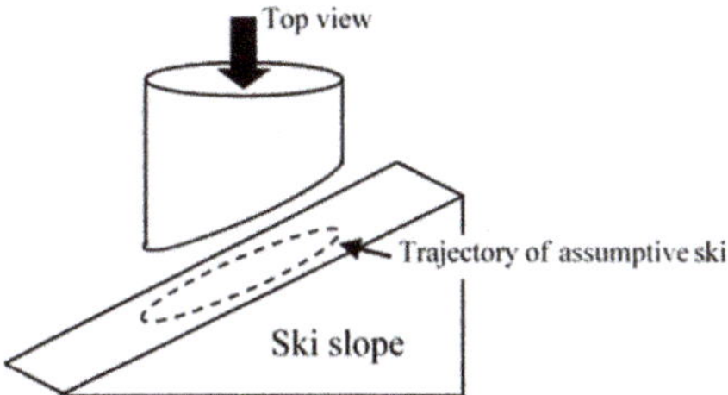

Figure 4. Simple assumption of the skiing robot turn.

Figure 5 was made to show perspectives from directly above a ski slope and directly from the side, devoting attention to the body, a left leg and a right leg. Conditions for depiction were assumed as follows.

- A 35 deg edge angle was used with a 25 deg ski slope.
- Side curve radii R of the skis are 400 mm.
- Speed at the turn is constant.
- Leg width is 80 mm.
- Body height is 40 mm.
- The COG is the body center.

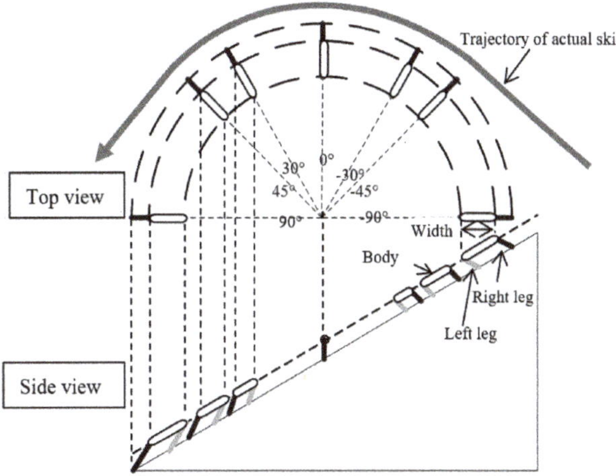

Figure 5. Simple assumption of turn of the skiing robot.

Results show the skiing robot turn motion as the following. Here, the angles in Figure 5 show the ski angle direction as ω. The slope along the Y-axis has $\omega = 0°$. As portrayed in Figure 5, first, the right foot in the upper part of the ski slope changes as a turn progresses to become different at the bottom of the ski slope. At about $\omega = 30°$, the body support changes from the left foot to support by the right foot: the angle of leaning of the body changes. The skiing robot is turning continuously while repeating this motion in alternate directions.

3.4. Experimental Method

For experiment using the passive-turn type of ski robot, we took measurements using three CCD cameras and performed three-dimensional motion analyses. Figure 6 shows the experiment setup. The X-, Y- and Z-axes are also presented in Figure 6. We placed one camera at the front part of the ski slope. The other two were placed respectively at 45 deg left and 45 deg right from the front part. The slope angle was set to 25 deg.

We chose to use a carpet (0.9 m × 1.8 m) as a ski slope, with its fluff height sufficient for the skis to retain contact when a ski edge was angled (fluff height > t in Figure 1). Stainless steel tape was placed on the lower surfaces of the skis to facilitate sliding. Figure 5 shows that the ski angle of direction ω was set.

During the experiments, the ski trajectory, the ski speed and the COG shift timing were observed. The data were verified by making the ski robot turn one direction after another continually while trying to use body heights of 40 mm or 50 mm and its skis with side curve radii R of 400 mm or 800 mm. Initially, the skis were oriented to $\omega = 0°$. The results obtained experimentally using the ski robot were based on averages of five trials.

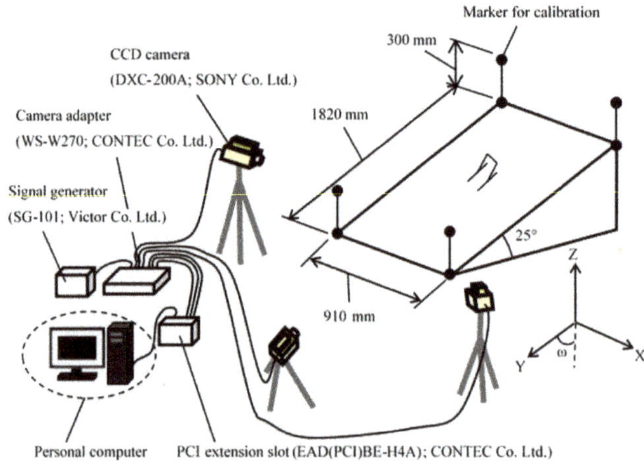

Figure 6. Experimental setup.

3.5. Body Height Difference Effects on Turning

For these experiments, robots of two types with lumbar heights of 40 mm and 50 mm were slid at an initial velocity of 0 m/s. Then the trajectories of the respective turns were analyzed.

With the start of the left ski running as the origin 0, the X-axis coordinate (positive) is taken in the left direction; the Y-axis coordinate (positive) is taken in the down direction as viewed from the front. The coordinates of the trajectory of the left ski are represented numerically. The results are presented in Figure 7. ω in Figure 7 represents the azimuth angle of the robot: is the angle between the ski orientation and the straight line in the downward slope direction. The radius of the arc formed by connecting the three points of the start point, the top point and the finish point of the turn of the ski robots did not differ significantly. However, results demonstrated that the point at which the robot starts to switch the turn to the opposite side is slower and that the cycle of the turn is larger as the lumbar region is positioned higher. The "cycle of the turn" means the duration of a continuous turn in the left or right direction.

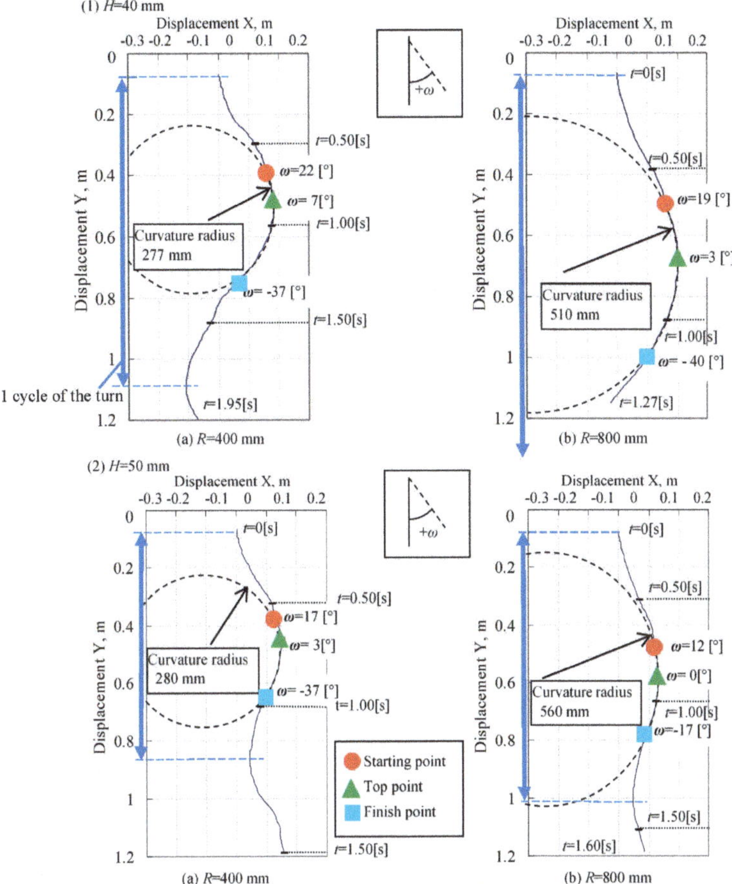

Figure 7. Experimental results related to the difference of the body height.

4. Ski Robot Using Variable Height Mechanism of COG

4.1. Composition of Ski Robot Incorporating a Variable Height Mechanism

During sliding of the passive turn type ski robot, the cycle of the turn also differed depending on the leg length difference. Therefore, attention is devoted to the influence of the turn because of the difference in the COG height. Results verified whether the robot ski turn can be controlled by changing the COG height during sliding. For moving the COG to the right and left, which is the basic operation of the ski robot, a passive turn ski robot is applied. We used a four-node link that operates only by gravity, as shown in Figure 8. A mechanism was added to change the leg length to control the robot's ski turn with the change in the COG height during sliding. This mechanism is a small servomotor (S3156; Futaba Corp., Chiba, Japan). The leg length changes as the two plates of the leg portion slide, as shown in Figure 9. Its operation combines batteries (NR5U50; Futaba Corp.) and transceivers (T6EX, R6004F; Futaba Corp.) to enable wireless radio-controlled radio. A link for moving the COG to the left and right and a mechanism for expanding and contracting the leg are provided. Therefore, a universal joint is used for the rotation axis of the mechanism to expand and contract the leg.

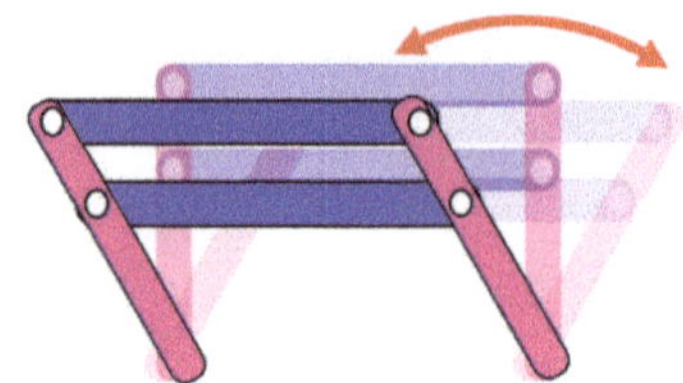

Figure 8. Rectangular linkage for center of gravity shift.

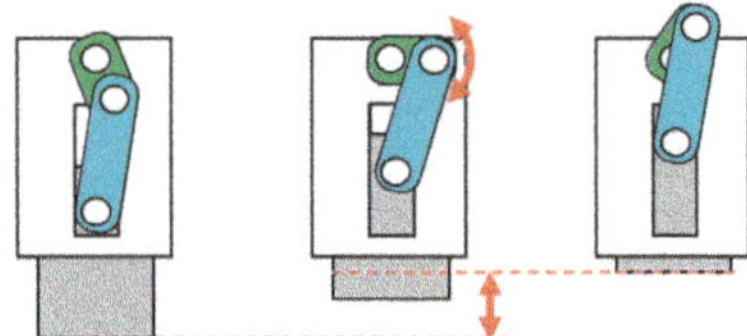

Figure 9. Variable height mechanism of the foot length.

Figure 10 shows a ski robot with an additional variable COG height mechanism. The robot leg length was a minimum of 50 mm; the length of extension of the legs was a maximum of 15 mm. The width of both legs was 110 mm. The side curve radius of the ski was 400 mm. The edge angle range was −35° to 35°. The total weight was 263 g.

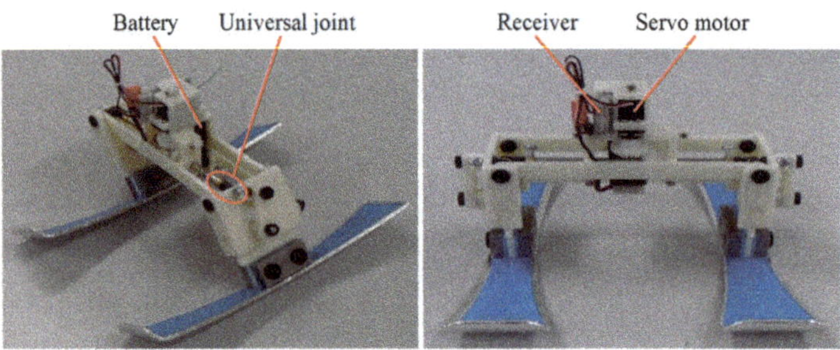

Figure 10. Ski robot with variable height mechanism.

4.2. Servomotor Response Characteristics

Servomotors and transmitters/receivers are used as robot mechanisms. The legs are extended and contracted by operation of the radio controller. Therefore, a delay occurs between the transmitter operation and the servomotor operation. To assess this delay, the respective operations of the transmitter and the servomotor were photographed using a high-speed camera (420 fps) using the S3156 servomotor and the T6EX and R6004F transceivers used for the robot. Then the reaction time of the servomotor was measured from the video. Dartfish software (Dartfish Japan Co., Ltd.) was used for the measurement. Results confirmed that a lag of about 0.06 s occurred between the transmitter and servomotor operations.

4.3. Switching Time in the COG Height

When the robot changes the COG height, it takes some time from the operation of the servomotor until the COG height changes. Therefore, change time for the case where the leg length was increased from 50 mm to 65 mm and for the case where the leg length was decreased from 65 mm to 50 mm ware examined. Similar to the servomotor response characteristics, the state of the change in the COG was photographed using a high-speed camera (420 fps); it was also measured using Dartfish software.

Measurement results confirmed that about 0.29 s was necessary for the leg length to change from 50 mm to 65 mm; about 0.21 s was necessary for the leg length to change from 65 mm to 50 mm.

4.4. Experimental Method

Measurement fields such as the measurement range and the slope angle are the same as those shown in Figure 6. As an experiment method, the position at which the distance from the origin in the Y-axis direction is 1500 mm distance is marked. This mark is used as a point for changing the COG height of the robot. We assessed sliding at an initial velocity of 0 m/s from the leg lengths of 65 mm and 50 mm and change of the leg length as it passes by the point. The legs were operated to have lengths of 65 mm to 50 mm and 50 mm to 65 mm, respectively. The cycle of the turn was compared in cases where the COG height was not operated. The results obtained experimentally using the ski robot were based on averages of five trials.

4.5. Experimental Results

Figure 11 shows the trajectory of both skis during the turn. The red lines in Figure 11b,d respectively show the points at which the COG changes. Results show that the trajectory of the turn thereafter also changes when the COG position is lowered from a high state to a low state. Moreover, the cycle of the turn is more reduced than in the case in which the COG height is not changed. In the case in which the position of the COG is raised from the low state to the high state, similarly, the subsequent trajectory changes; the cycle of the turn increases.

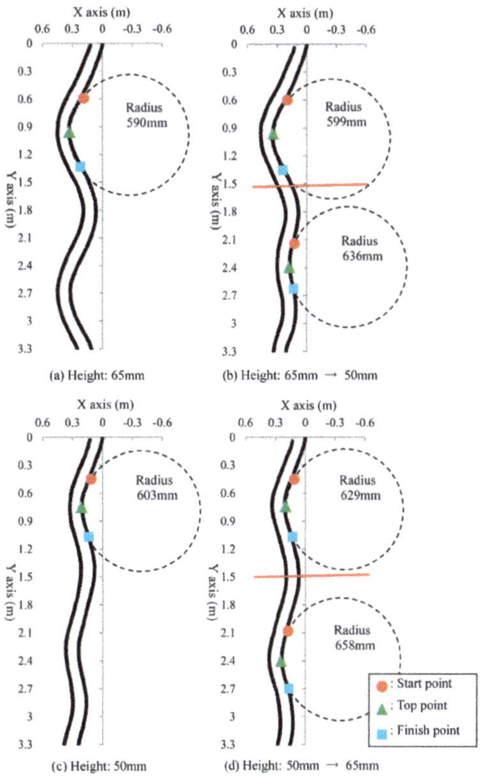

Figure 11. Trajectories of ski robot with variable gravitational center height mechanism.

However, no significant difference was found in the radius of curvature of the turns before and after the change in the COG in either case. The experimentally obtained results confirmed that the switching point of the turn, which is the movement of the COG in the left and right direction, was changed by changing the timing of raising and lowering the COG height.

5. Ski Robot Using Variable Edge Angle Mechanism

5.1. Composition of Ski Robot Using Variable Edge Angle Mechanism

In actual skiing, there is a difference in the position of the senior and the lower in turning. One of them is the difference in edge angle and the senior person performs the turn while making the edge large at the time of the turn. Therefore, attention is paid to the influence of the turn due to the difference in the edge angle and it is verified whether or not the ski turn of the robot can be controlled by changing the edge angle during the sliding.

For moving the center of gravity to the left and right, which is the basic operation of the ski robot, a four-node link capable of operating only by gravity is used, as in a passive turn type ski robot having a center of gravity variable mechanism. In addition, a mechanism that can change the edge angle has been added to control the robot's ski pattern by changing the edge angle during sliding. In this mechanism, a small servomotor (S3156; Futaba Electronics Industry) is installed on both leg portions and the edge angle is changed by rotating a small rod supporting the lumbar up and down as shown in Figure 12. Also, batteries (NR-4QC; Futaba Electronics) and transceivers (T6EX, R6004F; Futaba Electronics) can be combined to enable wireless-controlled operation.

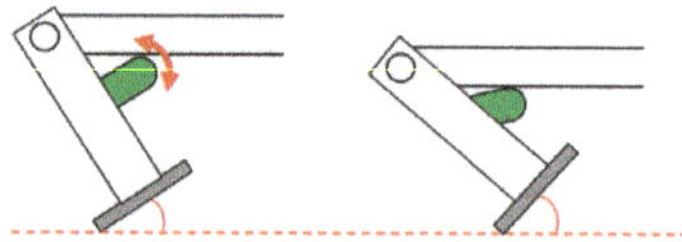

Figure 12. Variable mechanism of the edge angle.

The servomotors of both leg portions are connected to the two-forked cord. Therefore, they are mutually interlocked to perform the same degree of angulation on the left and right sides. Figure 13 shows a ski robot with an added edge angle variation mechanism. The edge angle can be varied from $\pm 30°$ to $\pm 35°$. The robot leg length was 55 mm. The width of both legs was 110 mm. The side curve radius of the ski was 400 mm. Its total weight was 238 g.

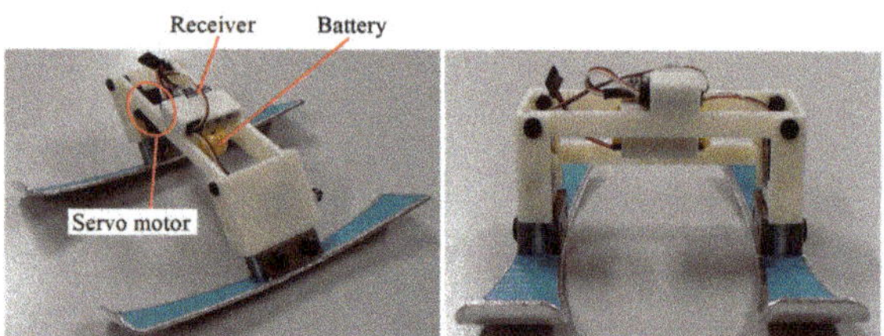

Figure 13. Ski robot with variable mechanism of the edge angle.

5.2. Switching Time in the Edge Angle

When the robot changes the edge angle, it takes some time from the operation of the servomotor until the edge angle changes. Therefore, the change times in the cases of increasing the edge angle 30° to 35° and of decreasing the edge angle from 35° to 30° were examined. The edge angle change was photographed using a high-speed camera (420 fps) and was measured using Dartfish software. Results confirmed that 0.14 s was necessary for the edge angle to change from 30° to 35°; about 0.16 s was necessary for the edge angle to change from 35° to 30°.

5.3. Experimental Method

Experiments were conducted to determine whether the cycle of the turn could be controlled by changing the edge angle during the sliding. The measurement fields such as the measurement range and the angle of the slope are the same as those shown in Figure 6.

As an experimental method, the position at which the distance from the origin in the Y-axis direction is the 1500 mm distance is marked. This mark is used as a point at which the edge angle of the robot is changed. The edge angles were operated to be 35° to 30° and 30° to 35°, respectively and the cycles of the turns were compared with those in the case where the edge angles were not operated.

5.4. Experimental Results

Figure 14 shows the trajectories of both skis during the turn. The red lines in Figure 14b,d represent points that change the edge angle. In Figure 14c,d, the "start, top and finish point" markers move from left side because the turn cycle is shorter than those shown in Figure 14a,b. The turn posture at 1500 mm of the switching point is not the same. When the edge angle is reduced from 35 deg to 30 deg, the trajectories of the subsequent turns are also changed. Results demonstrate that the radius of curvature of the turns is greater than in the case in which the edge angle is not changed. Similarly, when the edge angle is increased from 30° to 35°, the subsequent trajectory changes. The radius of curvature of the turn decreases. The experimentally obtained results confirmed that the magnitude of the curve of the turn was altered by changing, increasing or decreasing, the edge angle. The experimentally obtained results for the ski robot were based on the averages of five trials.

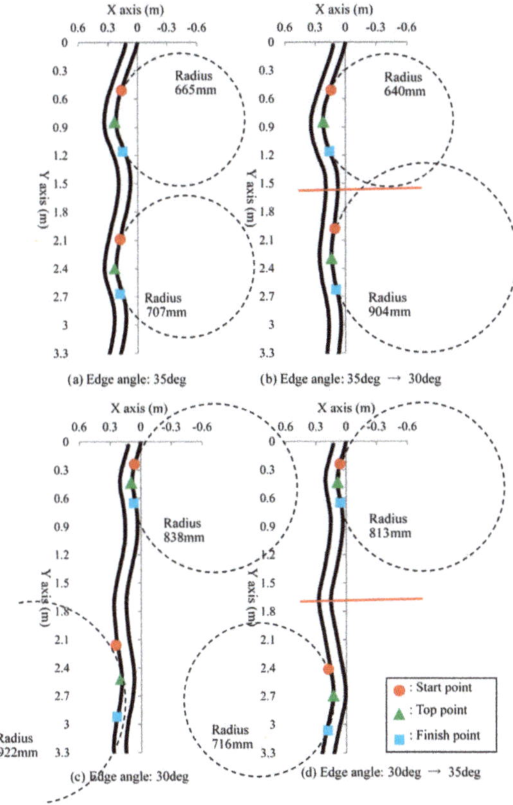

Figure 14. Trajectories of ski robot with variable edge angle mechanism.

6. Ski Robot Using Variable Mechanism of Ski Flexure

6.1. Ski Robot Incorporating a Mechanism of Variable Ski Flexure

For turns during skiing, the ski deflection is a major factor affecting motion. If the ski itself is straight and does not flex, then it cannot be turned well by raising the edge. The skier intentionally loads the ski plate surface when turning and turns while flexing the ski to create a curved surface. Therefore, attention must be devoted to the influence of the turn because of the presence or absence of ski deflection. Results verified whether the ski turn of the robot can be controlled by deflecting the ski during sliding. The ski has a side curve. Therefore, it is possible to turn the ski even in a state in which the ski is not bent.

For moving the COG to the left and right, which is the basic operation of the ski robot, a four-node link capable of operating solely by gravity is used as in a passive turn type ski robot with a variable mechanism of the COG (edge angle). In addition, a mechanism has been added to flex the plate to control the robot's ski turn by flexing the plate during sliding. This mechanism is powered by two small servomotors (S3156; Futaba Corp.). The ski can be flexed by simultaneously rotating the two plates and pulling both ends of the ski up with yarn, as portrayed in Figure 15. Its operation combines batteries (NR5U50; Futaba Corp.) and transceivers (T6EX, R6004F; Futaba Corp.) to enable wireless radio-controlled radio. The two servomotors are interlocked because they are connected to the two-forked cord. A link for moving the COG to the left and right and a mechanism for deflecting the ski are provided. Therefore, a universal joint is used for the rotation axis of the mechanism for deflecting the ski.

Figure 15. Variable mechanism of the ski flexure.

Figure 16 shows a ski robot with an additional variable ski deflection mechanism. The ski changes from a flat state to a deflected state. The robot leg length was 55 mm. The width of both legs was 110 mm. The side curve radius of the ski was 400 mm. The edge angle range was $-35°$ to $35°$. The total weight was 318 g.

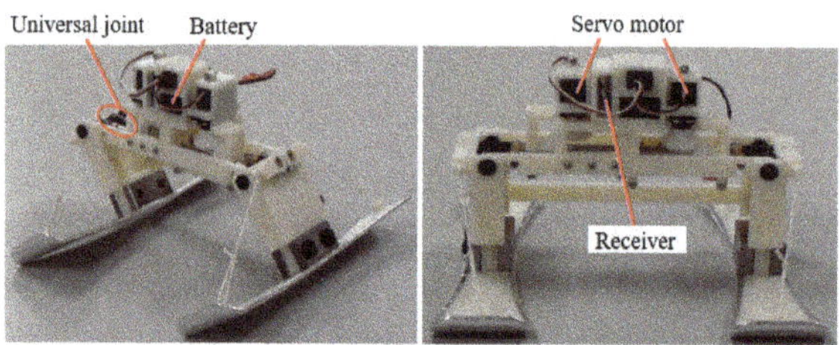

Figure 16. Variable mechanism of the ski flexure.

6.2. Switching Time in the Ski Flexure

When the robot deflects the ski, it takes some time after the servomotor operates until the plate deflects. Therefore, the change time in the case where the ski was bent from the flat state and in the case where it was returned from the bent state to the flat state was examined. The ski changes are photographed with a high-speed camera (420 fps) and measured by Dartfish software. From the measurement results, it can be confirmed that about 0.13 s is required for the ski to deflect from the flat state and about 0.10 s is required for the ski to return from the bent state to the flat state.

6.3. Experimental Method

Experiments were conducted to ascertain whether the cycle of turn can be controlled by varying the ski deflection during sliding. Measurement fields such as the measurement range and the slope angle are the same as those shown in Figure 6.

With the start of sliding of the left ski as the origin 0, the X-axis coordinate (positive) is taken in the left direction, the Y-axis coordinate (positive) is taken in the down direction as viewed from the front and the coordinates of the trajectories of the left ski and the right ski are represented numerically. As an experiment method, the position at which the distance from the origin in the Y-axis direction is the 1500 mm distance is marked. This mark is used as a point for changing the deflection of the ski of the robot. Sliding of the ski from flexed and flat at an initial velocity of 0 m/s is done to change the ski deflection as it passes by the point. The ski was manipulated from a deflected state to a flat state and from a flat state to a deflected state, respectively. The cycle of the turn was compared with the case in which the plate deflection was not manipulated. The results obtained experimentally using the ski robot were based on averages of five trials.

6.4. Experimental Results

Figure 17 shows the trajectory of both skis during the turn as the ski robot slid. The red lines in Figure 17b,d show points that change the deflection of the ski plate. When the ski is made to slide while bent, the radius of curvature of the turn is smaller. Also, the turn is deeper than when the ski is made to slide in a flat state. For the turn in which the ski plate deflection was changed in the middle of the sliding, although the curvature radius after the change could not be calculated because the number of turns within the measurement range was small, the change of the turn trajectory in both cases can be confirmed. Experiment results suggest that a turn with a flexed ski can be done with a smaller radius of curvature and that the subsequent trajectory can also be changed by changing the flexion of the ski during sliding. Therefore, the curvature of the turn can be changed by deflecting the plate during turning.

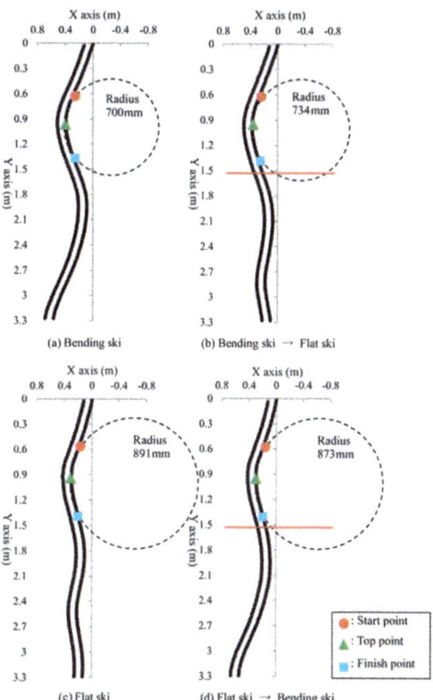

Figure 17. Trajectories of ski robot with variable ski flexure mechanism.

7. Conclusions

For this study, a passive turn ski robot was developed to improve the competitive ability of alpine skiers. Subsequently, experiments were conducted to ascertain whether mechanisms for varying the COG height, edge angle and ski deflection could be added to the passive turn type ski robot to control turns solely by these changes. The results of the respective investigations are summarized below.

1. A simple passive-turn type ski robot was fabricated to assess effects of differences of the robot construction and COG height on turning. The use of a hinge to connect the lumbar and leg allows the robot to repeat adduction and abductor movements of the hip joint using gravity exerted during sliding and to turning the hip joints continuously. In addition, when comparing the sliding trajectories of robots having different COG heights, the higher the COG is, the later the timing of switching to the opposite side turn becomes. For that reason, the cycle of the turn becomes larger.
2. To clarify the importance of the timing of changing the COG height in the turn by noting the influence on the turn caused by the COG height difference, a mechanism for varying the COG height was added to the passive turn ski robot. For turns in which the COG height was manipulated during running, the radius of curvature was almost identical. Only the cycle of the turn changed. Results show that the timing of raising and lowering the COG height changed the switching point of the turn, which is the movement of the COG in the left and right directions.
3. To clarify the importance of the edge angle in the turn by noting the influence on the turn by the difference in the edge angle, which is one factor affecting the turn, a mechanism for varying the edge angle was added to the passive turn type ski robot. Results show that the radius of curvature of the turn was changed during turning in which the edge angle was manipulated during sliding. Moreover, the depth of the arc of the turn was altered by the timing of the edge angle increase or decrease.
4. A mechanism for varying the ski deflection was added to the passive turn ski robot to clarify the importance of the ski deflection during turning by elucidating the influence on turning of the ski deflection the skier exerts during turning. Results demonstrated that the turn can be performed with a small curvature by deflecting the ski and show that the turn trajectory was altered by the plate deflection during sliding.
5. No consideration of friction exists in this study. Therefore, we would like to defer to future work our planned comparison of the running experiment results and a theoretical model incorporating the contact area of the snow in an alpine ski model.

Author Contributions: T.S. wrote the paper, performed the experiments and analyzed the data; N.S. conceived the method and helped to modify the paper. All authors have read and approved the final manuscript.

Funding: This research received no external funding.

Acknowledgments: The authors would like to thank K. Kono for technical assistance with the experiments.

Conflicts of Interest: The authors declare no conflict of interest.

References

1. Hosokawa, K.; Kawai, S.; Sakata, T. Improvement of damping property of skis. *Sports Eng.* **2002**, *5*, 107–112. [CrossRef]
2. Khmelev, V.N.; Levin, S.V.; Tsyganok, S.N.; Shalunov, A.V.; Chipurin, E.V. The New Technology of Sliding Ski Surface Covering. In Proceedings of the 6th Annual 2005 International Siberian Workshop and Tutorials on Electron Devices and Materials, Erlagol, Altai, Russian Federation, 1–5 July 2005; pp. 86–89.
3. Shimizu, S.; Hasegawa, K.; Nagasawa, T. Alpine skiing robot. *Adv. Robot.* **1992**, *6*, 375–376. [CrossRef]
4. Kawai, S.; Otani, H.; Sakata, T. Coupled Motion of Ski and Elastic Foundation Under Ski Control. *JSME Int. J. Ser. C* **2003**, *46*, 614–621. [CrossRef]

5. Sakata, T.; Tsukiyama, M. Effects of Position of Shoe Center on Ski Turn. *JSME Int. J. Ser. C* **1999**, *42*, 922–929. [CrossRef]

6. Tada, N.; Hirano, Y. In Search of the Mechanics of a Turning Alpine Ski Using Snow Cutting Force Measurements. *Sports Eng.* **2002**, *5*, 15–22. [CrossRef]

7. Sahashi, T.; Ichino, S. Coefficient of Kinetic Friction of Snow Skis during Turning Descents. *Jpn. J. Appl. Phys.* **1998**, *37*, 720–727. [CrossRef]

8. Morawski, J.M. Control System Approach to a Ski-turn Analysis. *J. Biomech.* **1973**, *6*, 267–279. [CrossRef]

9. Sakata, T.; Ito, T. Simulation of Ski Turn. In Proceedings of the 2nd International Conference the Engineering of Sports, Sheffield, UK, 8 July 1998; pp. 361–368.

10. Kagawa, H.; Yoneyama, T.; Okamoto, A.; Komatsu, H. Development of a Measuring System for Joint Angles of a Skier and Applied Forces during Skiing. *JSME Int. J. Ser. C* **1998**, *41*, 214–219. [CrossRef]

11. Kawai, S.; Yamaguchi, K.; Sakata, T. Ski Control Model for Parallel Turn Using Multibody System. *JSME Int. J. Ser. C* **2004**, *47*, 1095–1100. [CrossRef]

12. Hasegawa, K.; Shimizu, S. Dynamics of Repeated Parallel Turns of Snow Ski. *J. Jpn. Soc. Sports Ind.* **1995**, *5*, 9–18. (In Japanese) [CrossRef]

13. Sahashi, T.; Ichino, S. Carving-turn and edging angle of skis. *Sports Eng.* **2001**, *4*, 135–145. [CrossRef]

14. Nemec, B.; Lahajnar, L. Control and Navigation of the Skiing Robot. In Proceedings of the 2009 IEEE/RSJ International Conference on Intelligent Robots and Systems, St. Louis, MO, USA, 10–15 October 2009; pp. 2321–2326.

15. Yoneyama, T.; Kagawa, H.; Unemoto, M.; Iizuka, T.; Scott, N.W. A ski robot system for qualitative modeling of the carved turn. *Sports Eng.* **2009**, *11*, 131–141. [CrossRef]

16. Shimizu, S.; Doki, H.; Nojiri, N. Robot Models for Passive Dynamic Skiing—A Skidding-turn Model of Abduction and Adduction with Inner Rotation of the Hip Joints. *J. Ski Sci.* **2012**, *9*, 29–33. (In Japanese)

17. Shimizu, S.; Doki, H.; Yamane, M.; Nojiri, N. Robot Models for Passive Dynamic Skiing—A Snowplow Model of Rotation Involving the Hip Joints around the Femur Axes. *J. Ski Sci.* **2013**, *10*, 13–18. (In Japanese)

18. Shimizu, S.; Doki, H.; Yamane, M.; Sakatani, M.; Nojiri, N. Robot Models for Passive Dynamic Skiing—A Skidding-turn Model of Inward Lean with Inner Rotation of the Hip Joints. *J. Ski Sci.* **2014**, *11*, 13–18. (In Japanese)

19. Kono, K.; Saga, N. Experimental Study on Alpine Skiing Turn Using Passive Skiing Robot. In *Intelligent Robotics and Applications*; Lecture Notes in Artificial Intelligence; Springer: Berlin/Heidelberg, Germany, 2009; Volume 5928, pp. 1044–1050.

20. Kono, K.; Saga, N. Development of a passive turn type skiing robot with variable height mechanism of gravitational center. *J. Robot. Mechatron.* **2012**, *24*, 372–378. [CrossRef]

Article

Implementation of Explosion Safety Regulations in Design of a Mobile Robot for Coal Mines

Petr Novák [1,*], Tomáš Kot [1,*], Jan Babjak [1], Zdeněk Konečný [1], Wojciech Moczulski [2] and Ángel Rodriguez López [3]

[1] Department of Robotics, Faculty of Mechanical Engineering, VŠB-Technical University of Ostrava, 70833 Ostrava, Czech Republic; jan@babjak.cz (J.B.); zdenek.konecny@vsb.cz (Z.K.)
[2] Instytut Podstaw Konstrukcji Maszyn, Konarskiego 18A, 44-100 Gliwice, Poland; wojciech.moczulski@polsl.pl
[3] Department of Systems Engineering and Automation, Universidad Carlos III de Madrid, 28911 Madrid, Spain; angrodri@ing.uc3m.es
* Correspondence: petr.novak@vsb.cz (P.N.); tomas.kot@vsb.cz (T.K.)

Received: 18 October 2018; Accepted: 13 November 2018; Published: 19 November 2018

Featured Application: The mobile robot described in the article is used for reconnaissance and inspection of underground coal mines after a catastrophic event. Knowledge from the article can be used as guidelines and aid for design of a mechatronic system with explosion safety for Group I (underground mines), Category M1 (presence of an explosive atmosphere).

Abstract: The article focuses on specific challenges of the design of a reconnaissance mobile robotic system aimed for inspection in underground coal mine areas after a catastrophic event. Systems that are designated for these conditions must meet specific standards and regulations. In this paper is discussed primarily the main conception of meeting explosion safety regulations of European Union 2014/34/EU (also called ATEX—from French "Appareils destinés à être utilisés en ATmosphères Explosives") for Group I (equipment intended for use in underground mines) and Category M1 (equipment designed for operation in the presence of an explosive atmosphere). An example of a practical solution is described on main subsystems of the mobile robot TeleRescuer—a teleoperated robot with autonomy functions, a sensory subsystem with multiple cameras, three-dimensional (3D) mapping and sensors for measurement of gas concentration, airflow, relative humidity, and temperatures. Explosion safety is ensured according to the Technical Report CLC/TR 60079-33 "s" by two main independent protections—mechanical protection (flameproof enclosure) and electrical protection (automatic methane detector that disconnects power when methane breaches the enclosure and gets inside the robot body).

Keywords: mobile robot; coalmine; exploration; robotics; ATEX; safety; methane

---

1. Introduction

Despite the constant improvement of mining technology and ever more comprehensive knowledge of the geological composition of coal resources, there are still catastrophes that happen to underground coal mines. Thousands of miners die from mining accidents each year, especially from underground coal mining [1,2]. Underground coal mining is considered to be much more hazardous than hard rock mining due to flat-lying rock strata, the presence of methane gas, and coal dust.

The focus of the project "System of the mobile robot TeleRescuer for inspecting coal mine areas affected by catastrophic events" (supported by European Commission research fund Coal and Steel) was the development and realization of a system for virtual teleportation (virtual immersion) of rescuers to the underground areas of a coal mine that have been closed due to a catastrophic event within them [3]. It was an international project managed by a consortium composed of the Silesian University of Technology (Gliwice, Poland), the VSB—Technical University of Ostrava (Ostrava, Czech Republic), the Universidad Carlos III de Madrid (Madrid, Spain), COPEX (Katowice, Poland), Simmersion GmbH (Groß-Siegharts, Austria), and Skytech Research (Gliwice, Poland) during years 2014–2017. All authors of this article are members of the consortium team.

The inspection with use of a mobile robot should take place primarily in situations and places where the presence of a rescue team is absolutely precluded (e.g., after a decision to withdraw the team). In some situations, the robot could for a long time (e.g., a few or dozen days) remain in the danger zone as a remote measurement observatory. It seems to be recommended that the robot could be remotely controlled from a safe room (rescue base). It should be equipped with a set of cameras and a set of sensors for the analysis and recording of physical parameters of the mine and composition of the mine atmosphere. One should also consider the possibility of using a robot vehicle as a means of transport to provide e.g., a specialized equipment to/from the rescue base to/from the rescue team present in the danger zone (going to the zone or returning) or to the crew waiting for help.

A mobile robot that is designed for harsh conditions must be able to properly operate in such conditions [4]. This includes not only a heavy-duty construction and good driving abilities, but primarily the robot must not make the situation even worse by, for example, causing a methane explosion. To secure this, most countries adopted certain regulations and standards for all devices that are intended for areas with potential risk of explosion, and these regulations must be indispensably followed.

This article describes a practical application of said regulations for the mobile robot TeleRescuer. After the initial overview of the related legislation in the main world regions and analysis of existing mobile robots for similar tasks, TeleRescuer is introduced by a brief description of individual subsystems. Then follows the main part of the article—implementation of explosion safety regulations that begins by selecting the overall concept of protection, which is then described in detail (separation of electrical components into galvanic isolated subsystems; flameproof enclosure with an example of the performed stress analyses; automatic safety gas detector; and, other protections).

2. State of Art

2.1. Legislation Overview

When designing the robot in the underground coal mine environment, it is necessary to take into account the requirements for safety in potentially explosive atmospheres, based on the standards in force in the country of use.

In the European Union, the legislation is based on the European Commission Directive 2014/34/EU (also called ATEX), which sets requirements for manufacturers and operators of equipment that is designed to work in potentially explosive atmospheres [5]. The requirements in this document result from national standards adopted by individual countries. In the EU, these national standards are harmonized with the IEC 60079 Series Explosive Atmosphere Standards [6].

In China, the GB standard—Guobiao system is applied (Chinese national standards issued by the Standardization Administration of China), together with standards GB3836, which are identical to IEC 60079 [7].

In the United States of America, this issue is addressed by the legislation of Hazardous Locations (abbreviated to HazLoc), which aims to control the risks that are related to the explosion in certain environments [8].

In the Russian Federation and several neighbouring countries, the document "Technical Regulations CU TR 012/2011 on the safety of equipment in explosion hazardous environments" [9] is in use.

In Australia, this is solved by the NSW Coal Mine Health and Safety Regulation [10]. In Brazil, the INMETRO Regulation "Portaria 83:2006" states the requirements for electrical equipment for use in explosive atmospheres of vapours and gases [11].

It should be mentioned that all of the above-mentioned documents approach the problem more or less equally—they classify environments into several levels of risk and for each level offer specific ways to achieve the required safety.

2.2. Existing Mobile Robots

There are a number of projects related to problems of mobile robots in underground coal mines [12–14]. One of the most important differences between these robots when compared to the "normal" field mobile robots should be the ability to work in the potentially dangerous environment of coal mines by fulfilling the corresponding directives.

An example of a mobile robot that is designed for usage in coal mines is the Mine Rescue Robot (MINBOT) [15]. Its second generation—MINBOT-II—is developed based on the experiences learnt from the applications and experiments of the first generation (MINBOT-I) shown in Figure 1. The robot is controlled remotely by the operator via optical fibre. Unlike the previous version, MINBOT-II has its own power supply. The most interesting information—compliance with explosion safety regulations—is not mentioned.

The mobile robot Numbat (CSIRO—Division of Exploration and Mining, Kenmore, Australia) shown in Figure 1 is a mine reconnaissance robot designed in the 1990s by the Australian Commonwealth Scientific and Industrial Research Organization. The Numbat is an eight-wheeled mobile platform with an onboard gas analysis package to provide information on the environmental conditions within the mine [16].

Figure 1. Minibot-II robotic platform (left), Robot Numbat (right).

Another example of a mobile robot for underground mines is Wolverine, as developed by Remotec (Oak Ridge, TN, USA)—Figure 2. Originally a military robot, which used to serve as a traditional bomb squad robot, has been made mine permissible [17]. It weighs over 550 kg and it is driven by explosion-proof motors and rubber tracks. It is equipped with navigation and surveillance cameras, lighting, atmospheric detectors, night vision capability, two-way voice communication, and a manipulator arm. The robot is operated remotely from a safe location and has the capability of exploring up to 1.5 km, communicating vital information about the conditions in the mine over a fibre optic cable. The operator can see real-time information, including video and concentrations of combustible and toxic gasses.

Figure 2. Wolverine V-2 robot developed by Remotec (left) and Gemini-Scout (right).

The Gemini-Scout (Sandia National Laboratories, Albuquerque, NM, USA)—Figure 2 is fully equipped with cameras and sensors, enabling it to provide feedback on environmental and structural conditions and can serve as a two-way communications device with trapped miners, providing critical life-saving information. The weight of the robot is about 90 kg. Explosion safety is solved as explosion-proof housing (thus the robot cannot work when methane is present—it has only the M2 category implemented.) [18].

A similar conception of the chassis and the method of explosion safety as Gemini-Scout robot is used on the MPI robot by Emag-Piap consortium (Warsaw, Poland)—Figure 3. The robot is aimed at support for the teams of mine rescuers [19]. The robot is supposed to be certified for Group 1, category M1 (protection by explosion-proof housing plus protection of overpressure), but this combination is arguable. It weighs about 1100 kg, maximal velocity 0.7 m/s, distance range 1 km, length 240 cm, width 115 cm, height 180 cm, supply 42 VDC. It is not possible to move the robot through the 80 cm diameter hole in dams.

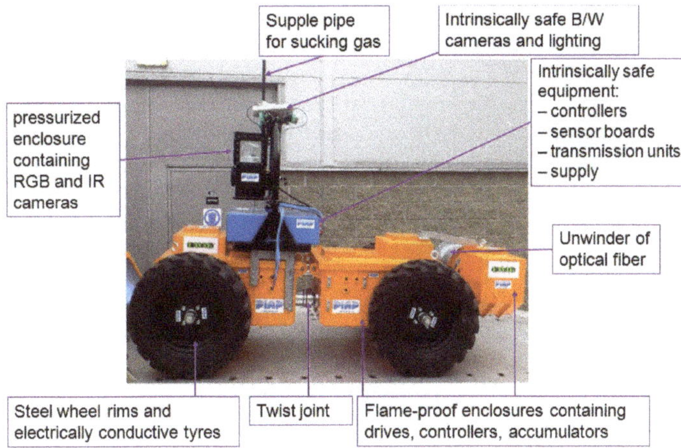

Figure 3. MPI robot by Emag-Piap consortium (Poland).

The mobile robots for coal mines described above have some weaknesses like their large size (cannot go through the fire-dam tube), teleoperation only (no autonomy), no ability to create a three-dimensional (3D) map of the surroundings and—most probably—problems with meeting the actual explosion safety directive requirements. Other serious problems include: communication distance is shorter than required, ability to overcome obstacles is low, and autonomous movement ability is weak or non-existent. Some tracked robots are not suitable for crossing rough surfaces that are caused by an explosion in a coal mine.

The goal of the TeleRescuer project was to deal with all of these problems and design a mobile robot that would be fully applicable and useful in the mentioned situations.

2.3. Requirements for the Robot

Required functionality and parameters of the robot TeleRescuer were specified based on the analysis of existing robots and a survey made in the Central Mine Rescue Station (Bytom, Poland).

The proposed unmanned vehicle should have a compact structure, small size, and high stability and mobility. Its dimensions cannot exclude the possibility of transport through a fire-dam tube (Ø 800 mm) in an anti-explosion dam. The device should also have as low weight as possible in order to enable manual handling (additional transport handles would be useful). Uncomplicated control shall be performed remotely—from the rescue base. Instrumentation (sensors, cameras) should be protected from possible damage.

The main obstacles and hindrances which the robot can encounter during the inspection and which should be dealt with include:

- significant reduction or total lack of visibility,
- high temperature (up to 60 degrees Celsius) and humidity (up to 100%),
- difficult terrain, i.e., significant excavation slope, uneven ground, water spills of different depths,
- reduced cross-sectional area of mining working,
- numerous obstacles specific to cave-ins and related to stored improperly or scattered material, and
- technological obstacles: structures of conveyors, conveyor drives, excavation protection structures and their intersections, hydraulic or wood racks, railroad tracks, turnouts, loading ramps, winches, transformers, switchgear or single switches, pumps, hoses, drainage, sheet, elements of concrete, machine constructions and their fixing—beam, struts, chains, wire ropes, tubes, pipes, cables, ventilation fans, and lutes.

As far as the sensory system is concerned, the device should be able to measure temperature, relative humidity and the four major gases (O_2, methane—CH_4, CO, CO_2). Beneficial could also be the ability to measure the air velocity and temperature of selected elements of the robot body. The exact scope and frequency of measurement should always be programmed after consultation with the head of the rescue operation. There also must be equipment for recording and transferring images to the operator (colour cameras operating in the visible light spectrum supported with additional lighting and infrared cameras), together with a 3D mapping functionality (not critical).

The respondents considered that the optimum working time for the robot would be:

- about 3–4 h of work, and
- from several hours to several days in idle mode.

3. Description of the Mobile Robot Telerescuer

The TeleRescuer robot (Figure 4) consists of the main chassis with four independent tracked arms (eight motors, gears, motor controllers, batteries, and the main control system are placed in a flameproof housing), a sensory arm with a sensory head, a 3D laser scanner unit, and a mote deploying subsystem (motes are small Wi-Fi repeater modules) [20,21]. Every subsystem has its own independent power supply.

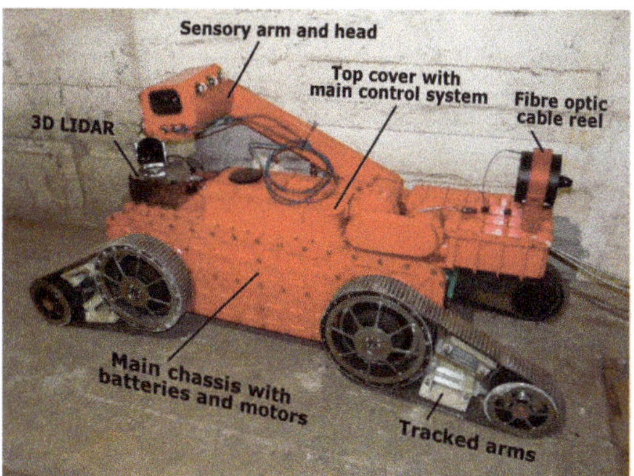

Figure 4. TeleRescuer—main subsystems.

3.1. Main Robot Chassis and Control System

The main robot chassis contains the motion subsystem, the main control system (MCS), the communication subsystem and power supply. The motion subsystem is based on four identical independent flipper arms with tracks; each of the arms contains two brushless DC motors.

The MCS is responsible for motion control, management of communication between all subsystems, autonomous behaviour, 3D map building, collision prevention, etc. This requires high computational power while keeping low power consumption, it was thus decided to use the IPC (Industrial PC) architecture [22,23].

The control system software is based on the Robotic Operating System (ROS). The system is modularly divided into several parts (ROS nodes) that are responsible for individual logical tasks (motion, sensors, autonomy, communication, 3D map building, etc.). The software architecture and implementation does not affect explosion safety, thus it is not described in detail here. More information can be found in [22].

3.2. Sensors, 3D Mapping and Autonomy

A very important part of the mobile robot is the sensory head located on the top of the tiltable sensory arm (see Figure 5). The sensory head contains five cameras (two for stereoscopic view, one with a wide field of view, one for rear view and one thermal camera), LED lighting, various gas sensors, and an inertial measurement unit. Elevation and rotation of the sensory head and lifting of the additional methane arm (this arm is part of the main sensory arm) are realized by only one DC motor with four electromagnetic clutches for selection of the type of movement. Detail description is beyond the scope of this paper.

The mapping subsystem is intended for 3D map building during robot movement in a coal mine [24]. This system contains a Sick LMS111 two-dimensional (2D) laser scanner mounted on a rotating axis adding the third dimension to scanning. Using a special visualization part of the operator control system [25–27], the rescuers can inspect the mine and plan their intervention. Mapping can also be used for regular inspections of coal mine areas—the system compares the actual map with the previous one and can report unexpected changes in the tunnel shape (a part of the tunnel is starting to collapse, etc.).

The second use of this subsystem is to provide real-time information about robot surroundings for the autonomy control. Autonomy is used for the automatic return of the robot in the case of losing communication with operator. An example of autonomous navigation control can be found in [28].

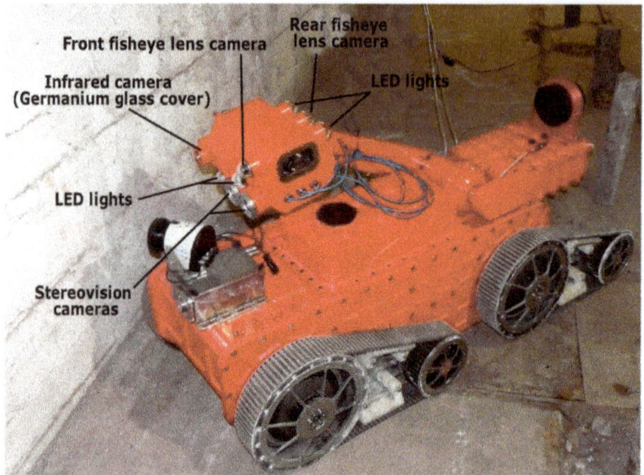

Figure 5. TeleRescuer—cameras and lights on the sensory head.

3.3. Operator—Robot Communication

A reliable system has been designed for communication between the operator and the robot. The main communication channel is based on an optical fibre cable. In case the cable is broken, a backup wireless communication system is activated automatically. The wireless network is built during robot motion by units called motes that act as repeaters to achieve hundreds of meters wirelessly. The motes are located on the rear part of the robot and they are automatically dropped depending on the intensity of the wireless signal.

3.4. Technical Data

The most important technical data of the mobile robot TeleRescuer include:

- weight: 590 kg,
- width: 741 mm,
- length: 2100 mm (tracks horizontally), 1540 mm (tracks vertically),
- height: 500 mm (minimal height, tracks horizontally), 920 mm (standing on tracks), additional +780 mm with arm in the top-most position,
- ability to drive through a tube with inner diameter 800 mm,
- driving speed: 0.5 m/s (software limited),
- battery capacity: approx. 2 h of operation,
- communication cable length: 2000 m, and
- pulling force: 1200 N (measured during tests).

4. Implementation of IEC 60079 for TeleRescuer

The robotic system TeleRescuer is intended for use in European countries, so the design was made according to the European Commission Directive 2014/34/EU [5] and the IEC 60079 Series Explosive Atmosphere Standards.

4.1. Classification

IEC 60079 classifies devices into two groups:

- Group I—equipment intended for use in underground mines and parts of surface installations of such mines, liable to be endangered by the explosion of methane and/or coal dust. Group I is further divided into Categories M1 and M2.
- Group II—equipment intended for use in other industries exposed to explosive atmospheres (further divided into Categories 1, 2, and 3).

The above-mentioned categories of devices define the required levels of security, namely in the underground mining area:

- Category M1—equipment designed so that it can safely operate in the presence of an explosive atmosphere. This is achieved through the use of integrated explosion protection measures selected, so that in the event of a failure of one of them, at least the second measure provides an adequate level of protection (two protections based on different principles); or, in case of two independent failures, an adequate level of protection is still assured (triple protection).
- Category M2—equipment designed to ensure a high level of safety under normal conditions, and in the case of severe operating conditions, resulting e.g., due to careless handling of the device or changing of environmental conditions.

One of the key requirements for the TeleRescuer system was that it should be approved for Group I, Category M1. This is the highest possible level and that poses a big challenge for the implementation of the robot.

4.2. Achieving ATEX Group I, Category M1

Proving the compliance with the essential safety requirements set out in the Directive is usually done by meeting the requirements of relevant ATEX standards. However, the high relative power used by drives, and the desire of using as many "common of the shelf" (COTS) components as possible, preclude the implementation of one of the protection modes (Ex ia I or Ex ma I) that would allow for achieving Category M1 directly; allowing only the use of those that give Category M2. But, for the Category M2, National Mining Regulations have the requirement to switch off power when the CH_4 concentration in the surrounding atmosphere exceeds some limit, usually between 1% and 2.5% *v/v*, using automatic meters. However, this is not acceptable for the intended TeleRescuer operational circumstances.

Directive 2014/34/EU offers two alternatives in this case (Annex I 1.1.a, Annex II 2.0.1): Either to apply two independent protection means, or to justify thoroughly that the required safety level is achieved. Some guidance on how to achieve this goal can be found in Technical Report CLC/TR 60079-33 "s" [29]; an IEC standard that was adopted by the EU as Technical Report or Recommendation. In Art 10.2.5 and 10.4. is open to the possibility of using a recognised (per standards) protection mode complemented by additional means of protection, which can be "innovative".

4.3. The Selected Solution for TeleRescuer

In TeleRescuer, the approach is using a recognised protection method (Flameproof, Ex d), which will give Category M2, combined with an automatic safety gas detector capable of tripping power to all non-Ex i_a electric devices in each Ex d enclosure. This solution is based on the Technical Report CLC/TR 60079-33 "s" mentioned above.

The innovation consists in placing the safety gas detector (with a trigger setpoint of 0.5% CH_4 *v/v*) inside the enclosure, which is made gas-tight using ad-hoc gaskets. In this way, under both normal and abnormal circumstances, even if CH_4 is present in the outer atmosphere, no gas would ingress into the enclosure, and the system will stay operational.

Only in the case of a failure in the sealing system, such ingress will happen, and power would be disabled. Setting a very low (0.5% *v/v*) trip point allows for avoiding the possibility of the inflammation of the inner atmosphere by the possible sparks that are generated when switching off power to internal devices.

Problematic is the 3D LIDAR used on the mobile robot. It uses the Sick LMS111 device, which has IP67 Ingress Protection but no level of explosion safety and no other laser scanner commercially available provides a sufficient protection. Thus, the 3D LIDAR module must be completely disconnected from power in environments with explosion risk. The 3D mapping functionality is not a crucial part of the whole system, so this solution is acceptable.

4.4. Separation of Subsystems

For increased safety, the mobile robot is divided into several galvanic isolated parts (depicted as grey boxes in Figure 6). Each of these subsystems has its own batteries and a safety methane detector (described in further chapters). The subsystems communicate over the optic fibre serial line (RS232) or Ethernet with galvanic isolated transformers.

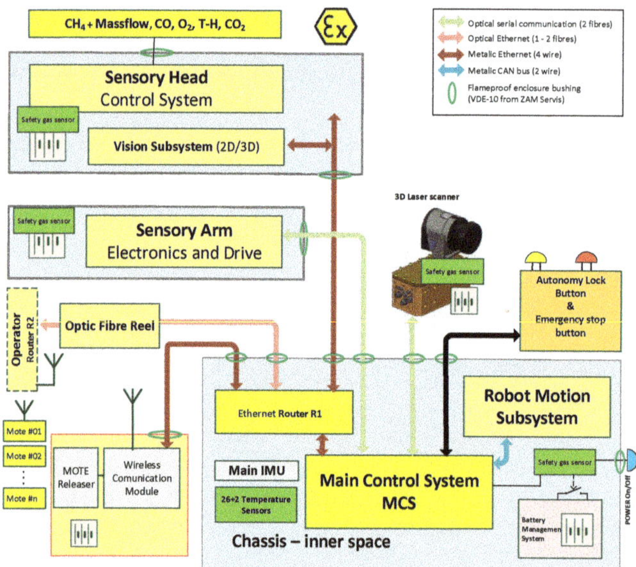

Figure 6. Separation of TeleRescuer subsystems.

4.5. Flameproof Enclosure

The requirements resulting from the standards place big demands on the design of the covers of the individual components of the robot. The most important is the encapsulation of the robot body, which houses eight robot motors, a battery subsystem, and the main control system. The design requirements are based on standard EN 60079-1—Explosive atmospheres—Part 1: Equipment protection by flameproof enclosure "d" [30], which specifies requirements for wall thickness, strength and resistance to the potential explosion of methane within the robot, contact surfaces of detachable parts, etc.

In designing the shape of the robot encapsulation, strength analyses of individual parts were performed continuously to achieve the optimal shape, strength, and weight ratio with respect to the potential pressure that could cause methane explosion inside the robot body. These analyzes were performed in the PTC Creo Simulate 3.0 CAD system, more details about the methodology can be found in [31]. The following example will demonstrate inspection and optimization of the top cover under which the robot control system is located inside the body (Figures 4 and 7).

In order to verify and optimize the top cover, it was necessary to create a computational model. The model contains a simplified assembly of the frame and the cover, with a contact between them. The frame is fixed and the cover is connected by screws, which are simulated as idealized Fastener elements (Figure 8). Based on the specification, the material "Stainless Steel 1.4462" was used for calculations (tensile strength Rm = 950 MPa, proof stress Re = 500 MPa).

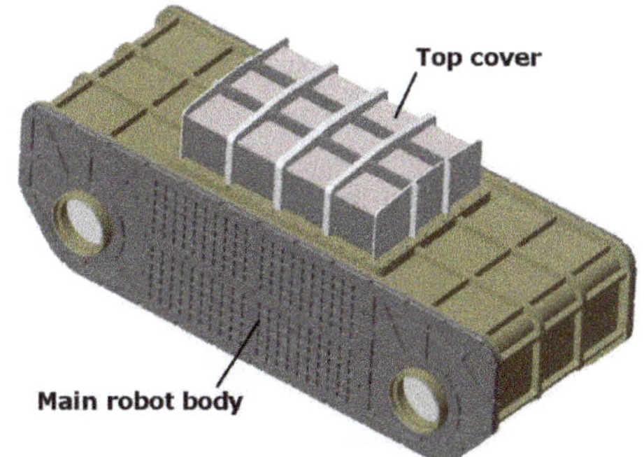

Figure 7. Main robot body with the top cover.

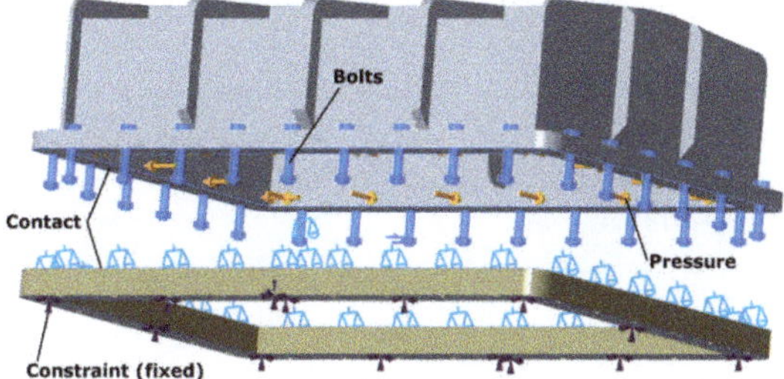

Figure 8. Computational model of the top cover assembly.

A pressure of 3 MPa was applied on the inner surfaces of the cover. This value simulates the explosion of methane inside the robot body and it is based on experiments from [32,33], increased by a safety factor. Results of the analysis are shown in the following figures.

Figure 9 shows the stress distribution on the cover. The red areas represent stress peaks reaching up to 1900 MPa, which means that the cover could be seriously damaged by the explosion and the flameproof enclosure protection could be broken. It was thus necessary to modify the design on the cover to lower the stress peaks.

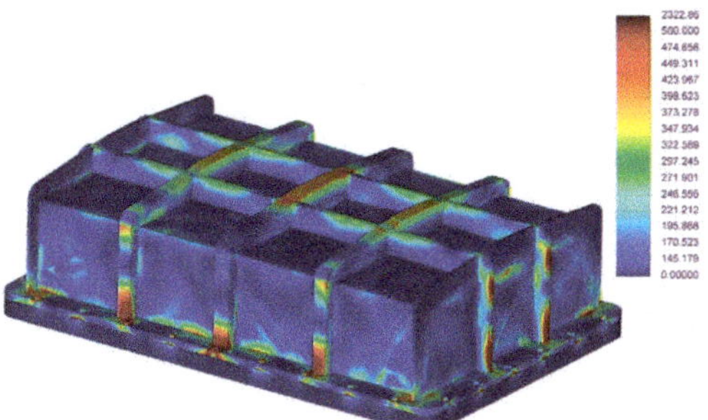

Figure 9. Stress distribution during an explosion simulation (initial design).

The final modified design of the cover is shown in Figure 10. The stress peaks are between 580 and 650 MPa, which does not exceed Rm and the cover would not be destroyed.

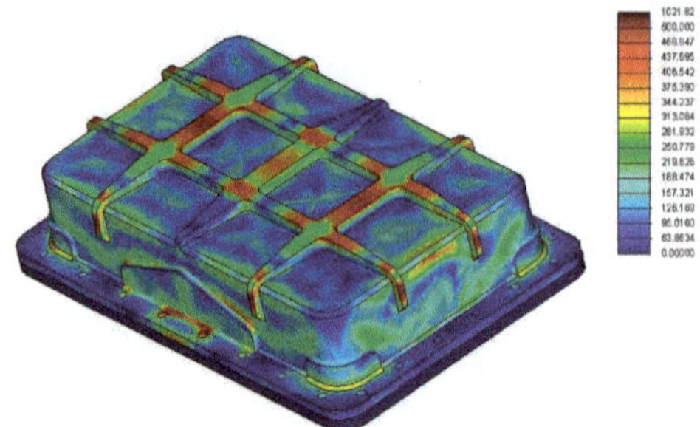

Figure 10. Stress distribution during an explosion simulation (final design).

Another very important test is for any possible gaps between individual parts of the enclosure caused by the inner explosion. The maximum gap size is controlled by EN 60079-1 ([29]: table "Minimum width of joint and maximum gap for enclosures of Groups I, IIA and IIB"), which in the case of a planar gap with the length bigger than 25 mm (the actual gap length is 30 mm, see Figure 11) and inner volume larger than 2000 cm^3 (the actual volume is approx. 40,000 cm^3) allows for a maximum width of the gap 0.5 mm for Group I.

The simulation results show that the contact surface between the top cover and the bottom frame deforms during the explosion and a gap appears. The width of this gap is different on the inner and outer edge of the cover and changes with the position along the edge (Figures 11 and 12), but it never exceeds the allowed limit (the maximum is 0.474 mm, which is less than 0.5 mm).

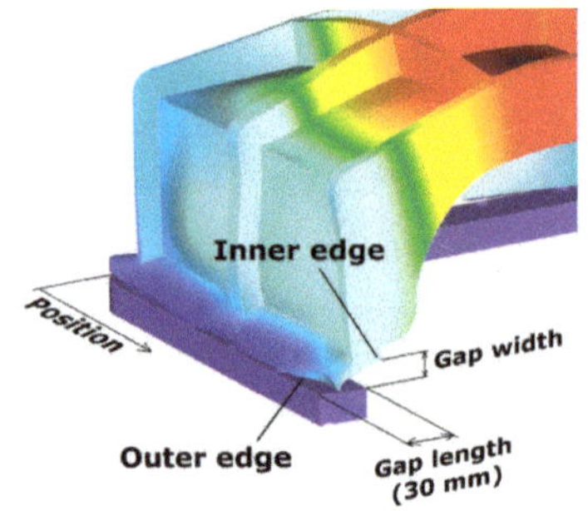

Figure 11. Top cover deformation.

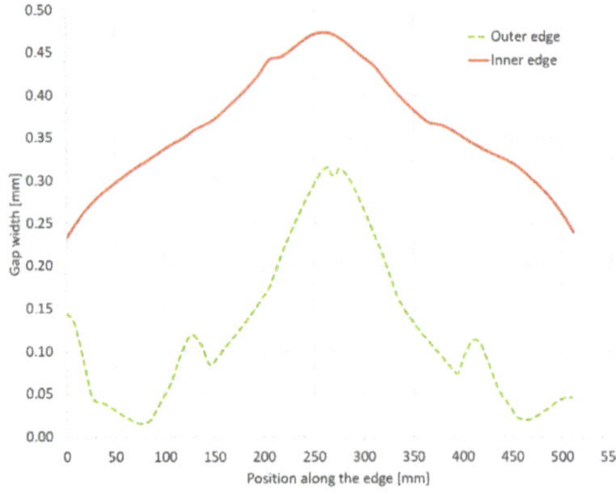

Figure 12. Gap width (for the inner and outer edge) in relation to a position along the edge (the local maxima correspond to ribs on the cover).

4.6. Automatic Safety Gas Detector

Even with the presence of methane in the atmosphere around the robot, methane should not get through the flameproof enclosure into the robot body. If, however, the enclosure is damaged, the second level of protection—automatic safety gas detector—prevents an explosion by turning the power of the whole robot off (except for the gas detector itself, which is designed with intrinsic safety) [34].

The design of the safety gas detector has a high safety level. It is purely hardware based (no microcontrollers and software), starting with an ATEX M1, SIL1, 0–5% *v/v* CH_4 sensor from Dynament. The output of the sensor is per the British Mining Standards, 0.4–2 V. Two independent under-voltage and over-voltage comparators are connected to sensor output through high-value resistors, to avoid crossed-comparator fails—see Figure 13. Each comparator energises a relay, and the contacts of these relays are connected in series. Under-voltage (V < 0.4 V) is interpreted as sensor failure. Over-voltage is interpreted as CH_4 > 0.5%. In both cases, power is disconnected by relays. Even if one comparator fails, the other will open the circuit. Intrinsically safe power supplies with appropriate voltage levels are included in the design of the safety gas monitor. The design is intentionally non-self-resetting. If the relays trip, it will remain de-energized until the arming (or re-arming) switch is operated. This feature is also used for avoiding draining the battery during long-term storage.

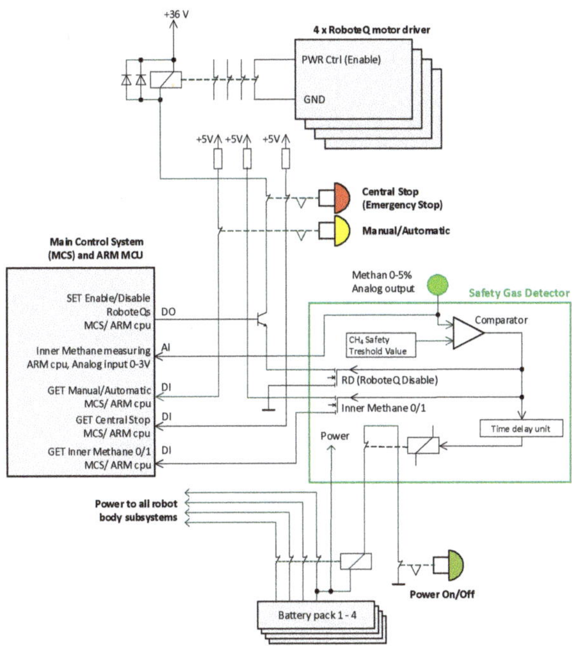

Figure 13. Implementation of safety gas detector into the robot subsystems—simplified diagram.

The detector also provides a logical signal for the main control system, which acts as a warning about an imminent power-down because of increased methane concentration. This allows for the control system to switch off in a controlled manner and primarily to disable DC motor drives to lower the currents for safer power-off switching. After a short delay, the logical signal is followed by power-down of the whole system.

The same automatic safety gas detector is installed separately in the main robot body, in the sensory arm and the sensory head; and a similar system is implemented also in the 3D LIDAR. The safety gas sensor is a part of the power management system, which distributes power from the batteries. The command for switching power off can come from several sources:

- manual control of the power of the whole system (the main power on/off button),
- manual activation of the safety central stop button,
- command from the operator control system (safety central stop button on the operator panel),
- the dangerous concentration of methane detected inside the subsystem, and
- activation of an independent watchdog monitoring the embedded control system.

The methane detector was tested in atmosphere that contained methane and other gases in well-known amounts in a special gas chamber. A calibrated gas sensor Draeger X-am 5000 (reference) and the methane detector were closed inside the chamber with gas entry for methane mixture and a small hole for cable harness and to allow a small airflow. Figure 14 shows one of the graphs that were obtained during the testing after calibration, where the measured values closely match values from the reference sensor.

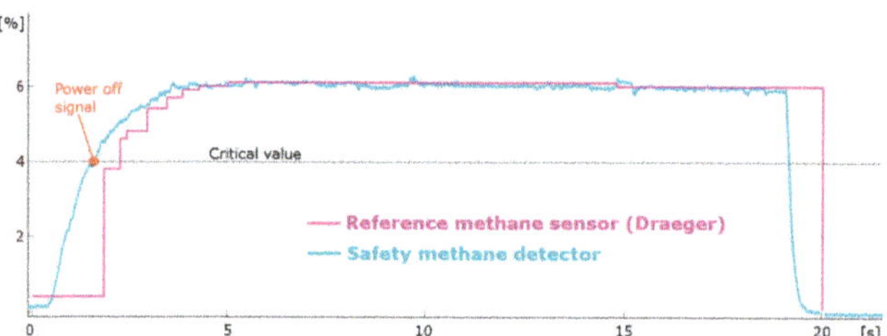

Figure 14. Safety methane detector testing (CH_4 concentration [%] over time [s] in a testing chamber).

4.7. Other Protections

The two main independent protections mentioned above (mechanical protection—flameproof enclosure; electrical protection—automatic methane detector) are supplemented by many different partial protections.

All electronic components of the main control system are sealed with a compound according to IEC 60079-18 (Explosive atmospheres—Part 18: Equipment protection by encapsulation "m") [35]. All cables leading outside of the flameproof enclosure are going through certified flameproof enclosure bushings.

In critical parts of the robotic system are installed thermometers that continuously monitor temperatures and allow for the control system to turn the robot off in case of unexpected overheating of some components.

The regulations also preclude the use of some types of materials for construction of the robot—for example, all light metal alloys (aluminium, titanium …). Plastic components pose a threat because of static charge and are thus allowed only with special precautions—the maximal continuous surface area of plastic without grounding is limited. All mechanical parts of the mobile robot TeleRescuer are made from steel, rubber (tracks), and glass (camera lenses covers).

5. Conclusions

A completely functional prototype of the reconnaissance mobile robotic system TeleRescuer has been built and thoroughly tested in various simulated and real conditions, including a training coal mine Queen Luiza in Zabrze, Poland—Figure 15. Tested were driving abilities (on various terrain material and quality, over obstacles of various sizes and shapes—perpendicular and at an angle, slalom, incline surfaces etc.), power abilities (pulling/pushing an obstacle), sensors accuracy, cameras placement, and image quality, etc. The tests showed some minor problems that should be improved in the following versions of the robot, for example, insufficient traction between the tracks and the ground, and complicated maneuverability in tight spaces during remote control based only on camera images. A detailed report from the tests is available in [36].

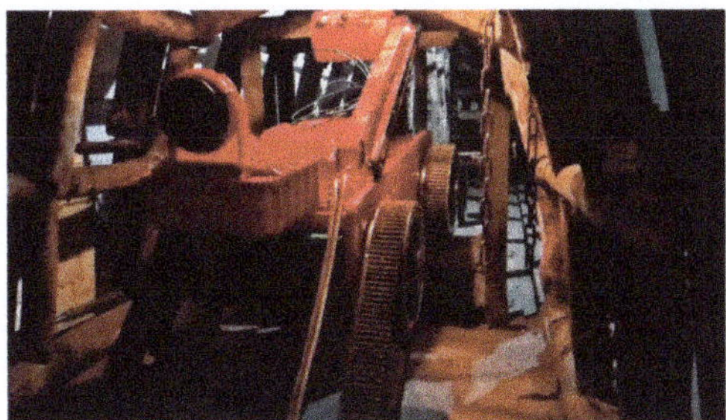

Figure 15. TeleRescuer—testing in a training coal mine Queen Luiza (March 2018).

Explosion safety is ensured by two independent protections according to Technical Report CLC/TR 60079-33 "s". The first protection is a flameproof enclosure that prevents methane from getting inside the body of the mobile robot. If this protection is damaged and a dangerous concentration of methane forms inside the body, the second protection (automatic gas detector) disconnects power from all electronic systems. This combination is valid for Group I, Category M1 because the whole system is fully operational even in environments with methane concentration and turns off only when the first protection is breached.

The proposed solution of explosion safety was evaluated by a specialized certification authority (OBAC Institute for Research and Certification Ltd., Gliwice, Poland) and several minor modifications were recommended in the evaluation report, but the overall concept was approved. The outcome of this process was, however, only a qualified opinion, the robot was not officially certified for explosion safety yet (the certification is very expensive and getting a full certification was not one of the goals of the project). Physical destructive tests of the flameproof enclosure (Section 4.3) were not performed at this stage.

The process of designing such a complex system (a reconnaissance mobile robot) in conformity with the very strict regulations that are related to explosion safety proved to be very difficult and demanding. It is highly recommended to discuss partial steps and decisions regularly during the process with a specialized authority.

Future work on the system will include implementation of the proposed minor modifications and improvements of construction of the robot, control system algorithms, and user interface regarding observations and feedback achieved during the final tests.

Author Contributions: Conceptualization, P.N., T.K. and J.B.; Data curation, Z.K.; Formal analysis, Z.K.; Funding acquisition, W.M.; Methodology, A.R.L.; Project administration, P.N. and W.M.; Software, T.K.; Supervision, P.N.; Validation, P.N.; Visualization, T.K.; Writing—original draft, P.N., T.K., J.B. and A.R.L.; Writing—review & editing, P.N. and T.K.

Funding: The project has been carried out in a framework of an EU programme of the Research Fund for Coal and Steel under the grant agreement No. RFCR-CT-2014-00002, by Polish Ministry for Science and Higher Education from financial sources that constitute the public aid in years, 2014–2017 assigned for an accomplishment of an international co-financed project; and partly supported by the European Regional Development Fund in the Research Centre of Advanced Mechatronic Systems project, project number CZ.02.1.01/0.0/0.0/16_019/0000867 within the Operational Programme Research, Development and Education.

Conflicts of Interest: The authors declare no conflict of interest.

References

1. Wikipedia. Mining Accident. Available online: https://en.wikipedia.org/wiki/Mining_accident (accessed on 3 July 2018).
2. The National Institute for Occupational Safety and Health. Coal Mining Disaster: 1839 to Present. Available online: https://www.cdc.gov/niosh/mining/statistics/content/coaldisasters.html (accessed on 3 July 2018).

3. Moczulski, W.; Cyran, K.; Januszka, M.; Novak, P.; Rodriguez, A. *System for Virtual Teleportation of Rescuer for Inspecting Coal Mine Areas Affected by Catastrophic Events*; Technical Annex of the Project Application; Silesian University of Technology: Gliwice, Poland, 2014.

4. Yang, D.; Wang, L.; Chen, J. The basic safety requirements of the coal mine rescue robot. *Coal Mine Saf.* **2009**, *42*, 104–107.

5. EUR-Lex. Directive 2014/34/EU of the European Parliament and of the Council. Available online: https://eur-lex.europa.eu/legal-content/En/TXT/?uri=CELEX:32014L0034 (accessed on 9 July 2018).

6. Czech Office for Standards, Metrology and Testing. *ČSN EN 60079-0 ed. 4. Explosive atmospheres—Part 0: Equipment—General Requirement*; Czech Office for Standards: Prague, Czech Republic, 2013; p. 100.

7. CNEX-Global. Chinese Ex Product Certification. Available online: http://www.cnex-global.com/services/chinese-ex-product-certification.html (accessed on 3 July 2018).

8. Petzl. Explosive Environments: HAZLOC Standard. Available online: https://www.petzl.com/BE/en/Professional/Explosive-environments--HAZLOC-standard (accessed on 3 July 2018).

9. Certification of Conformity & Industrial Safety. Technical Regulations CU TR 012/2011 on the Safety of Equipment in Explosion Hazardous Environments—Certification and Declaration of Conformity. Available online: http://www.ccis-expertise.com/en/technical-regulations-cu-tr-012-2011-on-safety-of-equipment-in-explosion-hazardous-environments (accessed on 3 July 2018).

10. New South Wales Government. Coal Mine Health and Safety Act 2002 No. 129. Available online: https://legislation.nsw.gov.au/#/view/act/2002/129 (accessed on 11 July 2018).

11. Compilation of Regulatory Approaches Used in Various Countries. Available online: https://www.unece.org/fileadmin/DAM/trade/wp6/SectoralInitiatives/EquipmentForExplosiveEnvironment/SIEEE-QuestionsRepliesE.pdf (accessed on 11 July 2018).

12. Ray, D.N.; Majumder, S.; Maity, A.; Roy, B.; Karmakar, S. Design and development of a mobile robot for environment monitoring in underground coal mines. In Proceedings of the 2015 Conference on Advances in Robotics, Goa, India, 2–4 July 2015; ISBN 978-1-4503-3356-6.

13. Gomathi, V.; Sowmeya, S.; Avudaiammal, P.S. Design of an Adaptive Coal Mine Rescue Robot using Wireless Sensor Networks. *Int. Comput. Appl.* **2015**, *2*, 8–11.

14. Ma, X.; Mao, R. Path Planning for Coal Mine Robot to Avoid Obstacle in Gas Distribution Area. *Int. J. Adv. Robot. Syst.* **2018**, *15*. [CrossRef]

15. Wang, W.; Dong, W.; Su, Y.; Wu, D.; Du, Z. Development of search-and-rescue robots for underground coal mine applications. *J. Field Robot.* **2014**, *31*, 386–407. [CrossRef]

16. Jonathon, C.R.; David, W.; Hainsworth, A. The numbat: A remotely controlled mine emergency response vehicle. In *Field and Service Robotics*; Springer: London, UK, 1998; pp. 53–59, ISBN 978-1-4471-1273-0.

17. United States Department of Labor. MSHA-Wolverine Robot. Available online: https://arlweb.msha.gov/sagomine/robotdetails.asp (accessed on 10 July 2018).

18. Sandia National Laboratories. Gemini-Scout Mine Rescue Vehicle. Available online: http://www.sandia.gov/research/robotics/unique_mobility/gemini-scout.html (accessed on 10 July 2018).

19. Kasprzyczak, L.; Szwejkowski, P.; Cader, M. Robotics in Mining Exemplified by Mobile Inspection Platform. *Min. Inf. Autom. Electr. Eng.* **2016**, *2*, 23–28.

20. TeleRescuer. Available online: http://www.telerescuer.polsl.pl (accessed on 3 July 2018).

21. Department of Robotics, TeleRescuer. Available online: http://robot.vsb.cz/telerescuer/ (accessed on 3 July 2018).

22. Babjak, J.; Novák, P.; Kot, T.; Moczulski, W.; Adamczyk, M.; Panfil, W. Control System of a Mobile Robot for Coal Mines. In Proceedings of the 2016 17th International Carpathian Control Conference (ICCC), Tatranska Lomnica, Slovakia, 29 May–1 June 2016; pp. 17–20, ISBN 978-146738606-7.

23. Novak, P.; Babjak, J.; Kot, T.; Bobovský, Z.; Olivka, P.; Moczulski, W.; Timofiejczuk, A.; Adamczyk, M.; Guzman, B.G.; Armada, A.G.; et al. Telerescuer—Reconnaissance mobile robot for underground coal mines. In Proceedings of the 2017 Coal Operators' Conference, Wollongong, Australia, 8–10 February 2017; The University of Wollongong Printery: Wollongong, Australia, 2017; pp. 332–340, ISBN 978-1-74128-261-0.

24. Olivka, P.; Mihola, M.; Novák, P.; Kot, T.; Babjak, J. The 3D laser range finder design for the navigation and mapping for the coal mine robot. In Proceedings of the 17th IEEE International Carpathian Control Conference (ICCC), Tatranska Lomnica, Slovakia, 29 May–1 June 2016; pp. 533–538, ISBN 978-146738606-7.

25. Kot, T.; Novák, P.; Babjak, J. Visualization of Point Clouds Built from 3D Scanning in Coal Mines. In Proceedings of the 17th IEEE International Carpathian Control Conference (ICCC), Tatranska Lomnica, Slovakia, 29 May–1 June 2016; pp. 372–377, ISBN 978-146738606-7.
26. Kot, T.; Novák, P.; Babjak, J.; Olivka, P. Rendering of 3D Maps with Additional Information for Operator of a Coal Mine Mobile Robot. In *Modelling and Simulation for Autonomous Systems MESAS 2016*; Lecture Notes in Computer Science; Springer: Berlin, Germany, 2016; Volume 9991, pp. 214–225.
27. Kot, T.; Novák, P.; Babjak, J. Virtual Operator Station for Teleoperated Mobile Robots. In Proceedings of the International Workshop on Modelling and Simulation for Autonomous Systems (MESAS), Prague, Czech Republic, 29–30 April 2015; pp. 144–153, ISBN 978-3-319-22383-4.
28. Arena, P.; Fortuna, L.; Frasca, M.; Patané, L. Learning Anticipation via Spiking Networks: Application to Navigation Control. *IEEE Trans. Neural Netw.* **2009**, *20*, 202–216. [CrossRef] [PubMed]
29. IEC Webstore. IEC 60079-33:2012 Explosive Atmospheres—Part 33: Equipment Protection by Special Protection 'S'. Available online: https://webstore.iec.ch/publication/652 (accessed on 9 July 2018).
30. Czech Office for Standards, Metrology and Testing. *ČSN EN 60079-1 ed. 3. Explosive Atmospheres—Part 1: Equipment Protection by Flameproof Enclosure "d"*; Czech Office for Standards: Prague, Czech Republic, 2015; p. 68.
31. Konečný, Z. Evaluating Methods of the Strength Analyzes Results Realize in PTC Creo/Simulation. *Appl. Mech. Mater.* **2015**, *816*, 357–362. [CrossRef]
32. Górny, M. Explosion Pressure Inside Flameproof Electrical Motors in Low Temperatures. *Zeszyty Problemowe Maszyny Elektryczne* **2008**, *80*, 99–105.
33. Górny, M. Gas Explosion Propagation Through Flameproof Induction Motor's Air Gap. *Zeszyty Problemowe Maszyny Elektryczne* **2013**, *99*, 121–127.
34. Czech Office for Standards, Metrology and Testing. *ČSN EN 60079-11 ed. 2. Explosive Atmospheres—Part 11: Equipment Protection by Intrinsic Safety "i"*; Czech Office for Standards: Prague, Czech Republic, 2012; p. 132.
35. Czech Office for Standards, Metrology and Testing. *ČSN EN 60079-18 ed. 3. Explosive Atmospheres—Part 18: Equipment Protection by Encapsulation "m"*; Czech Office for Standards: Prague, Czech Republic, 2015; p. 36.
36. Kopex Group. Report on Field Tests of TeleRescuer System [Final Report]. 2018. Available online: http://www.telerescuer.polsl.pl/ (accessed on 2 November 2018).

Article

A Robotic Drilling End-Effector and Its Sliding Mode Control for the Normal Adjustment

Laixi Zhang [1], Jaspreet Singh Dhupia [2],*, Mingliang Wu [1] and Hua Huang [1]

[1] School of Mechanical & Electronical Engineering, Lanzhou University of Technology, Lanzhou 730050, China; laixi_zh@163.com (L.Z.); wml0757@163.com (M.W.); hh318872@126.com (H.H.)

[2] Department of Mechanical Engineering, University of Auckland, Private Bag 92019, Auckland 1142, New Zealand

* Correspondence: j.dhupia@auckland.ac.nz; Tel.: +64-9-9235915

Received: 18 September 2018; Accepted: 9 October 2018; Published: 11 October 2018

Featured Application: Robotic Drilling.

Abstract: A robotic drilling end-effector is designed and modeled, and a sliding mode variable structure control architecture based on the reaching law is proposed for its normal adjustment dynamic control. By using a third-order nonlinear integration chain differentiator for obtaining the unmeasurable speed and acceleration signals from the position signals, this sliding mode control scheme is developed with good dynamic quality. The new control law ensures global stability of the entire system and achieves both stabilization and tracking within a desired accuracy. A real-time control experiment platform is established in xPC target environment based on MATLAB Real-Time Workshop (RTW) to verify the proposed control scheme and simulation results. Simulations and experiments performed on the designed robotic end-effector illustrate and clarify that the proposed control scheme is effective.

Keywords: dynamical model; nonlinear differentiator; robotic drilling; sliding mode control; drilling end-effector

---

1. Introduction

Robots have undoubtedly demonstrated their value in the automotive industry, they have been regarded suitable for aerospace applications such as drilling and fastening due to the obvious low investment costs compared to bespoke industrial solutions [1]. The traditional application areas for industrial robots involve highly repetitive operations such as drilling. Robotic drilling system has made extensive application in both industry and research field. In the aircraft industry, where the drilling of many thousands of holes per aircraft is needed, the usually adopted solution is to stack the parts in a joint position and drill them in a single operation [2,3]. The use of industrial robots for drilling is gaining an increasing interest due to their flexibility and the comparatively low cost of industrial robot systems [4,5]. DeVlieg et al. [6] and Atkinson et al. [7] developed a drilling system for the skin to substructure joint on the F/A-18E/F Super Hornet wing trailing edge flaps (TEF) and for 737 Aileron, which utilizes a mass produced, high capacity industrial robot such as Kuka KR350/2 and KR360-2 as the motion platform for an automated drilling, countersinking, and hole inspection machine.

In general, a servo-controlled robotic drilling end-effector is coupled with an industrial robot to complete the drilling process. Webb et al. [8] developed a compact and innovative end-effector which is capable of performing drilling, countersinking, sealing and riveting operations and also contains a significant amount of process monitoring sensors to enable automated in-process checks and quality measurement. Hempstead et al. [9] developed an end-effector for use on a Kuka KR350 robot. The end-effector pushes up on a wing panel with programmable pressure, drills a hole with

---

a servo drill, inspects the hole with a servo ball-type hole gauge and then drives a pin-tail style lock bolt into the hole. Liang et al. [10] further designed a drilling end-effector to use for industrial robots. The real-time force feedback can detect dull or broken bits, drill to breakthrough, and plot thrust force while drilling. Devlieg [11] also developed a single process end-effector that performs all required functions, including one-sided pressure application, touch probing, barcode scanning, drilling/countersinking, measurement of hole diameter and countersink depth, and face milling. Automatic drilling requires the drilling bit to be perpendicular to the curved surface of fuselage. However, most of these references focus exclusively on all-in-one end-effector structure design and functions such as drilling, countersinking, inspecting or monitoring, and fastening [12], and a few on the normal adjustment dynamical model and control. The development of these robotic drilling systems has created a research need for the dynamical models and control of the normal adjustment of robotic drilling end-effector.

The dynamic performance of the normal adjustment of the robotic drilling end-effector will affect the drilling quality and productivity. Olsson [13] presented methods and systems for force-controlled robot drilling, based on active suppression of drill sliding through a model-based force control scheme. Tian et al. [14] proposed a novel approach based on the detection system to detect the normal vector to product surface in real time for the robotic precision drilling system in aircraft component assembly. Mei et al. [15] proposed a novel in-process robot base frame calibration method with a 2D vision system to measure and compensate the position errors of drilled fastener holes. Frommknecht [16] presented a multi-sensor measurement system for robotic drilling. The system enables a robot to measure its 6D pose with respect to the work piece and to establish a reference coordinate system for drilling. Shi et al. [5] developed a normal adjustment cell (NAC) in aero-robotic drilling by using an intelligent double-eccentric disk normal adjustment mechanism (2-EDNA), a spherical plain bearing, and a floating compress module with sensors to improve the quality of vertical drilling. Qin et al. [17] presented an approach for the acquisition of vibration signals of the end-effector in robotic drilling. With the aid of the rigid body kinematics, the vibration signals of the end-effector can be obtained and analyzed. Garnier et al. [18] proposed two robotic drilling models, in which the cutting process and the stiffness of the robot hand are taken into account, respectively. However, these efforts focus basically on kinematics analysis and static accuracy insurance, depending quite a bit on sensors. Though Zhang et al. [19] proposed an intelligent surface-normal adjustment system to deal with the normal adjustment that focused on calculating surface-normal vector to regulate the feed direction of the drilling bit in real-time, this works didn't involve the dynamic characteristics of the normal adjustment of the end-effector.

This paper develops a robotic drilling end-effector for industrial robot. Based on the dynamics modeling and analysis, a sliding mode variable structure control architecture with the constant plus exponential rate reaching law is proposed for the normal adjustment dynamic control. By using a third-order nonlinear integration chain differentiator for obtaining the unmeasurable speed and acceleration signals from the position signals with the minimum noise, this sliding mode control scheme is developed with good dynamic quality. The new control law ensures global stability of the entire system and achieves both stabilization and tracking within a desired precision. A real-time control experiment platform is established in xPC target environment based on MATLAB Real-Time Workshop (RTW) to verify the proposed control scheme and simulation results. Simulations and experiments performed on the designed robotic end-effector illustrate and clarify that the controller is effective and robust.

2. Dynamical Model for the Robotic Drilling End-Effector

2.1. The Structure of the Robotic Drilling End-Effector

The main functions of the designed robotic drilling end-effector include: measuring and calculating the outward normal of the fuselage surface at drilling point, adjusting spindle axis to coincide with the normal and drilling after the surface is reliably compressed. A diagram and a photograph of the designed robotic drilling end-effector are shown in Figure 1. Figure 1a shows a diagram of the end-effector structure, Figure 1b shows the specifics of the internal structure of the end-effector, and Figure 1c shows the prototype of the end-effector, in which the spindle and feed units fulfill spindle driving and feeding. The compressing unit compresses the surface to diminish the vibration during the drilling process. It also helps to eliminate the clearances between the layers of material in order to avoid chips going into layers and causing stress concentration. There are four displacement sensors arranged evenly in the compressing unit around the spindle for measuring four space coordinates on the surface around drilling area. Thus, by the proposed method of measuring and calculating the normal of curved surface [19], the outward normal of fuselage surface at drilling point is obtained. The normal adjusting unit is fixed on the frame, and two mutually perpendicular axes (Joints 1 and 2) are installed in a plane normal to the spindle. The spindle can rotate about these two axes, thus fulfilling the function of adjusting the spindle to normal of the surface. The frame links and supports above parts, and connects the end-effector to robot.

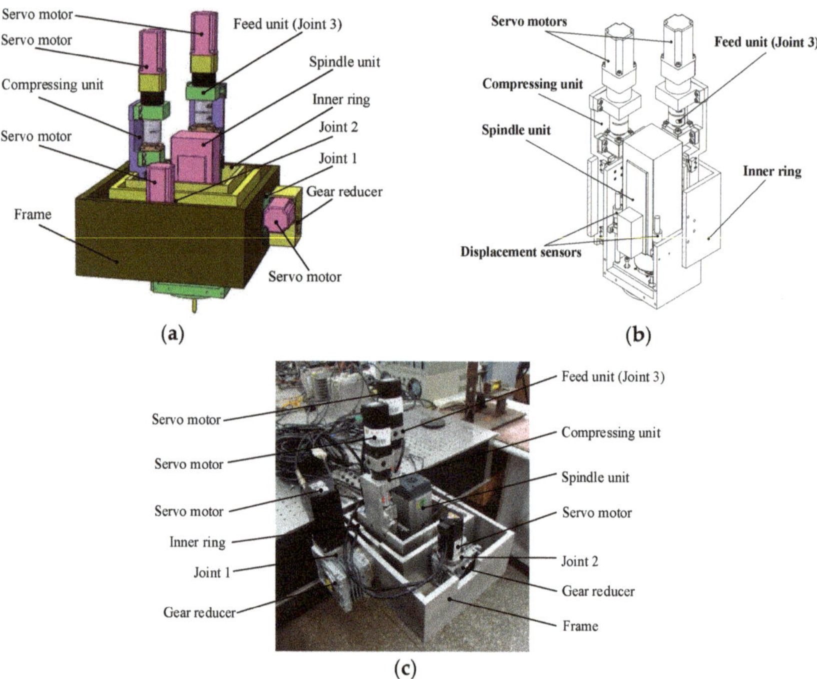

Figure 1. Structure and prototype of the designed end-effector: (**a**) the diagram of the end-effector structure; (**b**) the specific of the internal structure of the end-effector; and (**c**) the prototype of the end-effector.

2.2. Motion Equations of the Robotic Drilling End-Effector

The coordinate system of the designed robotic drilling end-effector is shown as Figure 2. The kinematic model is established using Denavit-Hartenberg (D-H) method. The base coordinate system is placed in rotating Joint 1 and concentric exactly with joint coordinate {1}. Because the normal adjustment unit of the end-effector consists of Joints 1 and 2, the world frame {0} is placed at the center of the Joint 1 and is concentric with the frame {1} for the sake of analysis.

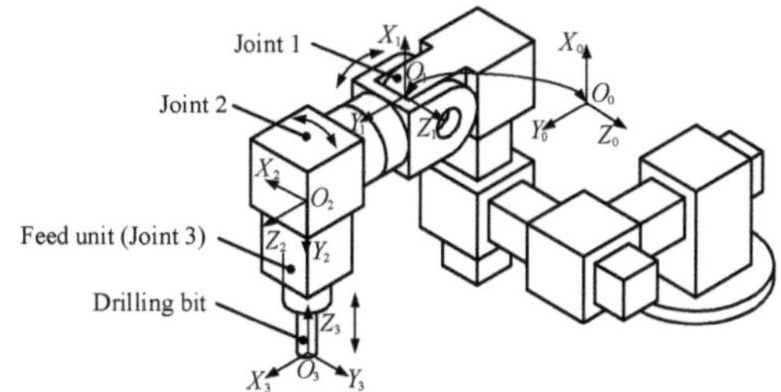

Figure 2. Coordination definition of the end-effector.

The transform matrixes in the coordinate system are calculated as:

$$
{}^0_1T = \begin{bmatrix} \cos\theta_1 & -\sin\theta_1 & 0 & 0 \\ \sin\theta_1 & \cos\theta_1 & 0 & 0 \\ 0 & 0 & 1 & 0 \\ 0 & 0 & 0 & 1 \end{bmatrix}
\tag{1}
$$

$$
{}^1_2T = \begin{bmatrix} -\sin\theta_2 & -\cos\theta_2 & 0 & 0 \\ 0 & 0 & 1 & 0 \\ -\cos\theta_2 & \sin\theta_2 & 0 & 0 \\ 0 & 0 & 0 & 1 \end{bmatrix}
\tag{2}
$$

$$
{}^2_3T = \begin{bmatrix} 0 & -1 & 0 & 0 \\ 0 & 0 & -1 & d_3 \\ 1 & 0 & 0 & 0 \\ 0 & 0 & 0 & 1 \end{bmatrix}.
\tag{3}
$$

Both spindle feeding unit and pressing unit will not move in the normal adjusting process of the end-effector. Thus, both their acceleration and velocity are 0.

A 3×3-order positive definite inertial matrix is calculated as:

$$
\mathbf{D}(\mathbf{q}) = \begin{bmatrix} D_{11} & D_{12} & D_{13} \\ D_{21} & D_{22} & D_{23} \\ D_{31} & D_{32} & D_{33} \end{bmatrix}
\tag{4}
$$

with

$$
\mathbf{q} = \begin{bmatrix} \theta_1 & \theta_2 & d_3 \end{bmatrix}^T
\tag{5}
$$

$$
\begin{aligned}
D_{11} &= I_{zz1} + I_{yy2} + I_{zz3} + I_{a1} + \left(I_{xx2} - I_{yy2} + I_{yy3} - I_{zz3} - 2m_3 d_3 c_{z3} + m_3 d_3^2\right)\cos^2(\theta_2) \\
&\quad + \left(I_{xy2} + I_{yz3} - m_3 d_3 c_{y3}\right)\sin(2\theta_2)
\end{aligned}
\tag{6}
$$

$$
D_{12} = \left(-I_{yz2} + I_{xz3} - m_3 d_3 c_{x3}\right)\sin(\theta_2) + \left(I_{xz3} - I_{xy3}\right)\cos(\theta_2)
\tag{7}
$$

$$
D_{13} = m_3 c_{x3} \cos(\theta_2)
\tag{8}
$$

$$
D_{21} = \left(-I_{yz2} + I_{xz3} - m_3 d_3 c_{x3}\right)\sin(\theta_2) + \left(I_{xz2} - I_{xy3}\right)\cos(\theta_2)
\tag{9}
$$

$$
D_{22} = I_{zz2} + I_{xx3} - 2m_3 d_3 c_{z3} + m_3 d_3^2 + I_{a2}
\tag{10}
$$

$$
D_{23} = -m_3 c_{y3}
\tag{11}
$$

$$
D_{31} = m_3 c_{x3} \cos(\theta_2)
\tag{12}
$$

$$
D_{32} = -m_3 c_{y3}
\tag{13}
$$

$$D_{33} = m_3 + I_{a3} \tag{14}$$

where θ_1 and θ_2 denote the rotation angles of the Joints 1 and 2, respectively, and d_3 denotes the linear displacement of the prismatic Joint 3. I_{abi} is the product of inertia of Joint i with respect to a pair of orthogonal axes a and b, a, b denote x, y, or z; m_i is the mass of Joint i. c_{ai} is the centroid coordinate of Joint i on axis a, a denote x, y, or z.

A 3×1-order vector of centrifugal and Coriolis force is calculated, regarding that $\dot{q}_3 = \dot{d}_3 = 0$, the result is:

$$\mathbf{H}(\mathbf{q},\dot{\mathbf{q}}) = \begin{bmatrix} h_1 & h_2 & h_3 \end{bmatrix}^T \tag{15}$$

with

$$\begin{aligned} h_1 &= \left[\left(-I_{xx2} + I_{yy2} - I_{yy3} + I_{zz3} - m_3 d_3^2 + 2m_3 d_3 c_{z3} \right) \sin(2\theta_2) + 2\left(I_{xy2} + I_{yz3} - m_3 d_3 c_{y3} \right) \cos(2\theta_2) \right] \dot{\theta}_1 \dot{\theta}_2 \\ &\quad + \left[\left(-I_{xz2} + I_{xy3} \right) \sin(\theta_2) + \left(-I_{yz2} + I_{xz3} - m_3 d_3 c_{x3} \right) \cos(\theta_2) \right] \dot{\theta}_2^2 \end{aligned} \tag{16}$$

$$h_2 = \left[\frac{1}{2} \left(I_{xx2} - I_{yy2} + I_{yy3} - I_{zz3} - 2m_3 d_3 c_{z3} + m_3 d_3^2 \right) \sin(2\theta_2) + \left(-I_{xy2} - I_{yz3} + m_3 d_3 c_{y3} \right) \cos(2\theta_2) \right] \dot{\theta}_1^2 \tag{17}$$

$$h_3 = m_3 \cos(\theta_2) \left[c_{y3} \sin(\theta_2) + (c_{z3} - d_3) \cos(\theta_2) \right] \dot{\theta}_1^2 + m_3 (c_{z3} - d_3) \dot{\theta}_2^2. \tag{18}$$

A 3×1-order gravity vector is calculated as:

$$\mathbf{G}(\mathbf{q}) = \begin{bmatrix} G_1 & G_2 & G_3 \end{bmatrix} \tag{19}$$

with

$$\begin{aligned} G_1 &= \left[m_1 c_{x1} \sin(\theta_1) + \left(m_1 c_{y1} + m_2 c_{z2} + m_3 c_{x3} \right) \cos(\theta_1) + \left(-m_2 c_{x2} + m_3 c_{y3} \right) \sin(\theta_1) \sin(\theta_2) \right. \\ &\quad \left. + \left(-m_2 c_{y2} + m_3 c_{z3} - m_3 d_3 \right) \sin(\theta_1) \cos(\theta_2) \right] g \end{aligned} \tag{20}$$

$$G_2 = \left[\left(-m_2 c_{y2} + m_3 c_{z3} - m_3 d_3 \right) \sin(\theta_2) + \left(m_2 c_{x2} - m_3 c_{y3} \right) \cos(\theta_2) \right] g \cos(\theta_1) \tag{21}$$

$$G_3 = m_3 g \cos(\theta_1) \cos(\theta_2). \tag{22}$$

According to Lagrange's equation, the dynamical equation of the end-effector is derived as:

$$\mathbf{D}(\mathbf{q})\ddot{\mathbf{q}} + \mathbf{H}(\mathbf{q},\dot{\mathbf{q}}) + \mathbf{G}(\mathbf{q}) = \boldsymbol{\tau} \tag{23}$$

or

$$\begin{bmatrix} D_{11} & D_{12} & D_{13} \\ D_{21} & D_{22} & D_{23} \\ D_{31} & D_{32} & D_{33} \end{bmatrix} \begin{bmatrix} \ddot{\theta}_1 \\ \ddot{\theta}_2 \\ \ddot{d}_3 \end{bmatrix} + \begin{bmatrix} h_1 \\ h_2 \\ h_3 \end{bmatrix} + \begin{bmatrix} G_1 \\ G_2 \\ G_3 \end{bmatrix} = \begin{bmatrix} \tau_1 \\ \tau_2 \\ f_3 \end{bmatrix} \tag{24}$$

where τ_1 and τ_2 denote the torque acting on Joints 1 and 2, respectively, f_3 denotes the force acting on prismatic Joint 3.

The joint parameters of the end-effector calculated by the model are listed in Table 1.

Table 1. Joint parameters of the end-effector.

Parameters	$i = 1$	$i = 2$	$i = 3$
I_{xxi} (m^4)	0.03	0.04	0.03
I_{xyi} (m^4)	0	0.01	0
I_{xzi} (m^4)	0	−0.01	0
I_{yxi} (m^4)	0	0.01	0
I_{yyi} (m^4)	0.09	0.11	0.04
I_{yzi} (m^4)	0	0	0
I_{zxi} (m^4)	0	−0.11	0
I_{zyi} (m^4)	0	0	0
I_{zzi} (m^4)	0.06	0.11	0.01
m_i (kg)	5.8	7.6	8.6
c_{xi} (m)	0.05	0.02	0
c_{yi} (m)	0	0.02	0
c_{zi} (m)	0	−0.01	0.14

3. Sliding Mode Controller Design

3.1. Friction Compensation

In order to research dynamical characteristic of the end-effector during normal orientation adjusting, considering that feed unit and compressing unit maintain resting state during the normal adjusting process, Equation (23) is rewritten as:

$$\mathbf{D}(\mathbf{\Theta})\ddot{\mathbf{\Theta}} + \mathbf{H}\left(\mathbf{\Theta},\dot{\mathbf{\Theta}}\right) + \mathbf{G}(\mathbf{\Theta}) = \mathbf{\tau} \tag{25}$$

where $\mathbf{\Theta} = \begin{bmatrix} \theta_1 & \theta_2 \end{bmatrix}^T$ and $\mathbf{\tau} = \begin{bmatrix} \tau_1 & \tau_2 \end{bmatrix}^T$.

Equation (25) is an ideal dynamical model, because it didn't take into account the effects of friction torque and disturbing torque on the joints. Friction is a phenomenon with obvious nonlinearity that remains ubiquitous in mechanical systems. It always causes untoward consequences such as system steady-state error, limit cycle, static-slip and oscillation. Therefore, the friction torque and disturbing torque should be considered to obtain comparative control quality. The dynamical model needs to be modified as:

$$\mathbf{D}(\mathbf{\Theta})\ddot{\mathbf{\Theta}} + \mathbf{H}\left(\mathbf{\Theta},\dot{\mathbf{\Theta}}\right) + \mathbf{G}(\mathbf{\Theta}) + \mathbf{F}\left(\mathbf{\Theta},\dot{\mathbf{\Theta}},\ddot{\mathbf{\Theta}}\right) + \mathbf{\tau_d} = \mathbf{\tau} \tag{26}$$

where $\mathbf{F}\left(\mathbf{\Theta},\dot{\mathbf{\Theta}},\ddot{\mathbf{\Theta}}\right)$ denotes friction torque while $\mathbf{\tau_d}$ denotes disturbing torque.

Therefore, in order to compensate the friction torque, it is necessary to establish a friction model to estimate the friction torque. Stribeck model is a classical friction model which has components that are either linear in velocity or constant and describes the mechanical characteristics of various friction stages. The resulting friction force F_f can be expressed as [20]:

$$F_f = \begin{cases} F(v) & \text{if } v \neq 0 \\ F_e & \text{if } v = 0 \text{ and } |F_e| < F_s \\ F_s \text{sgn}(F_e) & \text{other} \end{cases} \tag{27}$$

$$F(v) = F_c + (F_s - F_c)e^{-|v/v_s|^2} + F_v v \tag{28}$$

where v is motion velocity; v_s is named Stribeck velocity; F_e denotes driving force; F_s denotes maximum static friction force; F_c denotes Coulomb friction force; F_v denotes viscous frictional torque coefficient. It is easy to identify the expression for F_f by measuring friction force in condition of uniform motion.

3.2. Sliding Mode Control Based on Model

Utilizing controller decomposition, a dynamic control law combining the linearized feedback of joints' parameters and servo compensator is investigated. The feedback linearization method which is often called computed torque method is a kind of design method based on system dynamical model. Define the driving torque as Equation (29).

$$\boldsymbol{\tau} = \alpha\mathbf{u} + \boldsymbol{\beta} \tag{29}$$

with

$$\begin{aligned} \alpha &= \mathbf{D}(\Theta) \\ \boldsymbol{\beta} &= \mathbf{H}\left(\Theta, \dot{\Theta}\right) + \mathbf{G}(\Theta) + \mathbf{F}\left(\Theta, \dot{\Theta}, \ddot{\Theta}\right) + \boldsymbol{\tau_d} \end{aligned} \tag{30}$$

where $\mathbf{u}$ denotes control quality.

Substituting Equations (29) and (30) into Equation (26), we can obtain that:

$$\ddot{\Theta} = \mathbf{u} \tag{31}$$

A sliding mode control based on reaching law is designed according to the dynamical model. The switch function is taken as:

$$\mathbf{S} = \mathbf{Ce} + \dot{\mathbf{e}} \tag{32}$$

where $\mathbf{C} = \mathrm{diag}(c_1, c_2), c_1 > 0, c_2 > 0$, and $\mathbf{e} = \mathrm{diag}(e_1, e_2)$ is the position tracking error matrix, e_1 and e_2 are the tracking error of the Joints 1 and 2, respectively. Because $\mathbf{S} = \mathbf{0}$ denotes the sliding mode surface and the state of the phase trajectory moving along the sliding mode surface, only $\mathbf{S} \neq \mathbf{0}$ is considered.

To demonstrate the stability of the designed controller, a Liapunov's function is made as:

$$\mathbf{V} = \frac{1}{2}\mathbf{S}^{\mathrm{T}}\mathbf{S} > \mathbf{0}. \tag{33}$$

The first-order derivative of Equation (32) is calculated as:

$$\dot{\mathbf{V}} = \mathbf{S}^{\mathrm{T}}\dot{\mathbf{S}}. \tag{34}$$

Using switch law can ensure the dynamic quality of the phase trajectory during reaching segment by designing the change rate function of the switch function of reaching segment. This method makes the system reach the sliding surface quickly according to the dynamical characteristic of the change rate function $\dot{s}$.

The constant plus exponential rate reaching law combines the advantages of both constant rate reaching law and exponential rate reaching law and has good dynamic characteristics that make the system reach the sliding surface quickly and eliminate system chattering effectively. The constant plus exponential rate reaching law is expressed as [21]:

$$\dot{\mathbf{S}} = -\varepsilon\mathrm{sgn}(\mathbf{S}) - \mathbf{KS} \tag{35}$$

with

$$\begin{aligned} \varepsilon &= \mathrm{diag}(\varepsilon_1, \varepsilon_2), \, \varepsilon_1 > 0, \, \varepsilon_2 > 0 \\ \mathbf{K} &= \mathrm{diag}(k_1, k_2), \, k_1 > 0, \, k_2 > 0. \end{aligned} \tag{36}$$

By derivation of Equation (32) and then substituting Equation (31) into it, the change rate function can be acquired as:

$$\dot{\mathbf{S}} = \mathbf{C}\dot{\mathbf{e}} + \ddot{\Theta}_{\mathbf{d}} - \mathbf{u}. \tag{37}$$

By correlating Equations (35) and (37), the controlling quantity can be calculated as:

$$\mathbf{u} = \mathbf{C}\dot{\mathbf{e}} + \ddot{\boldsymbol{\Theta}}_{\mathbf{d}} + \varepsilon\mathrm{sgn}(\mathbf{S}) + \mathbf{KS}. \tag{38}$$

Substituting Equations (32) and (35) into Equation (34), we can obtain that:

$$\dot{\mathbf{V}} < \mathbf{0}. \tag{39}$$

Thus, the stability of the control law is verified. Through the aforementioned method of calculation, it is obvious that the control law expressed by Equation (38) is in accord with Liapunov stability theorem. The control block diagram of the sliding mode variable structure control based on the model with friction compensation and gravity compensation is shown as Figure 3. The output control quality **u** of a servo compensator with the sliding mode variable structure acts as the input of the linearizing compensator. The final output torque τ of the controller can be deduced by substituting Equation (38) into Equation (29) as:

$$\boldsymbol{\tau} = \mathbf{D}(\boldsymbol{\Theta}) \cdot \mathbf{u} + \mathbf{H}\left(\boldsymbol{\Theta}, \dot{\boldsymbol{\Theta}}\right) + \mathbf{G}(\boldsymbol{\Theta}) + \mathbf{F}\left(\boldsymbol{\Theta}, \dot{\boldsymbol{\Theta}}, \ddot{\boldsymbol{\Theta}}\right) + \boldsymbol{\tau}_{\mathbf{d}}. \tag{40}$$

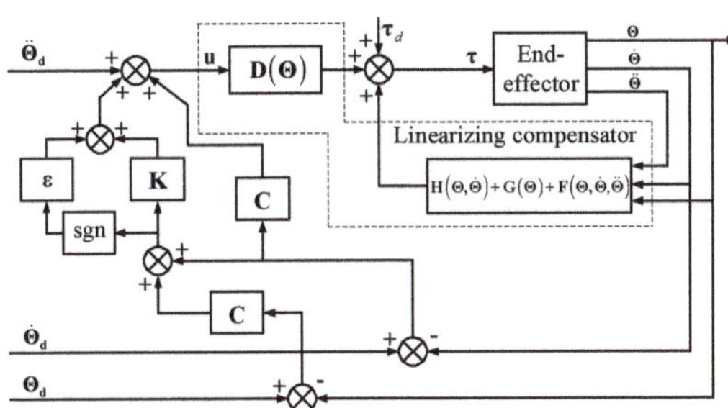

Figure 3. The sliding mode variable structure control block diagram.

3.3. Nonlinear Integration Chain Tracking-Differentiator

In order to realize the control law expressed by Equation (38), the position error change ratio of the joints, the speed error, and the acceleration of the joints must be known. In general, the photoelectrical encoders installed in the joints are displacement sensor. Theoretically, the speed signal can be obtained only by taking the first-order derivative of the displacement signal, and the acceleration signal by the second-order derivative. But only the signal expressed by analytic formula can be used for mathematically calculating its first or higher orders derivatives. The signal measured by sensor cannot be mathematically taking the derivatives because it has no analytic formula. In engineering, difference method is often used to approximate the derivatives of signal. Because the signal measured by sensor usually contains disturbances which will be further amplified after difference, the approximated first derivative signal usually intermingles with strong disturbances, and the approximated second derivative signal even can be submerged by disturbances. Much of the disturbances can be eliminated with digital filter such as Butterworth, Chebyshev, and Bessel et al. after difference, but this method unfortunately has obvious drawbacks such as time delay and nonlinear phase shift which will impact the control accuracy and even bring up system distortion or disability. The optimal estimation of the system state can be obtained from the observation equations using Kalman filter or other state observation methods such as time delay system observer [22], discrete-time system observer [23], sliding mode observer [24], and so on, but the observation equations are based on the object model, which limits its application. Extracting the derivatives of real-time signals is a common problem.

For the signals that are encountered in practical applications, developing the differentiator to take derivative is a realistic option [25]. The general structure of the differentiator is expressed as:

$$\begin{cases} \dot{x}_1 = x_2 \\ \dot{x}_2 = f(x_1 - v(t), x_2) \end{cases} . \tag{41}$$

Under the condition of existence of the solution of Equation (41), make x_1 converges to input signal $v(t)$ and x_2 converges to desired derivative $\dot{v}(t)$.

To solve the engineering questions, researchers proposed many practical differentiators such as linear high gain differentiator [26], sliding mode differentiator [27], hybrid differentiator [28], nonlinear integration chain differentiator [29], and so on. Adopted in this experiment is a third-order nonlinear integration chain differentiator which has multiple integration structure and high disturbance suppression performance and its form is expressed as Equation (42) [30].

$$\begin{cases} \dot{x}_1 = x_2 \\ \dot{x}_2 = x_3 \\ \dot{x}_3 = -\dfrac{1}{\varepsilon^3}\left[a_1 sig(x_1 - v(t))^{\alpha_1} + a_2 sig\left(\varepsilon^1 x_2\right)^{\alpha_2} + a_3 sig\left(\varepsilon^2 x_3\right)^{\alpha_3} \right] \end{cases} \tag{42}$$

where $x = \begin{bmatrix} x_1 & x_2 & x_3 \end{bmatrix}^T$ are the system state variables, and $sig(x)\alpha = |x|^{\alpha}\mathrm{sgn}(x)$. In the following experiments, x_1, x_2, and x_3 correspond to joint positions, velocities, and accelerations, respectively.

In the simulink model of the designed third-order nonlinear integration chain differentiator expressed by Equation (42), the parameters are set as $a_1 = 10$, $a_2 = 10$, $a_3 = 10$, $\alpha_1 = 1$, $\alpha_2 = 1$, $\alpha_3 = 1$, and $\varepsilon = 0.01$. A first order low pass filter $1/(0.02s + 1)$ is added after the second derivative output to eliminate the excessive amount of disturbances.

4. Simulations and Experiments Results and Discussion

4.1. Simulations Results and Discussion

The parameters of the sliding mode controller (SMC) are selected as $\mathbf{C} = \mathrm{diag}(20, 30)$, $\varepsilon = \mathrm{diag}(10, 20)$, and $\mathbf{K} = \mathrm{diag}(20, 60)$. Set the disturbing torques of Joints 1 and 2, respectively, as $\tau_{d1} = 0.5\sin(10\pi t)$ and $\tau_{d2} = 0.5\cos(10\pi t)$.

The parameters of the Stribeck friction model are selected as listed in Table 2.

Table 2. Parameters of the Stribeck friction model.

α	0.01
a_1	1
F_{s1}	0.1 N·m
F_{s2}	0.05 N·m
F_{c1}	0.04 N·m
F_{c2}	0.02 N·m
K_{v1}	0.02
K_{v2}	0.01

The simulation results of the control law of sliding mode controller expressed by Equation (38) with the above parameters on step input are shown in Figures 4 and 5 and on sinusoid input are shown in Figures 6 and 7. A classical PID (proportional-integral-derivative) controller was also designed for comparison. The values of proportional, integral, and derivative gains of Joint 1 were $P_1 = 102$, $I_1 = 1.2$, and $D_1 = 21.5$, and the control parameters of Joint 2 were $P_2 = 87$, $I_2 = 0.7$, and $D_2 = 15$. These values have been optimally tuned using a modified Ziegler–Nichols' (Z-N) tuning methods [31].

Figure 4 shows the step responses of Joint 1 with the proposed SMC and the PID controller. Figure 4a shows the position tracking with PID controller and the SMC. Figure 4b shows the phase trajectory diagram of the SMC. Figure 4c,d shows the tracking error with PID controller and the SMC, respectively. Figure 5 shows the step responses of Joint 2.

For step input, the adjusting times of both joints with the proposed SMC and the PID controller are only 0.2 s. However, the position steady state error of each joint with the SMC is far less than that with the PID controller. As shown in Figures 4c,d and 5c,d, the tracking errors generated by the PID controller oscillate more intensely than that by the SMC. This phenomenon implies that the SMC possess better control accuracy and stronger robustness under disturbances than the PID. The phase trajectory of the SMC rapidly reaches the sliding mode surface from the initial point and quickly converges along the sliding mode surface to the equilibrium point. This illustrates that the designed SMC has good dynamic quality and uniform stability.

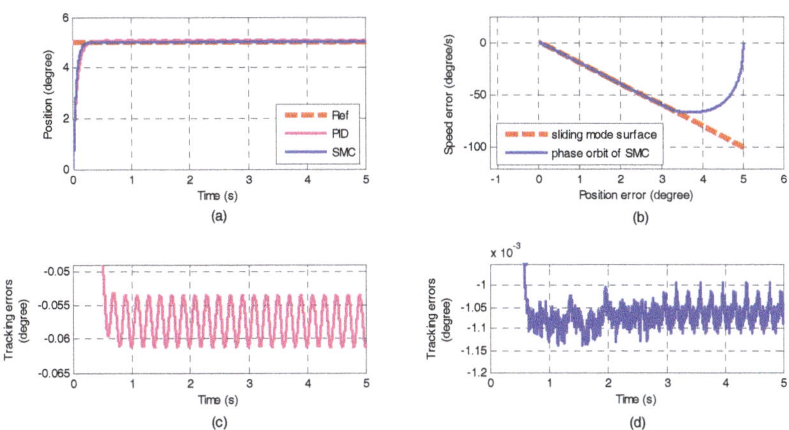

Figure 4. Simulation of step response of Joint 1: (**a**) Position tracking; (**b**) Phases trajectory diagram of SMC; (**c**) Tracking error of PID (proportional-integral-derivative); and (**d**) Tracking error of SMC (sliding mode controller).

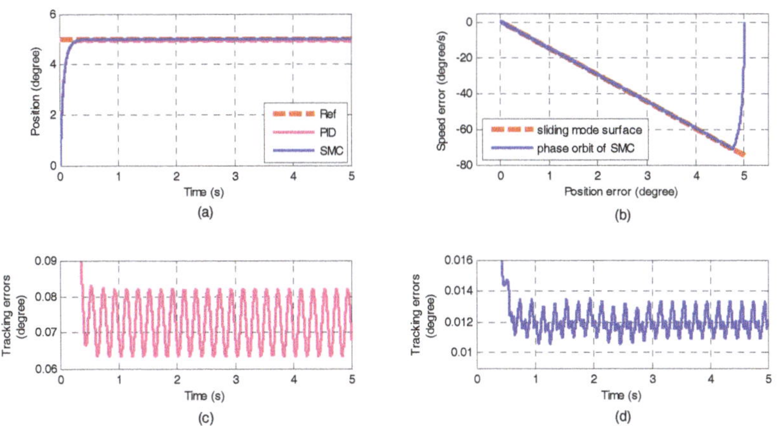

Figure 5. Simulation of step response of Joint 2: (**a**) Position tracking; (**b**) Phases trajectory diagram of SMC; (**c**) Tracking error of PID; and (**d**) Tracking error of SMC.

Figure 6 shows the sinusoidal responses of Joint 1 with the proposed SMC and the PID controller. Figure 6a shows the position tracking with PID controller and the SMC. Figure 6b shows the phase trajectory diagram of the SMC. Figure 6c,d shows the tracking error with PID controller and the SMC, respectively. Figure 7 shows the sinusoidal responses of Joint 2.

For sinusoidal input, as show in Figures 6 and 7, an obvious lag or steady error is presented in the trajectory produced by the PID controller, which suffers from some difficulties in obtaining accurate tracking. The position tracking error of each joint with the PID controller oscillates more greatly than that with the SMC. The existence of this phenomenon suggests again that the SMC has better dynamic and static characteristic than the PID, and has high accuracy and stronger robustness. As shown in Figures 6b and 7b, the phase trajectory rapidly reaches the sliding mode surface and quickly tends along the sliding mode surface to a tiny stable limit cycles around the equilibrium point. This illustrates that the designed sliding mode variable control has bounded stability. And the existence of the limit cycle implicates that the system has small steady state error.

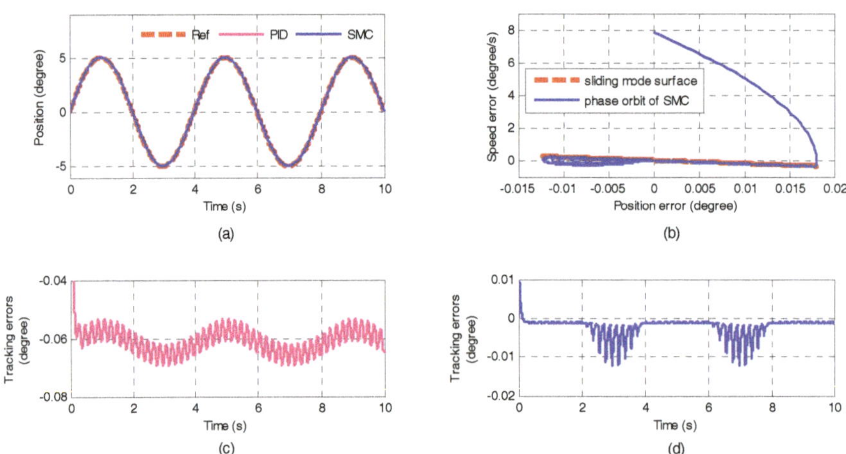

Figure 6. Simulation of sinusoidal response of Joint 1: (**a**) Position tracking; (**b**) Phases trajectory diagram of SMC; (**c**) Tracking error of PID; and (**d**) Tracking error of SMC.

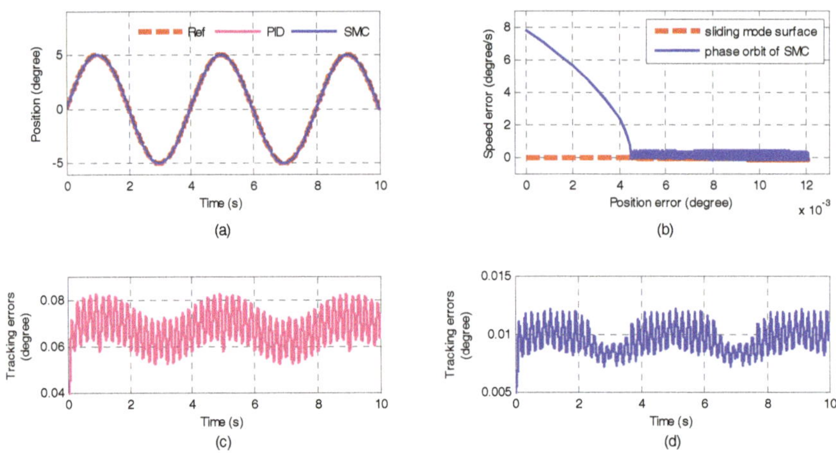

Figure 7. Simulation of sinusoidal response of Joint 2: (**a**) Position tracking; (**b**) Phases trajectory diagram of SMC; (**c**) Tracking error of PID; and (**d**) Tracking error of SMC.

The simulation results show that the influences of the friction torque and disturbances of the load torque and inertia are effectively suppressed by sliding mode variable structure control which has a strong adaptive ability to input signals.

4.2. Establishment of Real-Time Experiment Platform

The key to validation of the dynamical model and the designed controller is to establish a reliable real time control system which has the characteristics of rapid response speed, flexibility, and low hardware cost. MATLAB provides a real-time development environment to achieve system simulation and product rapid-prototype by adopting the RTW (Real-Time Workshop) toolbox. RTW can automatically transform the model into dynamic system model code running on the hardware to realize real-time simulation and control.

xPC target environment adopts the Host-Target pattern. The Host is used for running Simulink model, compiling and generating executable code and then downloading it to the Target, it can also monitor and control the Target. The target is used for running executable code by installing highly optimized real-time kernel of 32-bit protected mode. The Host communicates with the Target using RS232 serial port or ethernet [32,33]. The real-time control experiment platform based on RTW xPC target environment is shown in Figure 8.

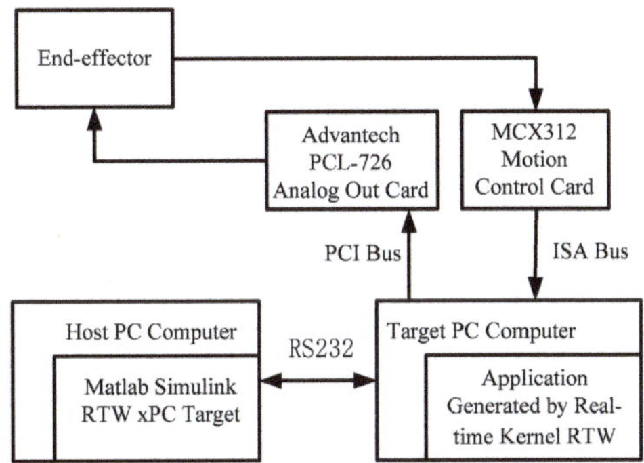

Figure 8. Real-Time Workshop (RT W) xPC target real-time control platform principle diagram.

In the experiment, an Advantech PCL-726 analog output card is used for output of the control signal, and an MCX312 motion control card is used for collecting the position signal from the encoder. PLC-726 can be directly driven by the corresponding module in the I/O module library of xPC target. MCX312 isn't included in the module library and its driver must be written. In the target system, the device driver must be written using C MEX S function which is a kind of S function written using C. Figure 9 shows the display of the RTW real-time experiment platform.

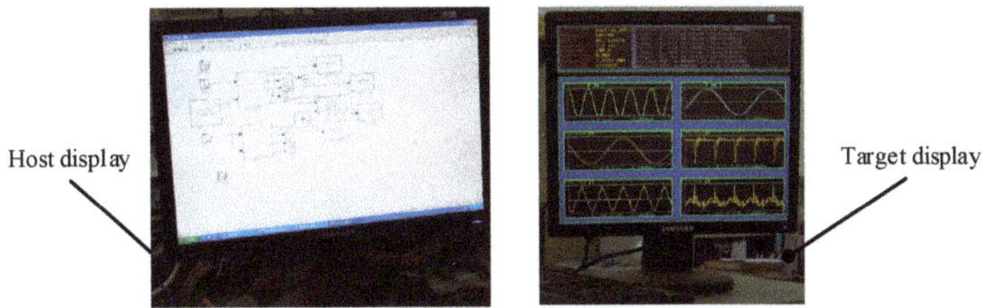

Figure 9. The display of the RTW real-time experiment platform.

4.3. Experiments Results and Discussion

The symbolic function in the variable structure control expressed by Equation (38) is replaced by the saturation function to design quasi-sliding mode controller for reducing the chattering which is brought naturally by the sliding mode variable structure control in the experimental system. The new control law is expressed as:

$$\mathbf{u} = \mathbf{C}\dot{\mathbf{e}} + \ddot{\mathbf{q}}_\mathbf{d} + \varepsilon sat(\mathbf{S}) + \mathbf{KS} \tag{43}$$

with

$$sat(s) = \begin{cases} 1 & s > \Delta \\ ks & |s| \leq \Delta \quad k = \frac{1}{\Delta} \\ -1 & s < -\Delta \end{cases} \tag{44}$$

where Δ is the boundary layer, and $\Delta = 0.05$.

Experiments are carried out with the PID and the SMC controller respectively by applying a five-degree step input and a sinusoid input $r = 5\sin(0.5\pi t)$ to both of the Joints 1 and 2 in the same time. Figures 10 and 11 show the step responses of the Joints 1 and 2, respectively. Figures 12 and 13 show the sinusoid responses of the Joints 1 and 2, respectively.

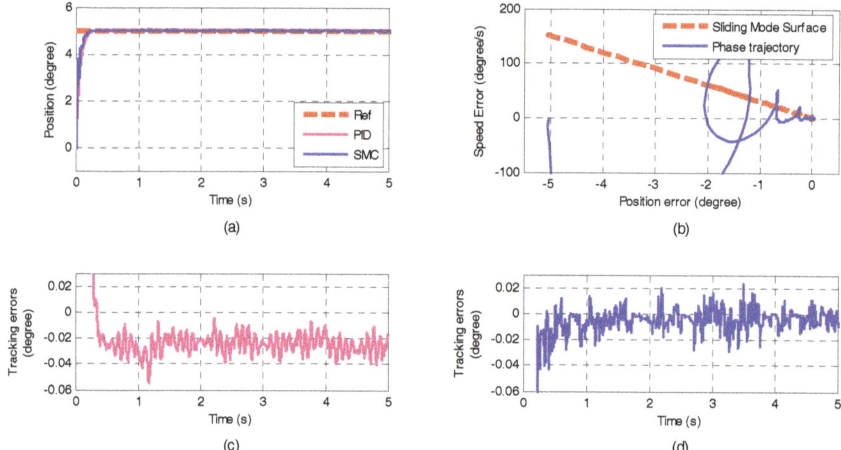

Figure 10. Experiment of step response of Joint 1: (**a**) Position tracking; (**b**) Phases trajectory diagram of SMC; (**c**) Tracking error of PID; and (**d**) Tracking error of SMC.

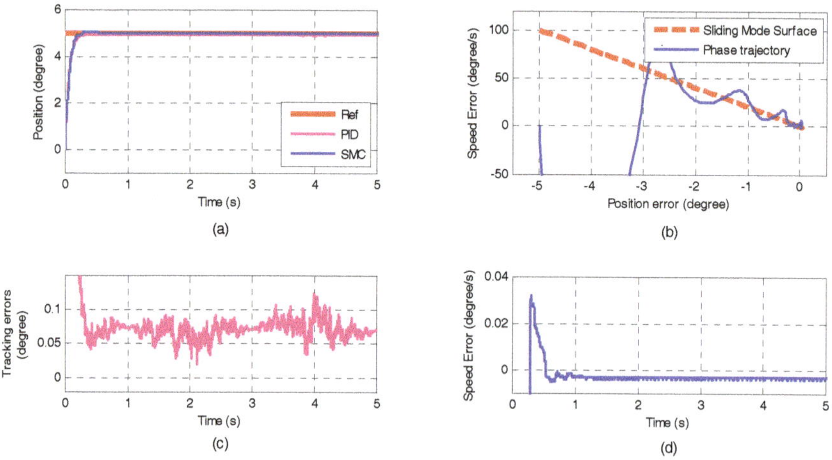

Figure 11. Experiment of step response of Joint 2: (**a**) Position tracking; (**b**) Phases trajectory diagram of SMC; (**c**) Tracking error of PID; and (**d**) Tracking error of SMC.

For step input, as shown in Figures 10a and 11a, the adjusting times of both joints with the proposed SMC and the PID controller are almost the same. However, the position steady state error of each joint with the SMC is far less than that with the PID controller, which is similar to the simulation results. As shown in Figure 10c,d, and Figure 11c,d, the tracking errors produced by the PID controller has higher oscillation amplitude than that by the SMC. This means that the SMC is more robust and accurate than the PID. As show in Figures 10b and 11b, both of the phase trajectories of the SMC converge to stable limit cycles, this implicates that the system yields bounded stability based on sliding mode variable structure control. In order to accelerate the trajectory reaching to the sliding mode surface and improve the system dynamic quality during the experiment, the values of the parameter **C** of the switch function and the proportional gain **K** are set bigger. This causes in initial response that the phase trajectory generates obvious chatter along the sliding mode surface.

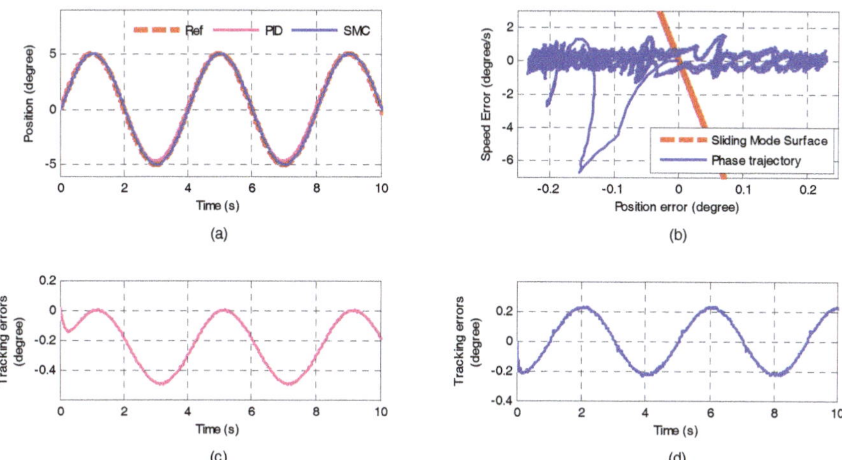

Figure 12. Experiment of sinusoidal response of Joint 1: (**a**) Position tracking; (**b**) Phases trajectory diagram of SMC; (**c**) Tracking error of PID; and (**d**) Tracking error of SMC.

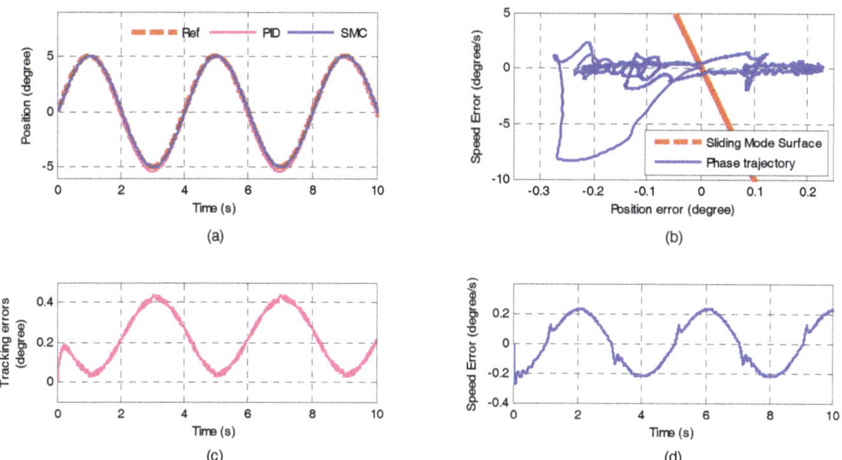

Figure 13. Experiment of sinusoidal response of Joint 2: (**a**) Position tracking; (**b**) Phases trajectory diagram of SMC; (**c**) Tracking error of PID; and (**d**) Tracking error of SMC.

For sinusoid input, as show in Figures 12 and 13, the position tracking experiment results of both of the Joints 1 and 2 with PID controller have obvious lag in contrast to that with the SMC, which suggests that the system with the SMC has higher tracking accuracy than that with the PID controller.

As shown in Figures 12b and 13b, both of the phase trajectories converge to limit cycles around the equilibrium point, this implicates that the system based on the SMC gains bounded stability and has certain steady state errors. For the reason, the x-axis and y-axis have different coordinate ranges, both of the limit cycles look tabular. Actually, if the x-axis and y-axis have the same ranges, both of the limit cycles will appear more slim, which implicates that the speed errors are larger than the position errors.

5. Conclusions

In robotic control systems, nonlinearity and disturbances such as friction and differential signal may severely limit the performance of control. In this paper, a sliding mode variable structure control architecture is proposed for the normal adjustment of a designed robotic drilling end-effector. By using computation torque control and by adopting a third-order nonlinear integration chain differentiator for obtaining the speed and acceleration signals from the position signals, this sliding mode control scheme is developed with good dynamic quality. The new control law ensures global stability of the entire system and achieves both stabilization and tracking within a desired precision. A real-time control experiment platform is developed in xPC target environment based on MATLAB RTW to verify the proposed control scheme and simulation results. The experimental results prove that the proposed sliding mode control strategy is effective and robust with regard to external disturbances. The proposed controller is successfully tested on the normal adjustment of the robotic drilling end-effector. This research provides a solution for the more accurate control of robotic drilling on the curved surface of an airplane.

Author Contributions: Conceptualization and methodology, L.Z.; formal analysis, L.Z, M.L.W., and H.H.; writing—original draft preparation, L.Z.; supervision, J.S.D.; Funding acquisition, L.Z.

Funding: This research was funded by National Natural Science Foundation of China, grant number 51765031 and Natural Science Foundation of Gansu Province, China, grant number 1508RJZA075.

Acknowledgments: The authors gratefully acknowledge the helpful comments and suggestions of the reviewers, which have improved the presentation.

Conflicts of Interest: The authors declare no conflict of interest.

Abbreviations

The abbreviations and symbols adopted throughout the paper are listed:

SMC	Sliding mode controller
PID	Proportional-Integral-Derivative
D	Positive definite symmetric inertia matrix
H	Centrifugal and Coriolis force vector
G	Gravity vector
θ_1	Rotation angle of joint 1
θ_2	Rotation angle of joint 2
d_3	Linear displacement of prismatic joint 3
I_{abi}	Product of inertia in respect to a pair of orthogonal axes a and b, a, b denote x, y or z, i denotes joint number
m_i	Mass of Joint i
c_{ai}	Centroid coordinate on axis a, a denotes x, y or z, i denotes joint number
τ_1	acts on joint 1
τ_2	Torque acts on joint 2
f_3	Force acts on prismatic joint 3
$\boldsymbol{\tau_d}$	Disturbance torque vector

τ	Computation torque vector
v	Motion velocity during friction
v_s	Stribeck velocity
F_f	Friction force
F_e	Driving force
F_s	Maximum static friction force
F_c	Coulomb friction force
F_v	Viscous frictional torque coefficient
u	Controlled quantity vector
V	Liapunov's function
S	Switch function
C	Proportional coefficient matrix in switch function
e	Position tracking error matrix
ϵ	Exponent matrix in reaching law
K	Proportional coefficient matrix in reaching law

References

1. Wilson, M. Robots in the aerospace industry. *Aircr. Eng. Aerosp. Technol.* **1994**, *66*, 2–3. [CrossRef]
2. Cirillo, P.; Marino, A.; Natale, A.; Marino, E.D.; Chiacchio, P.; Maria, G.D. A low-cost and flexible solution for one-shot cooperative robotic drilling of aeronautic stack materials. *IFAC-PapersOnLine* **2017**, *50*, 4602–4609. [CrossRef]
3. Olsson, T.; Haage, M.; Kihlman, H.; Johansson, R.; Nilsson, K.; Robertsson, A.; Björkmsn, M.; Isaksson, R.; Ossbahr, G.; Brogårdh, T. Cost-efficient drilling using industrial robots with high-bandwidth force feedback. *Rob. Comput. Integr. Manuf.* **2010**, *26*, 24–38. [CrossRef]
4. Schneider, U.; Drust, M.; Ansloni, M.; Lehmann, C.; Pellicciari, M.; Leali, F.; Gunnink, J.W.; Verl, A. Improving robotic machining accuracy through experimental error investigation and modular compensation. *Int. J. Adv. Manuf. Technol.* **2016**, *85*, 3–15. [CrossRef]
5. Shi, Z.; Yuan, P.; Wang, Q.; Chen, D.; Wang, T. New design of a compact aero-robotic drilling end effector: An experimental analysis. *Chin. J. Aeronaut.* **2016**, *29*, 1132–1141. [CrossRef]
6. DeVlieg, R.; Sitton, K.; Feikert, E.; Inman, J. ONCE (one-sided cell end effector) robotic drilling system. *SAE Tech. Pap.* **2002**. [CrossRef]
7. Atkinson, J.; Hartmann, J.; Jones, S.; Gleeson, P. Robotic drilling system for 737 aileron. *SAE Tech. Pap.* **2007**. [CrossRef]
8. Webb, P.; Chitiu, A.; Fayad, C.; Gindy, N.; Mckeown, C. Flexible automated riveting of fuselage skin panels. *SAE Trans.* **2001**, *110*, 218–221.
9. Hempstead, B.; DeVlieg, R.; Mistry, R.; Sheridan, M. Drill and drive end effector. *SAE Tech. Pap.* **2001**. [CrossRef]
10. Liang, J.; Bi, S. Design and experimental study of an end effector for robotic drilling. *Int. J. Adv. Manuf. Technol.* **2010**, *50*, 399–408. [CrossRef]
11. Devlieg, R. High-accuracy robotic drilling/milling of 737 inboard flaps. *SAE Int. J. Aerosp.* **2011**, *4*, 1373–1379. [CrossRef]
12. Gray, T.; Orf, D.; Adams, G. Mobile automated robotic drilling, inspection, and fastening. *SAE Tech. Pap.* **2013**. [CrossRef]
13. Olsson, T.; Robertsson, A.; Johansson, R. Flexible force control for accurate low-cost robot drilling. In Proceedings of the 2007 IEEE International Conference on Robotics and Automation, Roma, Italy, 10–14 April 2007; IEEE: New York, NY, USA, 2007.
14. Tian, W.; Zhou, W.; Zhou, W.; Liao, W.; Zeng, Y. Auto-normalization algorithm for robotic precision drilling system in aircraft component assembly. *Chin. J. Aeronaut.* **2013**, *26*, 495–500. [CrossRef]
15. Mei, B.; Zhu, W.; Yuan, K.; Ke, Y. Robot base frame calibration with a 2D vision system for mobile robotic drilling. *Int. J. Adv. Manuf. Technol.* **2015**, *80*, 1903–1917. [CrossRef]
16. Frommknecht, A.; Kuehnle, J.; Effenberger, I.; Pidan, S. Multi-sensor measurement system for robotic drilling. *Rob. Comput. Integr. Manuf.* **2017**, *47*, 4–10. [CrossRef]

17. Qin, C.; Tao, J.; Wang, M.; Liu, C. A Novel Approach for the Acquisition of Vibration Signals of the End Effector in Robotic Drilling. In Proceedings of the 2016 IEEE International Conference on Aircraft Utility Systems (AUS), Beijing, China, 10–12 October 2016; IEEE: New York, NY, USA, 2016.

18. Garnier, S.; Subrin, K.; Waiyagan, K. Modelling of robotic drilling. *Procedia CIRP* **2017**, *58*, 416–421. [CrossRef]

19. Zhang, L.; Wang, X. A novel algorithm of normal attitude regulation for the designed end-effector of a flexible drilling robot. *J. Southeast Univ.* **2012**, *28*, 29–34.

20. Olsson, H.; Åström, K.J.; Wit, C.C.D.; Gäfvert, M.; Lischinsky, P. Friction models and friction compensation. *Eur. J. Control* **1998**, *4*, 176–195. [CrossRef]

21. Hung, J.Y.; Gao, W.; Hung, J.C. Variable structure control: A survey. *IEEE Trans. Ind. Electron.* **1993**, *40*, 2–22. [CrossRef]

22. Wang, X.; Liu, J.; Cai, K. Tracking control for a velocity-sensorless VTOL aircraft with delayed outputs. *Automatica* **2009**, *45*, 2876–2882. [CrossRef]

23. Karafyllis, I.; Kravaris, C. On the observer problem for discrete-time control systems. *IEEE Trans. Ind. Electron.* **2007**, *52*, 12–25. [CrossRef]

24. Drakunov, S.; Utkin, V. Sliding Mode Observers Tutorial. In Proceedings of the 34th IEEE Conference on Decision and Control, New Orleans, LA, USA, 13–15 December 1995; IEEE: New York, NY, USA, 1995.

25. Ma, R.; Zhang, G.; Krause, O. Fast terminal sliding-mode finite-time tracking control with differential evolution optimization algorithm using integral chain differentiator in uncertain nonlinear systems. *Int. J. Robust Nonlinear Control* **2018**, *28*, 625–639. [CrossRef]

26. Vasiljevic, L.; Khalil, H. Differentiation with High-Gain Observers the Presence of Measurement Noise. In Proceedings of the 45th IEEE Conference on Decision and Control, San Diego, CA, USA, 13–15 December 2006; IEEE: New York, NY, USA, 2007.

27. Alwi, H.; Edwards, C. An adaptive sliding mode differentiator for actuator oscillatory failure case reconstruction. *Automatica* **2013**, *49*, 642–651. [CrossRef]

28. Wang, X.; Lin, H. Design and analysis of a continuous hybrid differentiator. *IET Control Theory Appl.* **2011**, *5*, 1321–1334. [CrossRef]

29. Listmann, K.D.; Zhao, Z. A Comparison of Methods for Higher-Order Numerical Differentiation. In Proceedings of the 2013 European Control Conference (ECC), Zurich, Switzerland, 17–19 July 2013; IEEE: New York, NY, USA, 2013.

30. Wang, X.; Chen, Z.; Yuan, Z. Design and analysis for new discrete tracking-differentiators. *Appl. Math. A J. Chin. Univ.* **2003**, *18*, 214–222. [CrossRef]

31. Meshram, P.M.; Kanojiya, R.G. Tuning of PID Controller Using Ziegler-Nichols Method for Speed Control of DC Motor. In Proceedings of the IEEE-International Conference on Advances in Engineering, Science and Management (ICAESM-2012), Nagapattinam, Tamil Nadu, India, 30–31 March 2012; IEEE: New York, NY, USA, 2012.

32. Low, K.H.; Wang, H.; Wang, M.Y. On the Development of a Real Time Control System by Using xPC Target: Solution to Robotic System Control. In Proceedings of the IEEE International Conference on Automation Science and Engineering, Edmonton, AB, Canada, 1–2 August 2005; IEEE: New York, NY, USA, 2005.

33. Grepl, R. Real-Time Control Prototyping in MATLAB/Simulink: Review of Tools for Research and Education in Mechatronics. In Proceedings of the 2011 IEEE International Conference on Mechatronics, Istanbul, Turkey, 13–15 April 2011; IEEE: New York, NY, USA, 2011.

Article

Prototype Design and Performance Tests of Beijing Astronaut Robot

Zeyuan Sun [1,2,3], Hui Li [1,2,3,*], Zhihong Jiang [1,2,3], Zhenzi Song [1,2,3], Yang Mo [1,2,3] and Marco Ceccarelli [2,4]

[1] School of Mechatronical Engineering, Beijing Institute of Technology, Beijing 100081, China; sunzeyuan222@foxmail.com (Z.S.); jiangzhihong@bit.edu.cn (Z.J.); songzhenzi1103@163.com (Z.S.); moyang602@163.com (Y.M.)
[2] Beijing Advanced Innovation Center for Intelligent Robots and Systems, Beijing 100081, China
[3] Key Laboratory of Biomimetic Robots and Systems, Ministry of Education, Beijing 100081, China
[4] LARM: Laboratory of Robotics and Mechatronics, DICeM, University of Cassino and South Latium, Via Di Biasio 43, 03043 Cassino FR, Italy; ceccarelli@unicas.it
* Correspondence: lihui2011@bit.edu.cn; Tel.: +86-186-1060-8649

Received: 2 July 2018; Accepted: 6 August 2018; Published: 10 August 2018

Featured Application: A novel chameleon-like astronaut robot that is designed to assist, or even substitute, an astronaut in a space station to complete dangerous and prolonged work.

Abstract: This paper proposes a novel chameleon-like astronaut robot that is designed to assist, or even substitute, a human astronauts in a space station to complete dangerous and prolonged work, such as maintenance of solar panels, and so on. The robot can move outside the space station freely via the hundreds of aluminum handrails, which are provided to help astronauts move. The robot weighs 30 kilograms, and consists of a torso, three identical 4-degree of freedom (DOF) arms, three end effectors, and three monocular vision system on each end effector. Via multi-arm associated motion, the robot can realize three kinds of motion modes: walking, rolling, and sliding. Numerous experiments have been conducted in a simulation environment and a ground verification platform. Experimental results reveal that this robot has excellent motion performance.

Keywords: space robot; hybrid bionic robot; chameleon; end effector

---

1. Introduction

With the development of space technology, a space station is a vital approach for human beings to study and research space, but its construction and maintenance require extravehicular activity (EVA) carried out by astronauts [1]. Since the space environment has the characteristics of high vacuum, large temperature differences, high radiation, and microgravity, which result in the failure of continuous working time and inefficiency, humans in space require a large and a complex life support system, an environmental control system, and life-saving system, and requires a lot of uninterrupted material supply [2]. Under such conditions, space operations of astronauts are not only dangerous but also costly. Therefore robots are used as assistants or replacements more and more frequently to reduce the risk and cost. In addition, space robots do better in load capacity, accurate positioning, adaption to environments, and durable work. Space robots can install precise devices, maintain the space station, and perform experiments in space without health concerns [3].

So far, a few robotic arms have been applied in the International Space Station which is being assembled in low Earth orbit, for example Canadarm2 [4] and Robonaut [5]. Canadarm2 was designed by the Canadian Space Agency and it is used mainly for payload handling, EVA support, and the international space station's assembly [6]. The Special Purpose Dexterous Manipulator (SPDM) is the third Canadian robotic arm used on the International Space Station (ISS), preceded by the Space Shuttle's Canadarm and the large Canadarm2 [7]. It is an extremely advanced, highly dexterous, dual-armed robot which can carry out delicate maintenance and servicing tasks on the ISS. Robonaut is a humanoid robot designed by the Robotic Systems Technology Branch at NASA's Johnson Space Center. Robonaut can reduce the burden of EVA on astronauts and also serve in rapid response capacities.

Canadarm2 and Robonaut are all well designed, but they can be considered as too complicated, and are designed only for specific activities, so they cannot work in all of the places outside the space station. There are many operational tasks in space, so the robot needs the ability of moving in a wide range and the ability of precise work. Moreover, the power supply on the space station is limited, so the path planning of the space robot should be power-saving [8,9].

Astronauts are weightless and floating in space. Thus, there are hundreds of aluminum handrails and poles on the extravehicular surface of the space station (as shown in Figure 1) for astronauts to get themselves to the expected position during spacewalks and operations. Most importantly, astronauts are fixed by these handrails so that they will not float away.

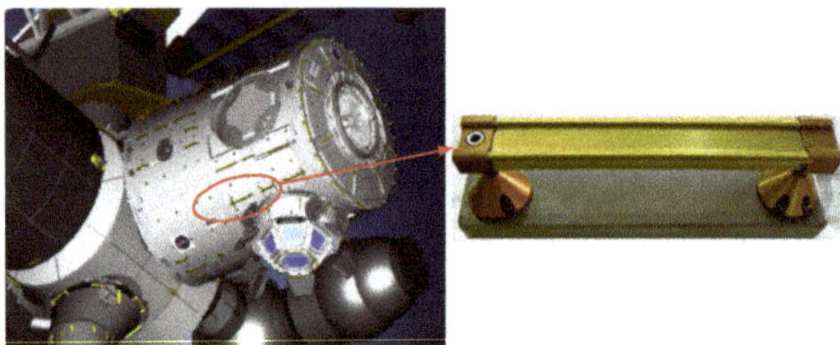

Figure 1. Handrails on extravehicular surface with appearance and dimensions.

A chameleon is a kind of arboreal reptile with a simple physical structure, but can move stably [10]. The conditions of the extravehicular surface are kind of similar to the structure of tree crown. Thus, referring to the simple physical structure and stable motion pattern, a new pattern of a hybrid bionic robot, called the Beijing Astronaut Robot, prototype is design is developed and performance tests are presented in this paper.

2. The Mechanical Structure and Test System Design of the Robot

2.1. Design Indicators

It is very difficult to transport large volume or heavy weight objects from the ground to a space station. The design of the space robot needs to meet the requirements of the operations on the space station and, at the same time, the mass and the volume should be as small as possible [11]. Parameters and functional indicators of the Beijing Astronaut Robot are as follows:

- The total mass of the robot is less than 30 kg;
- The total degrees of freedom (DOF) of the robot is no less than 12;
- The robot has at least three arms, each arm has no less than 4-DOF;
- The robot arm length is not more than 700 mm.

According to the structural environment outside the space station and the characteristics of the robot itself, the robot has three kinds of motion modes: walking, rolling and sliding. The motion index in each motion mode is as follows:

- Walking mode, the stride of the robot is not less than 30 cm, the robot movement speed is not less than 0.1 m/s;

- Rolling mode, the robot's stride is not less than 60 cm, the robot movement speed of not less than 0.2 m/s; and
- Sliding mode, the moving speed of the robot is not less than 0.5 m/s.

When the robot performs the motion and operation tasks outside the space station, the robot's visual measurement, end positioning and force control are supposed to have high precision. The specific indicators are as follows:

- Visual measurement accuracy not less than 1 mm in the range of 200 mm;
- End effector's positioning accuracy within 2 mm;
- End effector's force control accuracy within 2 N.

2.2. Structural Design and Key Technologies

The Beijing Astronaut Robot consists of three identical 4-DOF arms, three end effectors, and a torso, as shown in Figure 2.

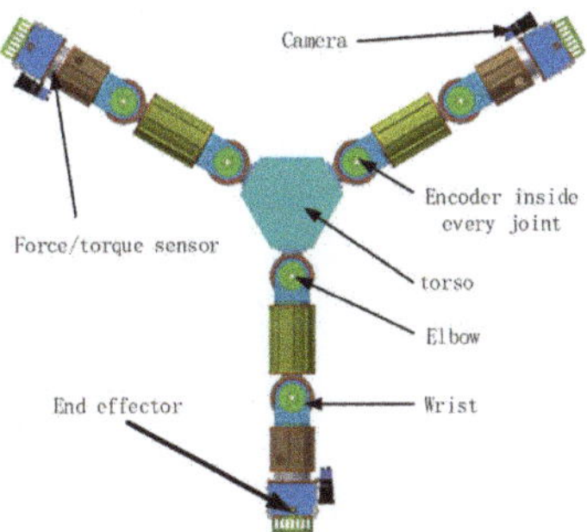

Figure 2. Overall structure of the robot.

The robot mimics the grasping of a chameleon, and its motion and operation mechanism is composed of three identical arms with end effectors, and has 15 degrees of freedom in total. Each arm consists of a wrist with three rotational degrees of freedom and an elbow with one rotational degree of freedom, a total of 3 × 4 = 12 degrees of freedom of motion. The end effectors have 3 × 1 = 3 degrees of freedom. In order to identify the environmental changes and determine its own state, the robot is equipped with a monocular vision camera, 6-dimensional force/torque and other sensors, like encoders in every joint, as shown in Figure 2. The overall scheme of the robot is shown in Figure 3.

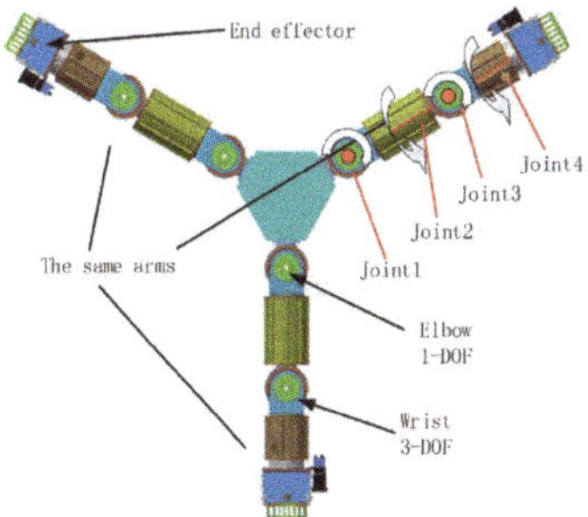

Figure 3. DOF (degree of freedom) configuration of the Beijing Astronaut Robot.

It is researched that the motion of the chameleon in a complex environment is very stable, this feature coincides with the robot's mobile demand [12]. Therefore, the design of our robot's bionic motion and operating mechanism draws on the characteristics of a chameleon. More specifically, the design of the robot arms' degrees of freedom and motion pattern draws on the skeletal structure of a chameleon [10,13]. The bionic robot arm and the end effector of the Beijing Astronaut Robot are shown in Figure 4.

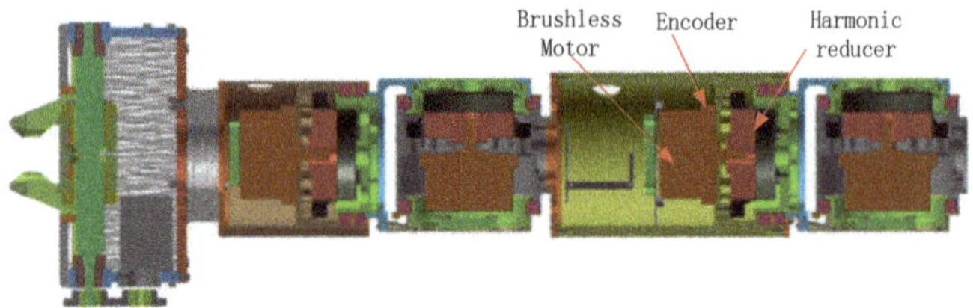

Figure 4. Bionic robot arm and end effector.

In order to make the robot end effector grasp the handrail and obtain sufficient force and torque to prevent the occurrence of sliding or rotation [1], we designed a claw driven by a screw (as shown in Figure 5). The rotation of the screw driven by a motor is converted into the opening and closing of the claw [14,15]. When the lead angle of the screw is smaller than the self-locking angle, the transmission mode has self-locking property, that is, the sliders can only be driven by the rotation of the screw, but the screw cannot be driven by stressing the sliders. Therefore, even if the motor's power is cut off, the gripper can still grasp the handrail to ensure the safe motion of the robot.

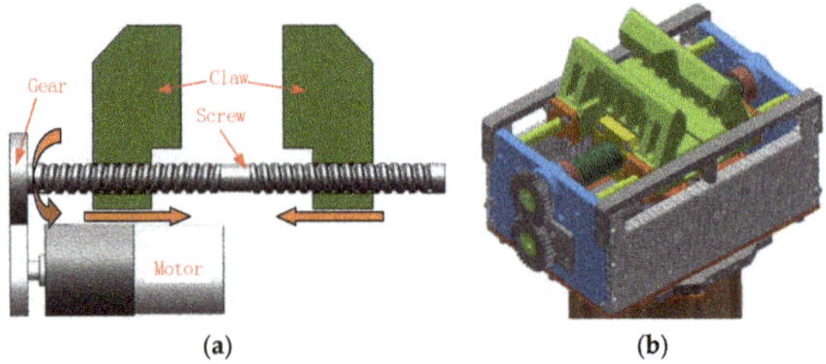

Figure 5. (**a**) The schematic diagram of mechanism of the claw; (**b**) The overall structure of the end effector.

2.3. Testing Layout

In order to test the motion performance of the robot, a control structure is designed, as shown in Figure 6.

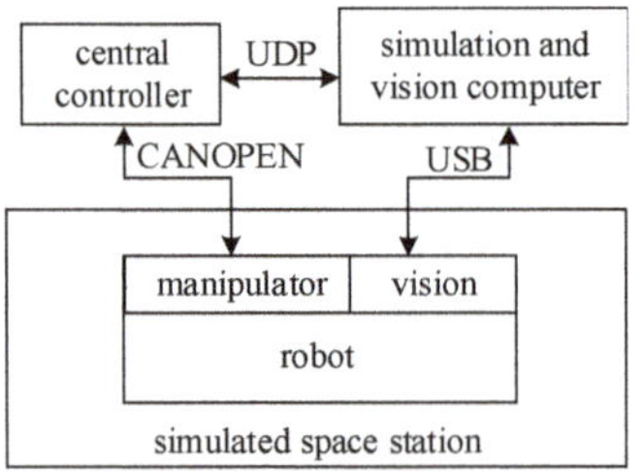

Figure 6. Control structure of the Beijing Astronaut Robot.

The motion test of the robot is done in the simulated space station. The Beijing Astronaut Robot can accomplish rolling, walking by grasping the handrails, and sliding driven by the rocker arm shown in Figure 7.

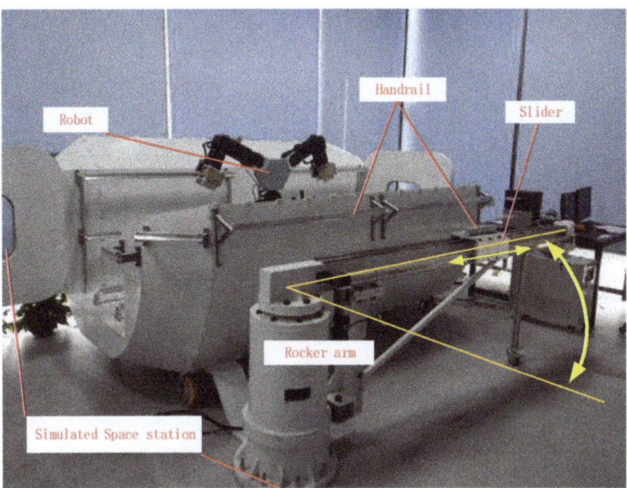

Figure 7. The actual test environment of the robot.

The central controller is in charge of the kinematics calculation, trajectory planning and sends motion instructions to every joint trough CANopen (a higher-layer protocol for the CAN bus) [16]. The simulation and visual processing computer completes 3D motion simulations, hand-eye vision processing, and task settings of the robot. The computer can capture images by communicating with hand-eye cameras through USB 2.0, and processes the images to obtain the pose of the target. The computer sends task and vision processing information to the central controller. Meanwhile, it receives the data of joint motion from the central controller through a UDP (User Datagram Protocol) network [17], so that the computer can preview the motion of the robot.

3. Testing Results

The performance tests include motion performance tests of a single arm, system simulation tests, and motion performance tests of the whole robot.

3.1. Motion Performance Test of A Single Arm

The motion performance of a single arm is the basis for the robot to complete all the tasks. Thus, we test the motion performance of a single arm to get the properties of the robot, like the motion accuracy, and the loading capability. We test the motion performance of a single arm under both no-load and load states by sending motion instructions and obtain the actual motion of joint from the encoder feedback.

The end effector of one arm holds the handrail while the other two are unmounted from the torso of the robot. We hope to test all joints at the same time, and all joint angles are set to the same value. Simple geometric analysis shows that when the joints of the single arm move from $0°$ to $30°$, the arm of the robot reaches a harsh condition, which can lay a foundation for the following load test. In the actual motion of the robot, we want the robot's displacement, velocity, and acceleration to change smoothly, and we can give the first and last values of the three. Thus, in the test, every joint rotates smoothly from using a fifth-order polynomial.

The robot finishes the planned motion under the no-load condition; the process is shown in Figure 8.

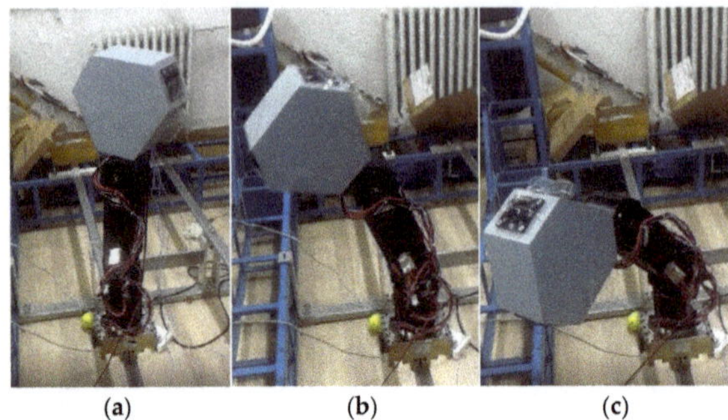

Figure 8. The motion of single arm under the no-load condition. (**a**) Motion start; (**b**) In the process of motion; (**c**) Motion complete.

In the process, the expected angles sent by the central controller and actual motion angles measured by the encoder of every joint is shown in Figure 9. The location of joints 1–4 are shown in Figure 3. It can be seen that the expected and actual angles are consistent, basically, and the maximal error is $0.03°$.

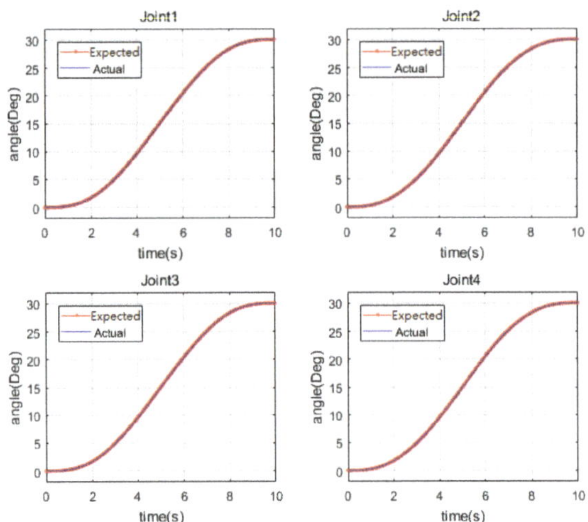

Figure 9. The motion of four joints under no load.

In order to check the performance of a single arm in the normal working condition, we add a 20 kg load, which is equivalent to the other two arms on the torso, and the process of motion is shown in Figure 10. Although the mass distribution is not exactly the same as the case where all three arms of the robot are installed, they are basically similar. Compared to the actual working conditions, such a loading method results in a slightly poor mass distribution causing a higher requirement for the single arm load capacity. Therefore, the load test can reflect whether the load capacity of each joint of one arm can meet the needs. It is proved that the arm can complete the planning motion well under the 20 kg load, and the maximal error is $0.04°$.

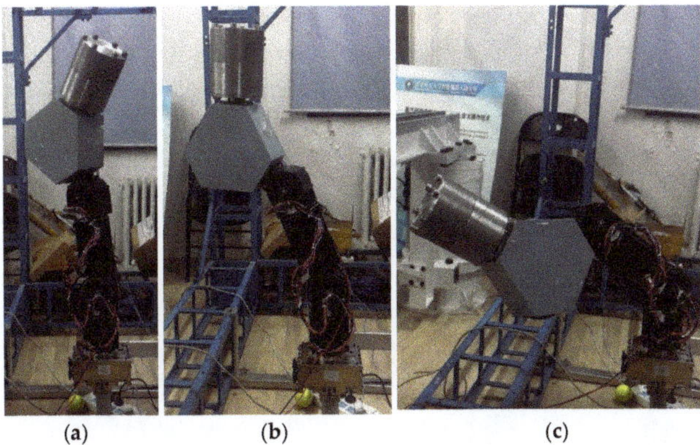

(a)   (b)   (c)

Figure 10. The motion of single arm under 20 kg load. (**a**) Motion start; (**b**) In the process of motion; (**c**) Motion complete.

3.2. System Simulation Test

Due to the large degree of freedom of the robot and the complex kinematics, the robot motion simulation system (as shown in Figure 11) is developed and becomes one of the ideal platforms for studying the kinematics of the robot. At the same time, the complexity of the space station structure has high requirements on the operation of the robot, resulting in a complexity of the robot's motion planning. Fortunately, the development of the simulation system can provide an auxiliary and verification platform for the motion planning of the robot.

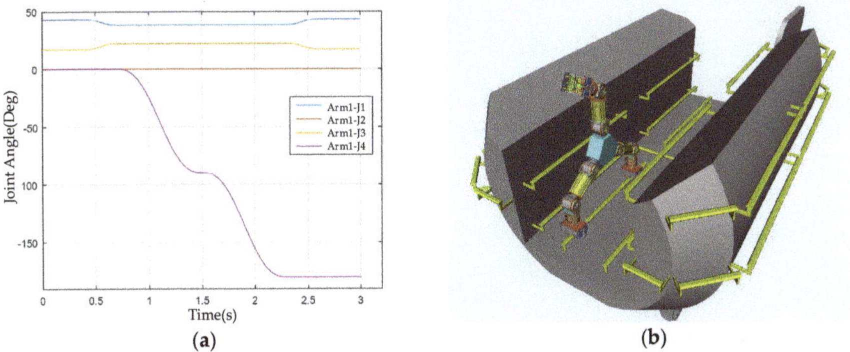

(a)   (b)

Figure 11. The simulation of the robot's motion. (**a**) The motion planning data; (**b**) Display interface of robot motion simulation system.

Before the motion performance test of the whole robot, we simulated the motion of robot with the simulation and visual processing computer. After the motion planning data has been entered into the simulation system, we can visually see the motion effect of the robot. With such a system, we can quickly check the robot's trajectory for obvious problems, ensure that the trajectory planning is correct, and then transfer it into the actual robot system, thus ensuring the safety of the robot motion.

3.3. Motion Performance Test of the Robot

After the motion performance test of a single arm and the system simulation test, we test the motion performance of the robot with the prototype. The central controller receives the revolving command from the simulation and visual processing computer, and send motion data of every joint to the prototype, so the robot revolves according to the data.

We tested three motion types in total: walking, rolling, and sliding, as shown in Figures 12–14.

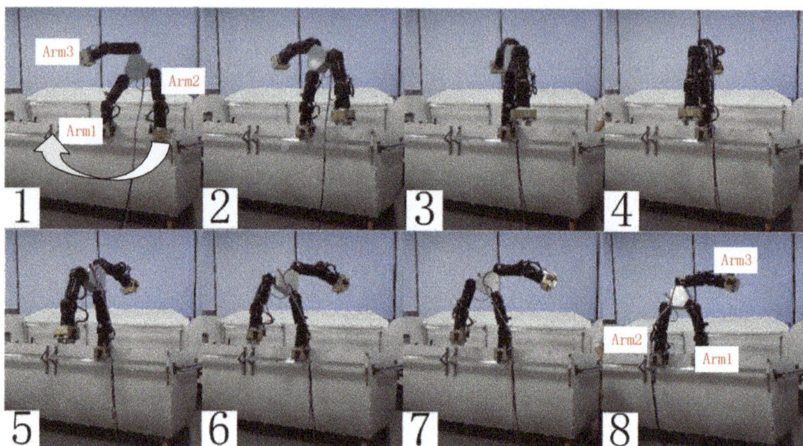

Figure 12. Walking mode. In this mode, arm 2 is initially located to the right of arm 1. During walking, arm 1 grasps the handrail, then arm 2 releases, the robot rotates 180° around arm 1. Then arm 1 reaches the left side of arm 2, and arm 2 grasps the armrest to complete one step of the robot. In the next step of walking, the robot rotates around arm 2 to complete the walking action. Arm 1 and arm 2 move alternately to achieve continuous walking of the robot.

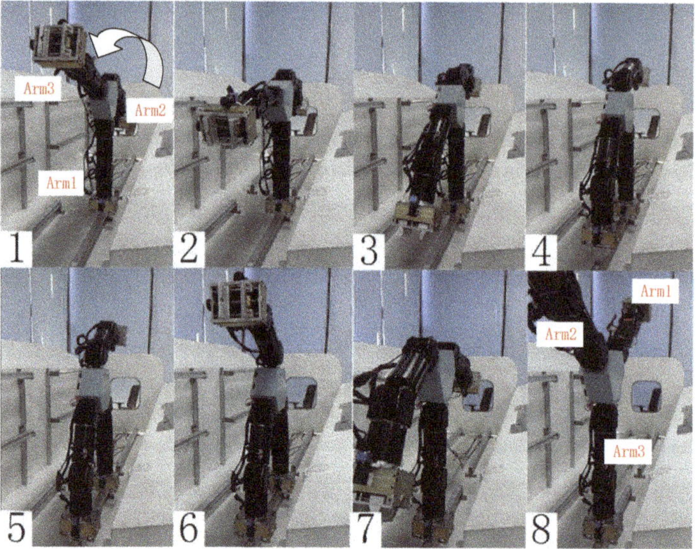

Figure 13. Rolling mode. In this mode, arm 1 grips the handrail, arm 2 rises, the center of gravity of the robot is adjusted, and then arm 3 drops and grasps the handrail. Finally, arm 1 rises to complete a tumbling motion. The three arms alternately run to achieve continuous rolling of the robot.

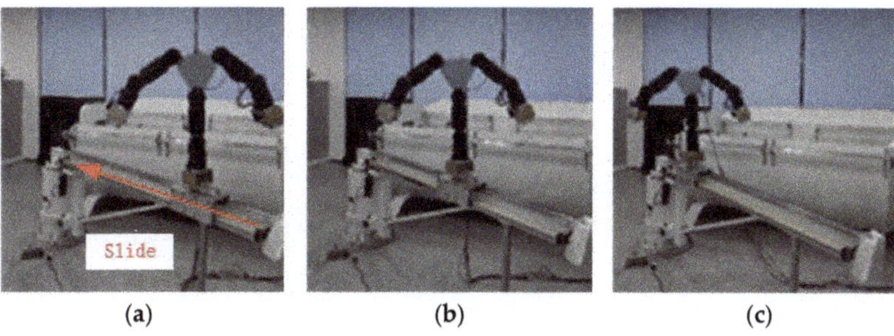

(a)   (b)   (c)

Figure 14. Sliding Mode. (**a**) Motion start; (**b**) In the process of motion; (**c**) Motion complete.

In the above modes, the joint most likely to get a large motion error is the joint 3 of arm 1 with the largest load in the rolling mode. The planned motion curve and the actual value read by the encoder are shown in Figure 15. The results show that the robot joint motion is stable, and its maximum error is 0.04°, which is consistent with the one-arm test results. The actual motion of the robot has indirectly demonstrated the accuracy of the joint motion, so it is unnecessary for the motion data for every joint to be listed.

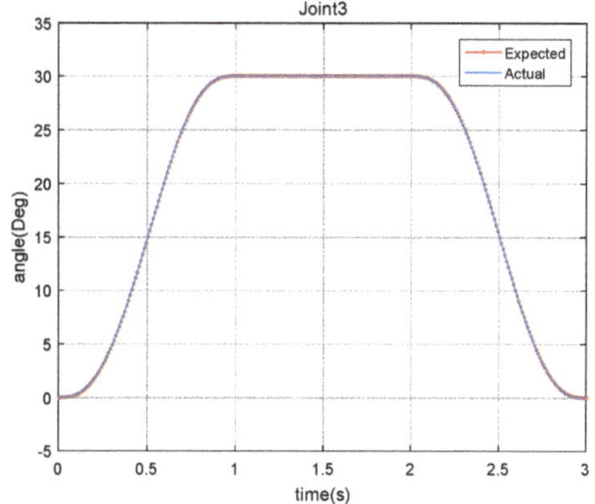

Figure 15. The motion of the root joint of arm 1.

In order to allow the robot to reach a specific working position outside the space station with the aid of a large space robot arm, an electric slide rail is arranged on the robot arm. The robot grasps the handrail on the slider driven by the electric slide rail to realize the sliding motion. In the test, a 2-DOF (yaw and telescopic) rocker arm is used to simulate the large space robot arm. The motion of the sliding mode is shown in Figure 14.

4. Discussion and Conclusions

Based on analysis of the features of the extravehicular environments, this paper learns from the mechanism of a chameleon, and creatively proposes a kind of astronaut robot that can help astronauts perform dangerous and prolonged work. The robot has three modular arms, each arm has four rotary joints, an end effector and a monocular vision system. The total mass of the robot is 30 kg, and each arm length is 700 mm, which meet the design requirements. The end effectors are designed with a self-locking ability to grasp the handrail firmly so the robot can move freely outside the space station via handrails. Numerous experiments, such as a single-arm test, simulation tests, and integrated ground verification, show that the astronaut robot has excellent motion performance. In the future, we will study how to save the power of our robot by drawing on the design of microrobots, so that it consumes less power from the space station power source [18].

Author Contributions: Conceptualization: H.L. and M.C.; structural design: Z.S. (Zeyuan Sun) and H.L.; software: Z.S. (Zhenzi Song) and Y.M.; funding acquisition: Z.J.

Funding: The authors would like to acknowledge the National Natural Science Foundation of China (61733001, 61573063, 61503029, U1713215) and Tianjin Science and Technology Plan Project (17YFCZZC00500) for their support and funding of this paper.

Conflicts of Interest: The authors declare no conflict of interest. The funders had no role in the design of the study; in the collection, analyses, or interpretation of data; in the writing of the manuscript, and in the decision to publish the results.

References

1. Thuot, P.J.; Harbaugh, G.J. Extravehicular activity training and hardware design consideration. *Acta Astronaut.* **1995**, *36*, 13–16. [CrossRef]
2. Macelroy, R.D.; Smernoff, D.T. Controlled Ecological Life Support System: Regenerative Life Support Systems in Space. Available online: https://ntrs.nasa.gov/search.jsp?R=19880002869 (accessed on 1 September 1987).
3. Zhang, T.; Chen, Z.; Wang, X.; Liang, B. Overview and Prospect of Key Technologies of Teleoperation of Space Robot. *Aerosp. Control Appl.* **2014**, *40*, 1–9.
4. Doetsch, K. Canada's role on space station. *Acta Astronaut.* **2005**, *57*, 661–675. [CrossRef] [PubMed]
5. Diftler, M.A.; Culbert, C.J.; Ambrose, R.O.; Platt, R.J. Evolution of the NASA/DARPA Robonaut control system. *IEEE ICRA* **2003**, *2*, 2543–2548.
6. Gibbs, G.; Sachdev, S. Canada and the international space station program: overview and status. *Acta Astronaut.* **2002**, *51*, 591–600. [CrossRef]
7. Abramovici, A. The Special Purpose Dexterous Manipulator (SPDM) Systems Engineering Effort—A successful exercise in cheaper, faster and (hopefully) better systems engineering. *J. Reduc. Space Mission Cost* **1998**, *1*, 177–199. [CrossRef]
8. Robotics, M.D. Mobile Servicing System—Data Sheet. MD Robotics, Brampton, Ontario, Canada 2002. Available online: http://www.spacenet.on.ca/data/pdf/canada-in-space/mss-ds.pdf (accessed on 23 January 2002).
9. Ni, W.; Zhang, B.; Yang, H.; Li, H.; Jiang, Z.; Huang, Q. Foot/hand design for a chameleon-like service robot in space station. In Proceedings of the 2013 IEEE International Conference on Robotics and Biomimetics (ROBIO), Shenzhen, China, 12–14 December 2013; pp. 215–220.
10. Fischer, M.S.; Krause, C.; Lilje, K.E. Evolution of chameleon locomotion, or how to become arboreal as a reptile. *Zoology* **2010**, *113*, 67–74. [CrossRef] [PubMed]
11. Coleshill, E.; Oshinowo, L.; Rembala, R.; Bina, B.; Rey, D.; Sindelar, S. Dextre: Improving maintenance operations on the international space station. *Acta Astronaut.* **2009**, *64*, 869–874. [CrossRef]
12. Prahlad, H.; Pelrine, R.; Stanford, S.; Marlow, J.; Kornbluh, R. Electroadhesive Robots—Wall Climbing Robots Enabled by a Novel, Robust, and Electrically Controllable Adhesion Technology. In Proceedings of the IEEE ICRA and Automation, Pasadena, CA, USA, 19–23 May 2008; pp. 3028–3033.
13. Hui, L.; Marco, C.; Qiang, H.; Giuseppe, C. A Chameleon-Like Service Robot for Space Station. In Proceedings of the International Workshop on Bio-Inspired Robots, Nantes, France, 6–8 April 2011.
14. Marco, C. *Grasping in Robotics*; Springer: Berlin, Germany, 2013; pp. 117–120. ISBN 978-1-4471-4664-3.
15. Sabatini, M.; Palumbo, N.; Gasbarri, P. Virtual and Rapid Prototyping of an Underactuated Space End Effector. *J. Robot. Autom.* **2017**, *1*, 10–21.
16. Farsi, M.; Ratcliff, K.; Barbosa, M. An introduction to CANopen. *Comput. Control Eng. J.* **1999**, *10*, 161–168. [CrossRef]

17. Postel, J. User Datagram Protocol. Available online: https://tools.ietf.org/html/rfc768 (accessed on 28 August 1980).
18. Baglio, S.; Castorina, S.; Fortuna, L.; Savalli, N. Modeling and design of novel photo-thermo-mechanical microactuators. *Sens. Actuators A* **2002**, *101*, 185–193. [CrossRef]

Article

Development of a Poppet-Type Pneumatic Servo Valve

Takahiro Kanno [1,†], Takashi Hasegawa [1,†], Tetsuro Miyazaki [1], Nobuyuki Yamamoto [2], Daisuke Haraguchi [2] and Kenji Kawashima [1,*]

1   Tokyo Medical and Dental University, 2-3-10, Kanda Surugadai, Chiyoda-ku, Tokyo 101-0062, Japan; kanno.bmc@tmd.ac.jp (T.K.); drivevolleyer@gmail.com (T.H.); tmiyazaki.bmc@tmd.ac.jp (T.M.)
2   RiverField Inc., Yotsuya Medical Bldg. 5th floor, 20 Samon-cho, Shinjuku-ku, Tokyo 160-0017, Japan; n-yamamoto.rfc@riverfieldinc.com (N.Y.); d-haraguchi.rfc@riverfieldinc.com (D.H.)
*   Correspondence: kkawa.bmc@tmd.ac.jp; Tel.: +81-3-5280-8163
†   These authors contributed equally to this work.

Received: 10 October 2018; Accepted: 30 October 2018; Published: 31 October 2018

Abstract: In pneumatic positioning and force-control systems, spool-type servo valves are widely used for obtaining quick responses and precise control. However, air leakage from these valves results in increased energy consumption. To address this problem, we developed a three-port poppet-type servo valve to reduce air leakage. The developed valve consists of a camshaft, two orifices, two metal balls, and a housing with two flow channels. The metal ball is pushed by fluid, and spring force closes the orifice. The port opens when the cam rotates and pushes the ball. The cam shape and orifice size were designed to provide the desired flow rate. The specifications of the DC motor for rotating the camshaft were determined considering the fluid force on the ball. Static and dynamic characteristics of the valve were measured. We experimentally confirmed that air leakage was 0.1 L/min or less. The ratio of air leakage to maximum flow rate was only 0.37%. Dynamic characteristic measurements showed that the valve had a bandwidth of 30 Hz. The effectiveness of the valve was demonstrated through experiments involving pressure and position control.

Keywords: servo valve; pneumatics; position control

---

1. Introduction

Recently, pneumatic servo systems have been widely adopted for use in robotic systems and vibration-isolation devices. The use of air in such applications has several advantages, including the nonmagnetism of air, the achievement of high power-to-weight ratios, and reduced heat generation. Pneumatic servo valves are a key element in pneumatic servo systems. The characteristics of these valves are a critical factor in achieving satisfactory performance [1–5]. There are mainly two types of pneumatic servo valve: the nozzle-flapper type, and the spool type. Nozzle-flapper type servo valves can directly control pressure. The pressure-control systems associated with the valves are generally approximated to a first-order lag system assuming an isothermal state change for the air in the load chamber [6,7]. These valves have both high responsiveness and high linearity. However, they require high exhaust flow rates in order to provide the precise control of pressure.

Spool-type servo valves can directly control flow rate. Because their exhaust flow rates are smaller than those of nozzle-flapper type valves, spool-type servo valves are widely used in pneumatically driven robot systems [8–10]. However, spool-type valves still experience air leakage owing to the presence of a gap between the spool and the sleeve. Usually, leakage flow is approximately 3% to 5% of the maximum flow rate. With the commercialization of servo valves, it is highly desirable to decrease exhaust flow rates in order to reduce energy consumption and operating costs.

To overcome this problem, poppet-type [11] and pinch-type valves [12–14] have been developed. Most of these valves only have two ports. The valve shown in Reference [14] has three ports, but the overlap condition should be improved. Three-port poppet-type servo valves had not yet been commercialized at the time of our investigation. A pneumatic cylinder can be controlled with a five-port servo valve, but two valves are required with a three-port valve. The use of three-port valves increases the cost. However, the use of two three-port valves becomes suitable for compliance control because each cylinder chamber can be individually controlled. A study [15] showed that the use of only one five-port servo valve may lead to very steep variation of the fluidic stiffness around zero velocity; independent pneumatic force control in each chamber, using two servo valves, surpasses this drawback. A zero-lap condition is desirable for precise control with three-port valves. Moreover, the dynamic response of servo valves is important for precise position control in a pneumatic servo system [16].

The purpose of this study was to develop a three-port poppet-type servo valve with high dynamic response and nearly zero lap. The target leak flow was 1% or less than the maximum flow rate. Further, to allow the valves to be used in soft robots controlled at low flow rates, maximum flow rate was designed to be less than 30 L/min. The characteristics of the valve were measured, and the effectiveness of the valve is demonstrated through pressure and position-control experiments.

The remainder of the paper is organized as follows: Section 2 describes the design of the proposed valve. The static and dynamic characteristics of the valve are discussed in Section 3, and Section 4 reports the results of the position-control experiments. Finally, the conclusions of this work are presented in Section 5.

2. Developed Valve

The proposed valve is developed to enable the control of pneumatically driven robotic forceps [10]. The required maximum flow rate is 25 L/min, as per the maximum speed of the forceps. To achieve good controllability, valve dynamics must be approximately the same as those of commercial valves. In particular, in surgical robots that must manipulate forceps for several hours during a surgical procedure, the valve leakage rate must be nearly zero in order to achieve energy savings.

2.1. Valve Structure

Figures 1 and 2 respectively show the schematic and image of the developed valve. The valve mainly consists of a camshaft, two orifices, two metal balls, and a housing with two flow channels. The valve has four ports: a supply port, an exhaust port, and two control ports. Two control ports connect to one flow path downstream of the housing. Thus, the proposed valve is functionally equivalent to general three-port valves, though it apparently has four ports. The two orifices are arranged in each flow channel. The balls are placed upstream of the orifices. When the input voltage to the motor to rotate the camshaft is zero, air flow is obstructed as the two balls are pushed into the orifices by upstream pressure and spring force. A spring with a spring constant of 0.02 N/mm was used. Two springs are placed between the ball and the washer as shown in Figure 1. The port is opened when the camshaft is rotated, and the cam at the downstream side of the orifice pushes the ball. The cost of the camshaft is about US$650, owing to the complex shape. The costs of two orifices are about US$300. The body of the valve is printed by a 3D printer. Other elements are relatively inexpensive. The cost of creating the valve prototype is about US$950. However, when the valve is mass-produced, its cost may decrease by half.

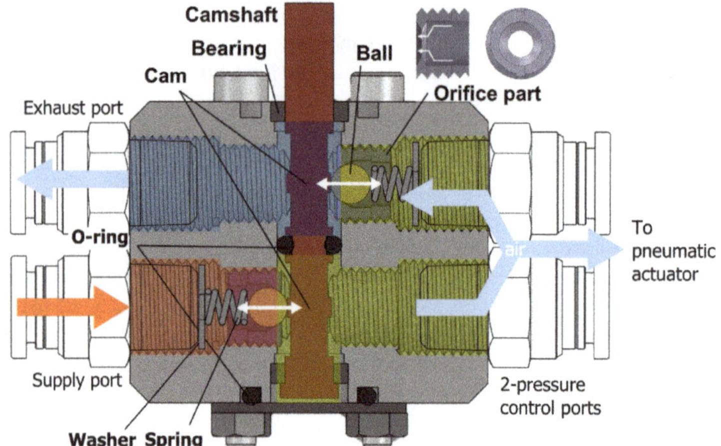

Figure 1. Schematic of poppet-type pneumatic servo valve.

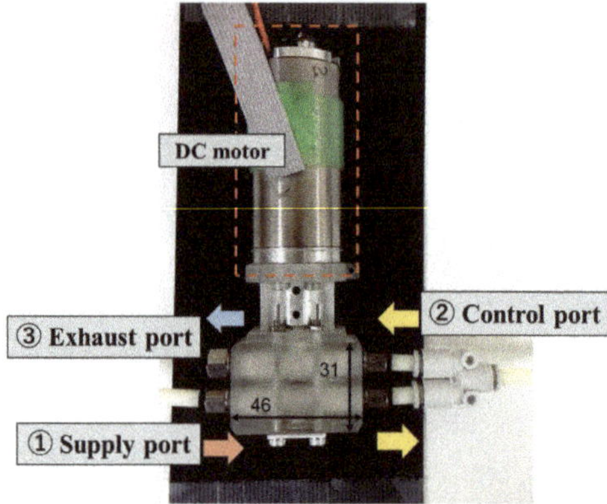

Figure 2. Photograph of the developed valve.

Figure 3 shows the arrangement of the camshaft, cams, and balls. The camshaft and cams are one element made of stainless steel. When the angle of the camshaft corresponds to the neutral position, the two balls come in contact with the orifices, thereby shutting off the flow. Thus, the initial condition of the valve is closed. Flow direction is determined by the rotational direction of the camshaft. Cams are placed on the camshaft at an initial offset angle of 30 degrees in order to prevent overlapping.

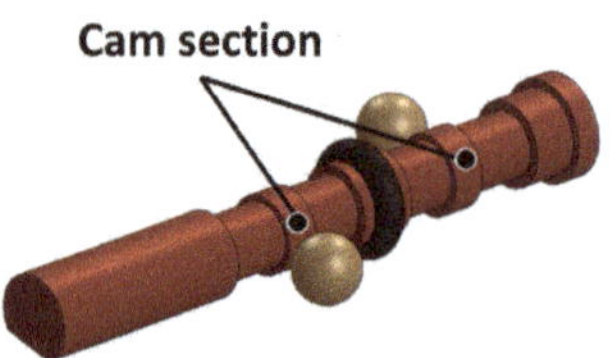

Figure 3. Arrangement of cams, camshaft, and balls.

Flow rate is controlled by the rotation angle of the camshaft. When the camshaft rotates counterclockwise, the ball at the control port is pushed to allow the exhaust port to open and discharge air. At this time, the supply port remains closed. When the camshaft rotates clockwise, the air flows toward the control port while the exhaust port is closed. The radius of the cam is designed to linearly change the effective area of the valve with respect to the cam angle. The position of the orifice can be adjusted with screws. Therefore, the lap condition of the valve can be changed by adjusting the position of the orifice.

2.2. Orifice Design

Flow rate through the orifice is represented in the following two formulas [17]. In the case of choked flow:

$$G = K_f S_e P_s \sqrt{\frac{T_0}{T_{air}}} \tag{1}$$

In the case of subsonic flow:

$$G = K_f S_e P_s \sqrt{\frac{T_0}{T_{air}}} \sqrt{1 - \left(\frac{\frac{P_c}{P_s} - b}{1 - b}\right)^2} \tag{2}$$

In Equations (1) and (2), G is the flow rate, K_f is the proportional constant, S_e is the effective area, T_0 is the temperature under normal conditions, T_{air} is the temperature of the air, P_s is the supply pressure, P_c is the downstream pressure, and b is the critical ratio.

The maximum flow rate at the choked condition was determined to be 26 L/min at a supply pressure of 500 kPa. Then, maximum effective area S_e was approximately 0.45 mm^2 as per Equation (1). Here, K_f = 11.1, T_0 = 273 K, and T_{air} = 293 K. Figure 4 shows the sectional view of the orifice. The upper and lower figures, respectively, show the cross section of the orifice when the valve is shut and opened by the cam. The sectional area of valve S can be obtained geometrically according to Figure 4.

$$S = \pi\{(r + b_{max})^2 - r^2\}\sin\varphi \tag{3}$$

Here, r, b_{max}, and φ, respectively, denote the radius of the ball, the gap between the orifice and the ball, and the slope angle of the orifice, as shown in Figure 4. Gap b_{max} can be obtained by the displacement of cam y_{st} as:

$$b_{max} = y_{st} \cos \varphi. \tag{4}$$

Then, the sectional area can be calculated from Equations (3) and (4) as follows:

$$S = (2ry_{st} \cos \varphi + y_{st}^2 \cos \varphi^2)\pi \sin \varphi \tag{5}$$

The ball surface must protrude from the orifice to be pushed by the cam. The displacement of the ball tip to orifice surface h_{out} can be described as:

$$h_{out} = r - r\cos \varphi - (r\sin \varphi - \frac{d_0}{2}) \tan \varphi. \tag{6}$$

Then, the diameter of orifice d_0 can be written as follows:

$$d_0 = 2(r\sin \varphi - \frac{r - r\cos \varphi - h_{out}}{\tan \varphi}) \tag{7}$$

The effective area of valve S_e can be obtained by multiplying Equation (5) by the contraction coefficient. The contraction coefficient of the orifices is approximately 0.85 [18]. Therefore, sectional area S becomes 0.53 mm^2 to satisfy the maximum flow rate of 25 L/min. First, radius r, angle φ, and the diameter of orifice d_0 were determined; they are listed in Table 1. Then, b_{max} was calculated from Equation (3), and y_{st} was obtained as 0.25 mm from Equation (4). Finally, the cam was designed to move the ball by 0.25 mm.

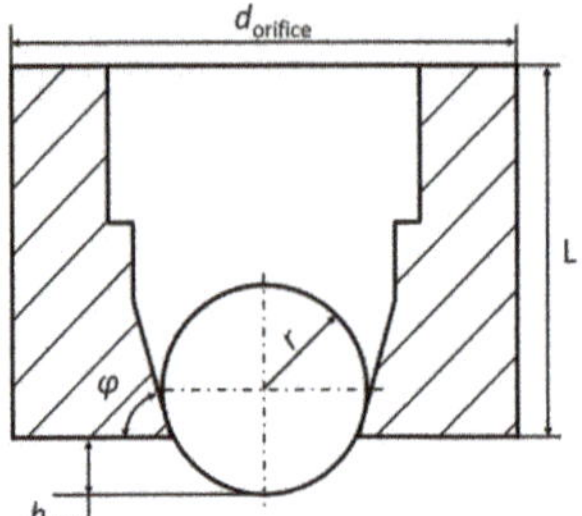 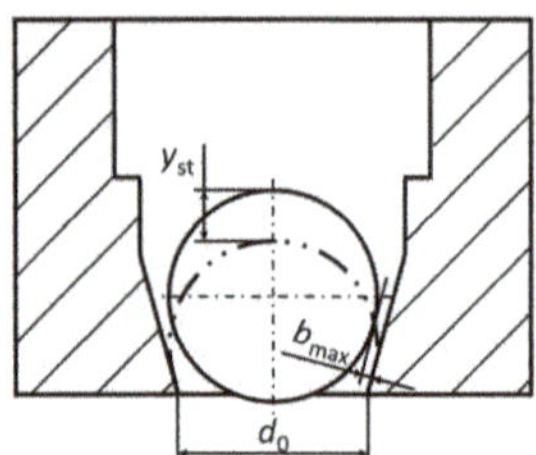

Figure 4. Sectional view of orifice part.

Table 1. Design variables of orifice parts.

Parameters	Value
Radius r	2 mm
Orifice angle φ	75 deg
Orifice diameter d_0	3.6 mm

2.3. Selection of Orifice Material

Because air viscosity is rather small compared to that of liquids, air leakage can easily occur between orifice and ball. Therefore, the selection of orifice material was an important consideration. We evaluated three materials for the orifice: aluminum, polypropylene, and polyacetal. The prototypes for the aluminum and polyacetal orifices were cut using a surface roughness of Ra 1.6. The polypropylene orifice was fabricated using a 3D printer. The air leakages at a supply pressure of 500 kPa were then measured using a flow meter.

The experimental results are shown in Table 2. Air leakage decreased when polyacetal was used owing to the elasticity of the material. Moreover, polyacetal is suitable for facilitating shutting off air flow. Therefore, polyacetal was selected as the orifice material.

Table 2. Air leakage with different orifice materials.

Materials	Value
Aluminum r	18.8 L/min
Polypropylene φ	1.0 L/min
Polyacetal. d_0	0.1 L/min

2.4. Cam Design

Figure 5 shows the designed cam shape. The left figure shows the cross-section of the cam, and the right figure shows the relation between the radius and angle of the cam. The difference between the maximum and minimum radii was set to 0.25 mm, as per the design described in Section 2.2. The cam was designed to linearly push the ball, as shown in the right figure of Figure 5. The controllable range of the cam was ±60 deg, as the cam was placed on the camshaft at an offset angle of 30 deg from the vertical axis, as shown in the middle figure of Figure 6, in order to prevent overlapping.

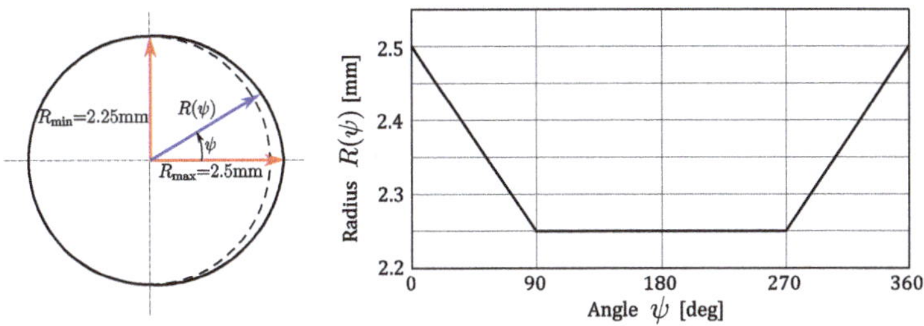

Figure 5. Designed cam shape.

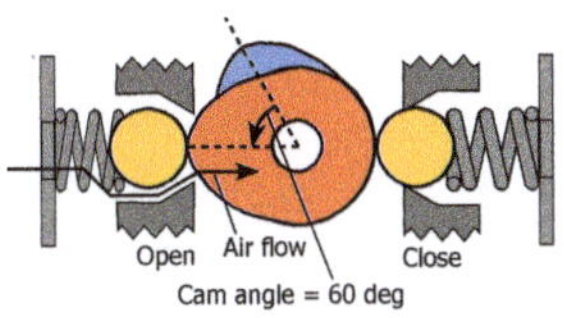
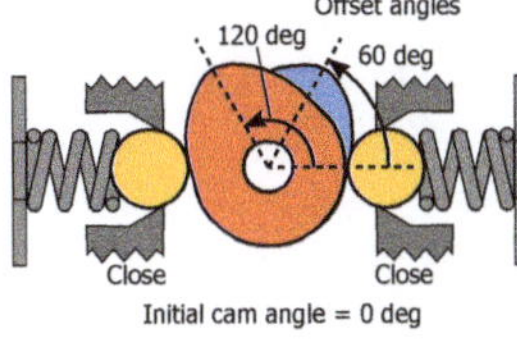
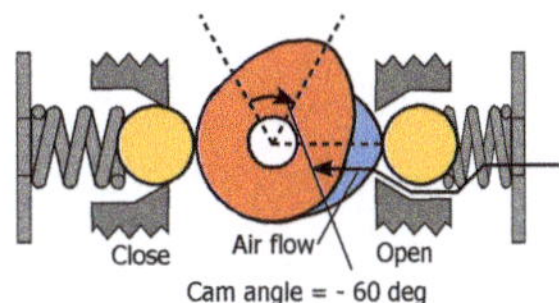

Figure 6. Cam arrangement.

2.5. Motor Selection

The camshaft must smoothly rotate to control the flow rate of the valve. The pushing and friction forces of the two cams and the O-ring act on the shaft. The torque required to push the cam, τ_c, is calculated by the principle of virtual work as follows:

$$\tau_c d\theta = (F_b + F_d)dx \tag{8}$$

Here, F_b is the pushing force of the ball, which is given as:

$$F_b = P\pi r^2. \tag{9}$$

F_b becomes 7.4 N at P = 500 kPa. F_d is the drag force on the ball, which is given as:

$$F_d = \frac{1}{2}\rho v^2 C_d \pi r^2. \tag{10}$$

Here, ρ is air density, v is air velocity, and C_d is the drag coefficient. Under the choked condition, velocity increases to the speed of sound. The Reynolds number is about 4600. Therefore, C_d is given as 0.47. Then, F_d becomes 0.6 N at p = 500 kPa. dx is the distance that the ball moves, and $d\theta$ is the rotation angle of the shaft. As a result, the torque τ_c required to push the cam for one ball is calculated as follows, as per the geometry from Figure 5, at P = 500 kPa.

$$\tau_c = \frac{(F_b + F_d)dx}{d\theta} = 0.02 \text{Nm} \tag{11}$$

The torque to rotate O-ring τ_o is given as follows:

$$\tau_o = \mu F_o r_o \tag{12}$$

Here, μ, F_o, and r_o are the friction coefficient, frictional force, and radius of the O-ring, respectively. The inner diameter and thickness of the O-ring were selected as 3.5 mm and 1.9 mm, respectively, as per the size of the camshaft. The stress of the O-ring (0.25 N/mm) was obtained from the O-ring catalog. Force F_o was obtained as 4.2 N with a compression of 0.1 mm. Then, τ_o became approximately 0.03 Nm with $\mu = 1.0$. A torque of at least 0.05 Nm was required to rotate the camshaft. Valve dynamics depend on the performance of the motor. We selected a DC motor (Maxon DCX 35L) to control the valve; this highly responsive motor, which can generate a torque of 0.12 Nm, has a mechanical time constant of 3.97 ms. The cost of the motor is about US$350.

3. Valve Characteristics

3.1. Static Characteristics

We measured the static characteristics of the developed valve. The camshaft was connected to the DC motor, and the angle of the camshaft was controlled using a microcomputer (Arduino). We measured the flow rates by changing the rotational angles of the cam. The supply and downstream pressures were set to 500 kPa and atmospheric-pressure values, respectively.

The experimental results are shown in Figure 7. The hysteresis of the flow rate to the cam angle was small. Flow rate changed linearly with respect to the cam angle. The maximum flow rate was 27 L/min, which was almost the same as the designed value. We confirmed that the value could achieve an almost zero-lap condition. Air leakage was 0.1 L/min or less, and the ratio of air leakage to the maximum flow rate was 0.37%. These values were fairly low compared to those of a spool-type servo valve (FESTO MPYE-5-M5-010-B), which exhibited air leakage of 1.9 L/min and a ratio of approximately 3.5%.

Then, we measured the flow characteristics of the valve using the experimental setup in accordance with the procedure recommended by ISO 6358. An experimental result with a cam angle of 40 deg is shown in Figure 8. As shown in the figure, the critical pressure ratio was 0.44 at a supply pressure of 500 kPa. The critical pressure ratios of commercially available valves are approximately 0.3 to 0.45 [17,19]. The path is geometrically symmetrical because the flow path is between the orifice and the ball. Therefore, the developed valve has a wide choke range compared with conventional valves.

Furthermore, we measured the pressure gain characteristics at zero flow. The pressure sensor was directly connected to the control port of the servo valve. The relation between the rotation angle of the cam and the pressure in the control port was measured. Figure 9 shows the results at a supply pressure of 500 kPa. We confirmed that the pressure gain was large because the valve leakage rate is small.

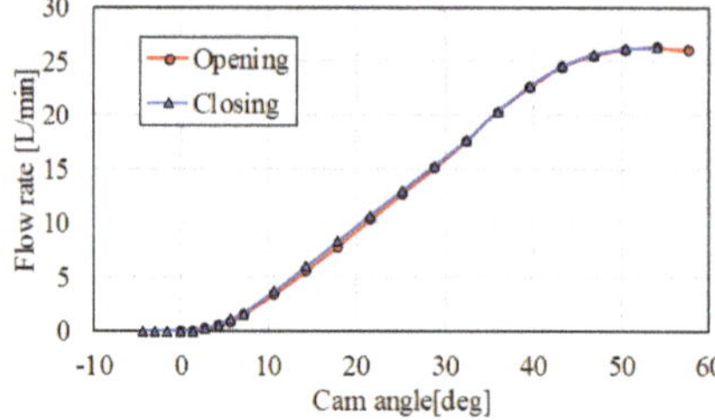

Figure 7. Static characteristics of developed valve.

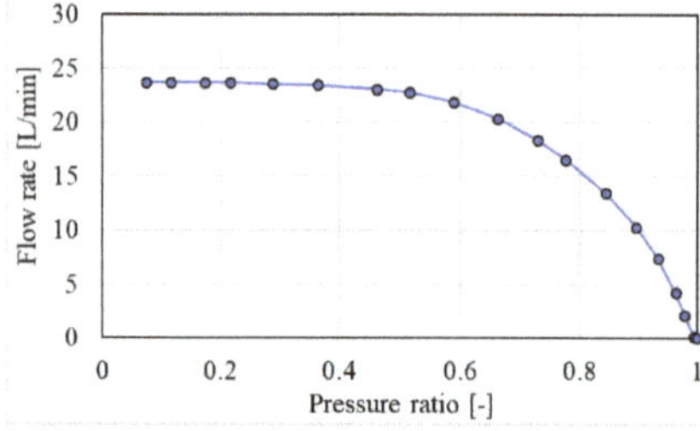

Figure 8. Flow characteristics of developed valve.

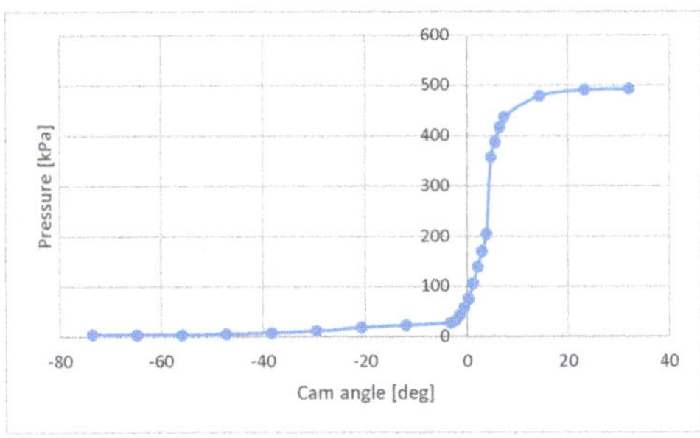

Figure 9. Pressure gain characteristics of developed valve.

3.2. Dynamic Characteristics

We executed a pressure-control experiment connecting a pressure sensor directly to the control port. During the experiment, the dead volume of 1.4×10^3 mm^3 remained even when directly connecting the sensor to the valve.

A PI feedback controller was used for pressure control. A PC was used as the controller; the measured pressure was entered into the PC by an AD converter. The control signal was provided to the DC motor through a DA converter. A block diagram of the pressure control is shown in Figure 10. The control parameters were determined by trial and error, and are listed in Table 3. The parameters were determined to be as large as possible to minimize offset during stable conditions. In this experiment, sinusoidal reference pressures with an amplitude of 20 kPa and an offset of 200 kPa were entered into the controller. The frequency of the reference input ranged between 0.1 and 30 Hz.

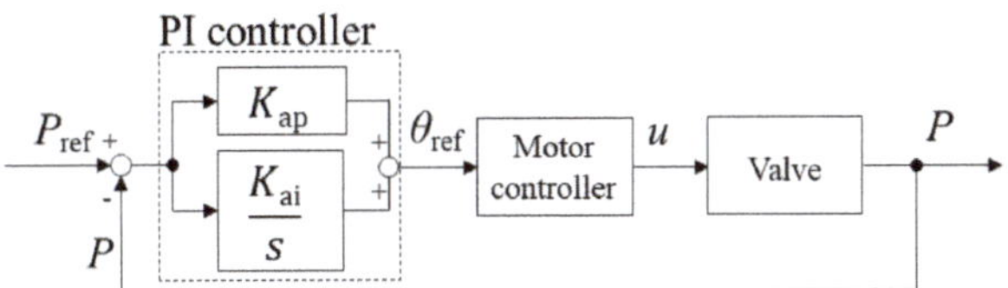

Figure 10. Block diagram of pressure control.

Table 3. Control gains of pressure control.

K_{ap}	4.0 V/kPa
K_{ai}	0.25 V/kPa s

The experimental results at frequencies of 0.1 and 5.0 Hz are shown in Figure 11. The upper figures show the pressure, and the lower figures show the cam angle. It is clear that the valve can control the pressure at a frequency of 0.1 Hz. A small phase delay is observed at a frequency of 5.0 Hz. The results were summarized in a Bode diagram, as shown in Figure 12. We confirmed that bandwidth was higher than 30 Hz when using the proposed valve. The phase lag became -180 deg at 40 Hz. This is considered to be a result of the motor dynamics. To enable comparison, we performed the same experiments using a commercially available spool-type servo valve (FESTO MPYE-5-M5-010-B), in which the maximum flow rate is approximately the same as that of the developed valve. The results for the spool-type valve are plotted in Figure 12. Up to 30 Hz, it is clear that the developed valve demonstrated almost the same performance as the spool-type servo valve. The dynamic response could be improved by changing the DC motor to one having a higher response.

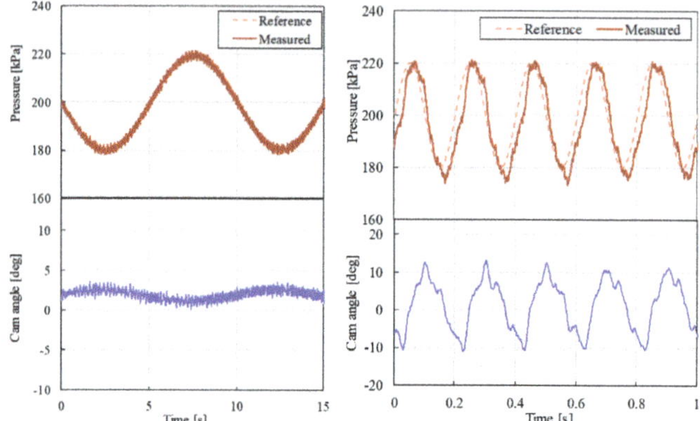

Figure 11. Results of pressure control (left: 0.1 Hz, right: 5.0 Hz).

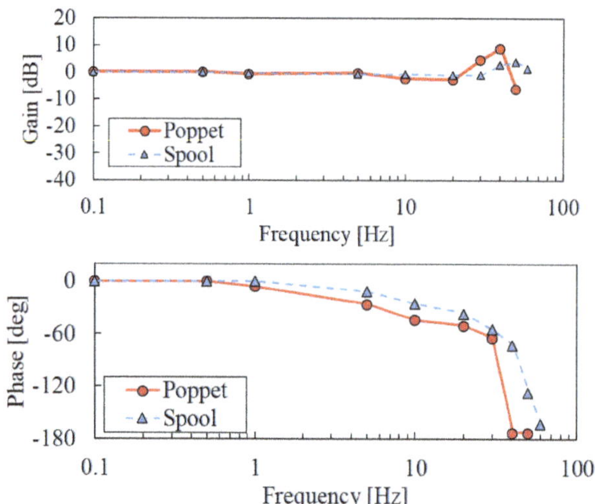

Figure 12. Frequency response of pressure control.

4. Position Control of Pneumatic Cylinder Using the Developed Valve

We executed a position-control experiment using a low-friction-type pneumatic cylinder. The experimental setup is shown in Figure 13. The DC motor was used to control the cam. The pneumatic cylinder used in the experiment had an inner diameter of 10 mm and stroke of 30 mm (SMC CJ2XB10-30Z). A load of 69.5 g was mounted on the rod tip of the cylinder. Supply pressure was set to 500 kPa. One chamber of the cylinder was kept at a constant pressure of 200 kPa, while the other chamber was controlled by the servo valve.

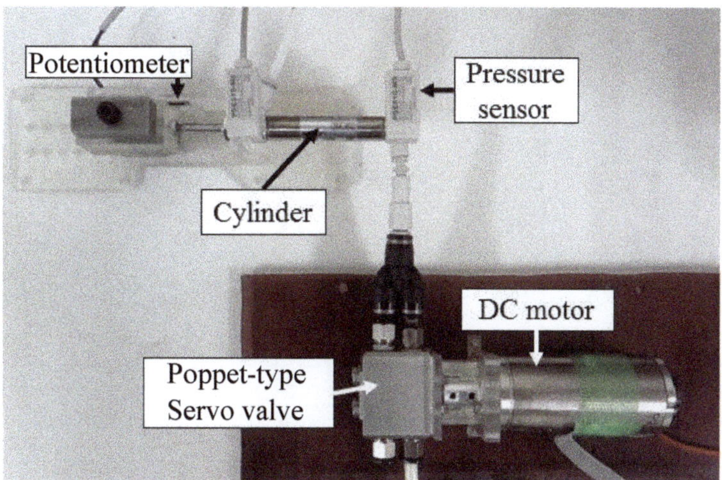

Figure 13. Experimental setup of position control.

A block diagram of the position control is shown in Figure 14. A cascade control was designed. The main loop was a PID position control, and the minor loop was a PI pressure control. The control parameters listed in Table 4 were determined experimentally.

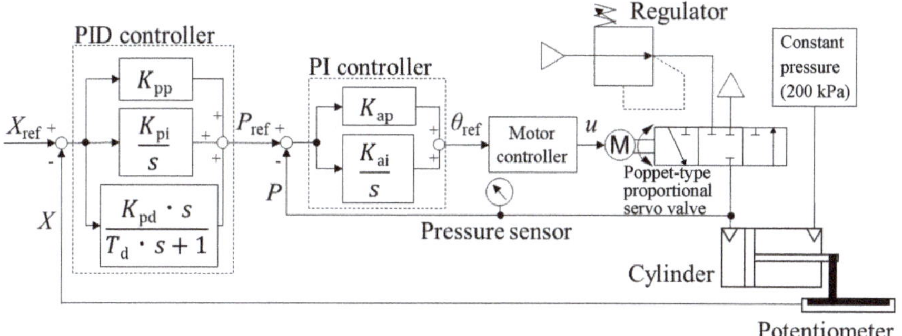

Figure 14. Block diagram of position control.

Table 4. Control gains of position control.

K_{pp}	38 kPa/mm
K_{pi}	1.5 kPa/mm s
K_{pd}	0 kPa s/mm
K_{ap}	0.25 V/kPa
K_{ai}	0.12 V/kPa s

Experimental results for sinusoidal input are shown in Figure 15, corresponding to frequencies of 0.5 and 2 Hz. The upper figure shows the position-tracking performance, the middle figure shows the pressure, and the lower figure shows the movement of the cam. We confirmed that the valve is effective for the position control of pneumatic cylinders.

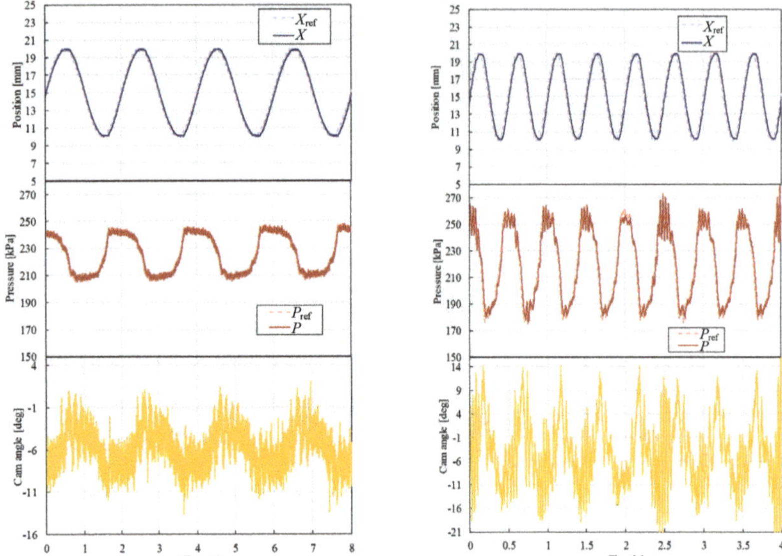

Figure 15. Position control results (left: 0.5 Hz right: 2.0 Hz).

5. Conclusions

We developed a three-port poppet-type servo valve for the purpose of reducing air leakage in pneumatic valves. In the valve, two resin orifices were placed in the flow channels of the valve, and a metal ball pushed to the orifice shut off air flow. The port was opened when the camshaft was rotated, and the cam at the downstream side of the orifice pushed the ball. Cam shape and orifice size were designed to have the desired maximum flow rate. A DC motor, having been selected based on the required torque, was employed to rotate the camshaft. The static and dynamic characteristics of the valve were measured. We confirmed from the flow characteristics that air leakage was 0.1 L/min or less. The ratio of air leakage to maximum flow rate was only 0.37% with the developed valve. Dynamic characteristics measurements showed that valve bandwidth was at least 30 Hz. Finally, the effectiveness of the valve was demonstrated through pressure and position-control experiments.

Future work will involve improving the dynamic response by using a high-response DC motor and by changing the O-ring. In addition, we intend to examine the durability of the valve.

Author Contributions: conceptualization, T.K. and K.K.; methodology, T.K., T.H., T.M. and K.K.; software, T.K., T.H., and T.M.; validation, T.H., T.M., and N.Y.; formal analysis, N.Y. and D.H.; investigation, T.M. and K.K.; resources, D.H. and K.K.; data curation, T.K.; writing—original draft preparation, T.K. and K.K.; writing—review and editing, T.K. and K.K.; visualization, T.K. and T.H.; supervision, K.K.; project administration, D.H. and K.K.; funding acquisition, K.K.

Funding: This research was funded by RiverField Inc.

References

1. Krivts, I.L. Optimization of performance characteristics of electro pneumatic two-stage servo valve. *J. Dyn. Syst. Meas. Control* **2004**, *126*, 416–420. [CrossRef]
2. Belforte, G.; Mauro, S.; Mattiazo, G. A method for increasing the dynamics performance of pneumatic servo systems with digital valves. *Mechatronics* **2004**, *14*, 1105–1120. [CrossRef]
3. Miyajima, T.; Fujita, T.; Sakaki, K.; Kawashima, K.; Kagawa, T. Development of a digital control system for high-performance pneumatic servo valve. *Precis. Eng.* **2007**, *31*, 156–161. 2006.05.003. [CrossRef]

4. Li, J.; Choi, J.; Kawashima, K.; Fujita, T.; Kagawa, T. Integrated control design of pneumatic servo table considering the dynamics of pipelines and servo valve. *Int. J. Autom. Technol.* **2011**, *5*, 485–492. [CrossRef]

5. Kawashima, K.; Youn, C.; Kagawa, T. Development of a Nozzle-Flapper Type Servo Valve Using a Slit Structure. *J. Fluid Eng.* **2007**, *129*, 573–578. [CrossRef]

6. Kagawa, T. Heat Transfer Effects on the Frequency Response of a Pneumatic Nozzle Flapper. *J. Dyn. Syst. Meas. Control* **1985**, *107*, 332–336. [CrossRef]

7. Kamali, M.; Jazayeri, S.A.; Najafi, F.; Kawashima, K.; Kagawa, T. Integrated nozzle-flapper valve with piezoelectric actuator and isothermal chamber: A feedback linearization multi control device. *J. Mech. Sci. Technol.* **2016**, *30*, 2293–2301. [CrossRef]

8. Liu, S.; Bobrow, J.E. An analysis of a pneumatic servo system and its application to a computer-controlled robot. *J. Dyn. Syst. Meas. Control* **1988**, *110*, 228–235. [CrossRef]

9. Pu, J.; Weston, R.H. A new generation of pneumatic servo for industrial robot. *Robotica* **1989**, *7*, 17–24. [CrossRef]

10. Haraguchi, D.; Kanno, T.; Tadano, K.; Kawashima, K. A Pneumatically-Driven Surgical Manipulator with a Flexible Distal Joint Capable of Force Sensing. *IEEE/ASME Trans. Mechatron.* **2015**, *20*, 2950–2961. [CrossRef]

11. Uehara, S.; Hirai, S. Unconstrained Vibrational Pneumatic Valves for Miniaturized Proportional Control Devices. In Proceedings of the 9th International Conference Mechatronics Technology, Kuala Lumpur, Malaysia, 5–8 December 2005.

12. Akagi, T.; Dohta, S.; Katayama, S. Development of Small-Sized Flexible Pneumatic Valve Using Vibration Motor and Its Application for Wearable Actuator. In Proceedings of the 15th International Conference Mechatronics Machine vision Practice, Auckland, New Zealand, 2–4 December 2008; pp. 441–446.

13. Nasir, A.; Akagi, T.; Dohta, S.; Ono, A. Analysis of Low-cost Wearable Servo Valve Using Buckled Tubes for Optimal Arrangement of Tubes. In Proceedings of the IEEE International Conference on Advaced Intelligent Mechatronics, Busan, Korea, 7–11 July 2015; pp. 830–835.

14. Morisaki, D.; Kanno, T.; Kawashima, K. Development of a Pinch-type Servo Valve Embedded in a Pneumatic Artificial Rubber Muscle. In Proceedings of the IEEE/SICE International Symposium on System Integration, Taipei, Taiwan, 11–14 December 2017.

15. Carneiro, J.F.; de Almeida, F.G. Using two servovalves to improve pneumatic force control in industrial cylinders. *Int. J. Adv. Manuf. Technol.* **2013**, *66*, 283–301. [CrossRef]

16. Kawashima, K.; Arai, T.; Tadano, K.; Fujita, T.; Kagawa, T. Development of Coarse/Fine Dual Stages using Pneumatically Driven Bellows Actuator and Cylinder with Air Bearings. *Precis. Eng.* **2010**, *34*, 526–533. [CrossRef]

17. Kawashima, K.; Ishii, Y.; Funaki, T.; Kagawa, T. Determination of Flow Rate Characteristics of Pneumatic Solenoid Valves Using an Isothermal Chamber. *J. Fluid Eng.* **2004**, *126*, 273–279. [CrossRef]

18. Andersen, B.W. *The Analysis and Design of Pneumatic Systems*; Krieger Publishing Company: Malabar, FL, USA, 2010.

19. Takosoglu, J.; Laski, P.; Blasiak, S.; Bracha, G.; Pietrala, D.; Zwierzchowski, J.; Nowakowski, L. Determination of flow-rate characteristics and parameters of piezo pilot valves. *EPJ Web Conf.* **2017**, *143*, 02126. [CrossRef]

Article

A Dual Stage Low Power Converter Driving for Piezoelectric Actuator Applied in Micro Mobile Robot

Chen Chen, Meng Liu and Yanzhang Wang *

College of Instrumentation & Electrical Engineering, Key Laboratory of Geo-exploration Instruments, Ministry of Education of China, Jilin University, Chang Chun 130026, China; cchen@jlu.edu.cn (C.C.); liumeng15@mails.jlu.edu.cn (M.L.)
* Correspondence: yanzhang@jlu.edu.cn; Tel.: +86-135-0442-0027

Received: 11 August 2018; Accepted: 11 September 2018; Published: 15 September 2018

Abstract: Piezoelectric actuators are widely utilized to convert electrical energy into mechanical strain with considerable potential in micro mobile robot applications. However, the use of Pb-based Lanthanumdoped Zirconate Titanates (PZTs) leads to two difficulties in drive circuit design, namely, high voltage step-up ratio and high energy conversion efficiency. When some devices driven by piezoelectric actuators are used in emerging technologies, such as micro mobile robot, to perform special tasks, low mass, high energy density, and high conversion efficiency are strategically important. When these demands are considered, conventional drive circuits exhibit the disadvantages of being too bulky and inefficient for low mass applications. To overcome the aforementioned drawbacks, and to address the need for a piezoelectric bimorph actuator, this work proposed a high step-up ratio flyback converter cascaded with a bidirectional half-bridge stage controlled, via a pulse width modulation strategy, and a novel control method. Simulations and experiments were conducted to verify the ability of the proposed converter to drive a 100 V-input piezoelectric bimorph actuator using a prototype 108 mg (excluding printed circuit board mass), 169 (13 × 13) mm^2, and 500 mW converter.

Keywords: piezoelectric actuator; high step-up ratio; high efficiency; small size; micro mobile robot

---

1. Introduction

Microrobots represent a new type of bionic robots inspired by insects to achieve ultracompact size, high mobility, and remote controllability; these robots have many applications, including those in bioengineering, disaster relief, microsurgery, and surveillance [1–3]. Among microrobots, micro mobile robots (MMRs) are widely used in environmental monitoring, rescue operations, and agricultural production, due to their high energy efficiency and high mobility. If the energy conversion efficiency of the frame, including the actuator, power circuit design, and mechanical transmission for stroke amplification can be improved, then the application prospects of such robots will be extended.

As a displacement device, actuators are widely used in many types of micromechanical devices like walking, swimming, jumping robots and so on [4–6]. Among them, especially, the actuator plays a critical role in obtaining sufficient lift force. Prior works on actuators in MMR applications include studies on an electromagnetic actuator, a shape-memory alloy actuator, an electrostatic actuator, and a dielectric elastomer actuator [7–10]. However, the disadvantages of low response rate, small mechanical displacement, and low conversion efficiency limit their applications to MMRs.

The breakthrough in piezoelectric bimorph actuators with high bandwidth and high power density achieved by Wood et al. in a multi-segmented centipede robot. Meanwhile, Wood et al. considered a complete fabrication solution for actuators, links, flexures, integrated wiring, and structural elements using high-performance materials; they then developed a highly efficient transmission link for a specific device [11]. However, a power electronic unit connected to a piezoelectric actuator faces two major challenges [12]. Firstly, high voltages within the range of 100–120 V are required to generate sufficient force and displacement. Most energy sources compatible with the weight budget of MMR applications have output voltages of 3–5 V. Consequently, power interface circuits require high step-up ratio. Secondly, a piezoelectric actuator provides only a fraction of the input energy required for effective displacement. In addition, prior work in dynamic performance analysis of piezoelectric actuator include Nabawy et al., which demonstrated comprehensive analytical model of both unimorph and bimorph structure [13,14]. Different drive frequencies and signal amplitudes can affect the electromechanical coupling factor and conversion efficiency of the piezoelectric actuator. Hence, the drive circuit must support drive signals with specific frequency and efficient bidirectional energy flow to fully utilize the unused energy stored in actuators, and consequently, maximize system efficiency.

Several researchers have discussed the miniaturization of voltage conversion circuits for microrobotic applications. A boost converter cascaded with a switched capacitor circuit was adopted by Steltz et al. to obtain high voltage [15]. However, this structure requires a large number of pump capacitors, which increases the size of the circuit and reduces its power density, particularly at a high output power. Chen proposed a piezoelectric transformer-based power amplifier with high voltage gain and power density [16]. Although the circuit topology can boost low voltage, the size of the structure faces difficulty in meeting the requirements of current manufacturing levels. Meanwhile, a piezoelectric transformer should operate close to the mechanical resonance frequency to obtain high voltage gain and efficiency. Other topologies with more components and a high step-up ratio are also suitable for driving high-voltage reactive loads; however, most efforts have focused on large-scale and high-power applications [17–19]. In another part of the circuit, a simple push–pull driver described in Reference [15] can generate a unipolar square wave voltage across the load without energy recovery. The majority of existing topologies focus on efficient piezoelectric driving for large-scale and high-power applications, which reduces the efficiency and power density of the power supply [20–25].

Harvard's team has previously produced a small-scale power conversion interface. However, as described in References [12], the boost tapped-inductors they use have difficulty in manufacturing compared to general flyback transformers. At the same time, in the direct current to alternating current (DC-AC) stage, the detection of current becomes quite difficult in the case of low power. This article is dedicated to improving and validating the short board mentioned above.

The main objectives of this study are as follows: (a) To determine the equivalent model parameters based on the impedance characteristics of piezoelectric bimorph actuators; (b) to introduce driving methods and requirements for piezoelectric actuators in MMRs; (c) to describe a circuit topology and its control strategy with low mass, high conversion ratio, high power density, and high efficiency; (d) to present a prototype that is capable of driving piezoelectric actuators; and (e) to observe and analyze the experimental results of the circuit output and displacement of material driving.

This paper is organized as follows. Section 1 introduces the electric properties and driving methods of piezoelectric actuators. Section 2 presents the overall circuit structure and control scheme. Sections 3 and 4 describe the simulation, fabrication specifications, and experimental realization of the circuit topology. Finally, Section 5 provides the conclusion of the study and directions for future research.

2. Driving Requirements and Strategies of Piezoelectric Actuator

2.1. Equivalent Model of a Piezoelectric Actuator

A bending actuator, called bimorph, with low cost and ultra-high energy density can be fabricated using generic piezo ceramics and high-performance composite materials coupled with the intelligent use of geometry and novel driving techniques. A good understanding of the electrical characteristics of each layer of a piezoelectric bimorph actuator is essential for designing and analyzing a drive circuit with good performance.

As a capacitive material, a piezoelectric element exhibits specific natural frequencies in different vibration modes. Many micro bio-inspired robots apply resonant body dynamics to obtain efficient locomotion. When a drive signal is applied to the resonant frequency of a piezoelectric element, the resulting vibration amplitude is considerably higher than those at other frequencies. For example, a piezoelectric bimorph is driven at the lowest resonant frequency in Harvard's microrobot [12]. To design an efficient drive circuit, a dynamic electrical model of the load (i.e., a piezoelectric bimorph driver) must be established. In the electrical domain, a piezoelectric element can be represented by an equivalent model, in which impedance is linked to the mechanical properties of the actuator and its load. The circuit model of Guan is adopted in this study, as shown in Figure 1 [26].

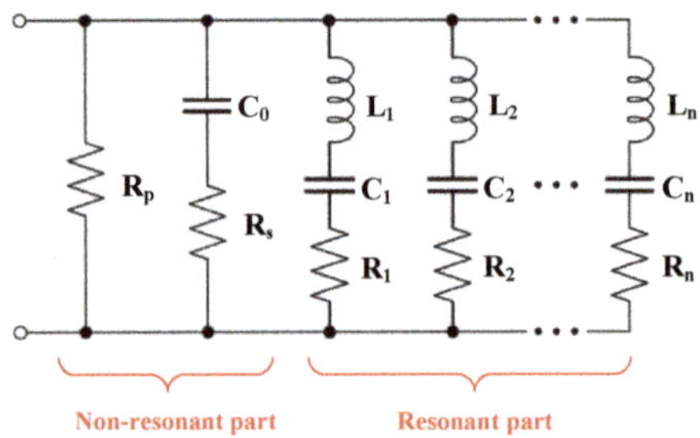

Figure 1. Electrical equivalent model of piezoelectric element.

The circuit model includes a static non-resonant part and a series of dynamic mechanical LCR (resistor R, inductor L, and capacitor C) branches. The static part represents the complex capacitive characteristics and loss mechanisms of piezoelectric actuators, which are made up of a parallel connection of series $C_0 R_s$, and R_p. Different mechanical resonance branches correspond to various mechanical resonance modes, where inductance L_n, capacitance C_n, and resistance R_n are related to the equivalent mass, compliance, and damping coefficient of each resonance mode, respectively. The impedance of the non-resonant part, every resonant shunt circuit, and total impedance are denoted as $Z_c(s)$, $Z_n(s)$, and $Z_t(s)$, respectively, and their corresponding formulas are as follows:

$$Z_c(s) = R_p \times [R_s + 1/(sC_0)] / [R_p + R_s + 1/(sC_0)], \tag{1}$$

$$Z_n(s) = R_n + sL_n + 1/(sC_n), n = 1, 2, 3, \ldots, \tag{2}$$

$$Z_t(s) = 1/[1/Z_c(s) + 1/Z_1(s) + 1/Z_2(s) + \ldots + 1/Z_n(s)], n = 1, 2, 3, \ldots. \tag{3}$$

The piezoelectric bimorph ceramic material (QDTE52-7.0-0.82-4, PANT, Suzhou, Jiangsu, China) used to test dynamic performance has the following parameters: 190 V highest driving voltage, 52 mm long, 7 mm wide, 0.82 mm thick, and 1.2 mm bidirectional displacement. The first resonant frequency of each bimorph layer is approximately 100 Hz close to the driving frequency of an insect-type robot, which makes the research practical.

The impedance characteristics of the material were measured using a 4294A impedance analyzer (Keysight Technologies, Inc., Santa Rosa, CA, USA). We can fit the experimental curve by choosing the values of the electronic components. Figure 2 shows the frequency response of one of the bimorph PZT layers and its equivalent model, in which the theoretical frequency response is calculated with MATLAB using the transfer function of the circuit model.

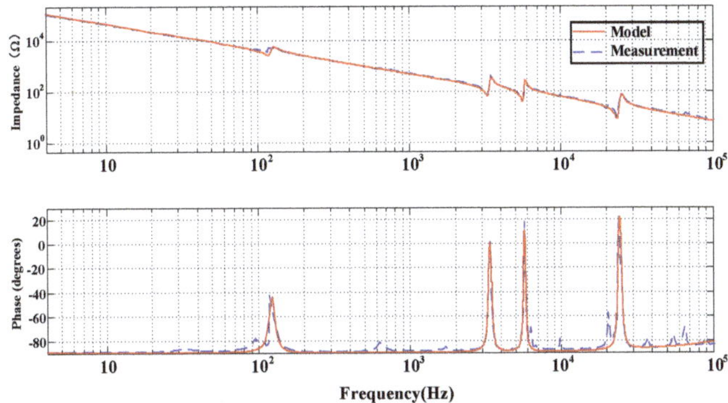

Figure 2. Impedance characteristic of one layer of bimorph.

Four prominent resonance modes are between 30 Hz and 30 kHz. Hence, the corresponding circuit model has four LCR branches. The parameters of the equivalent model are listed in Table 1.

Table 1. Equivalent parameters of the bimorph actuator.

Static Parameters	Value	Dynamic Parameters	Mode 1	Mode 2	Mode 3	Mode 4
C_0	24 nF	Ln	580 H	812 mH	422 mH	15 mH
R_s	1 Ω	Cn	3 nF	2.85 nF	1.9 nF	3 nF
R_p	10 MΩ	Rn	8.5 MΩ	7.5 MΩ	6.7 MΩ	5.2 MΩ

2.2. Drive Requirements

Piezoelectric bimorph actuators can be driven with bipolar or unipolar signals. In bipolar driving, the driving voltage varies between a positive voltage and a negative voltage, which causes the piezoelectric element to expand and contract, due to the converse piezoelectric effect. In unipolar driving, the voltage is only positive, which causes contraction. In general, piezoelectric actuators require considerable strain to provide power, and therefore, a high driving voltage is necessary to achieve large wing strokes. The drive signal must be unipolar, because a high-amplitude bipolar driving signal can cause depolarization of the piezoelectric element.

In the piezoelectric equivalent model, most of the energy is stored in C_0 and can be reused, thereby requiring the drive circuit to also complete the task of energy recovery. This energy part can be transferred to another layer of bimorph actuators or power supply to maximize the efficiency of the system.

2.3. Driving Methods

To avoid depolarization, the piezoelectric layers must be under unipolar driving (AC) signal condition in the polarization direction according to the characteristics of the piezoelectric actuator. Prior work about driving methods of piezoelectric actuator includes efforts by Shannon Rios, which made detailed analysis of the efficiency of several different configurations [27]. An available configuration of piezoelectric actuators is the "bimorph," which consists of two compliant electrodes on the top and bottom surfaces bonded to a compliant elastomer film inserted in the middle. One of the driving methods for the bimorph is called "alternating driving," which requires 2n drive stages per n bimorphs, as shown in Figure 3a. An optimized driving method, called "simultaneous driving," allows the sharing of a high-voltage bias among multiple bimorphs, thereby requiring n drive stages and one bias per n bimorphs, as shown in Figure 3b. A DC bias boost stage is connected to the two compliant electrodes, whereas a suitable unipolar drive stage is applied to the central compliant elastomer film.

The simultaneous driving architecture exhibits the advantage of using fewer components to reduce the size and mass of the converter when multiple bimorphs are required in a system [12].

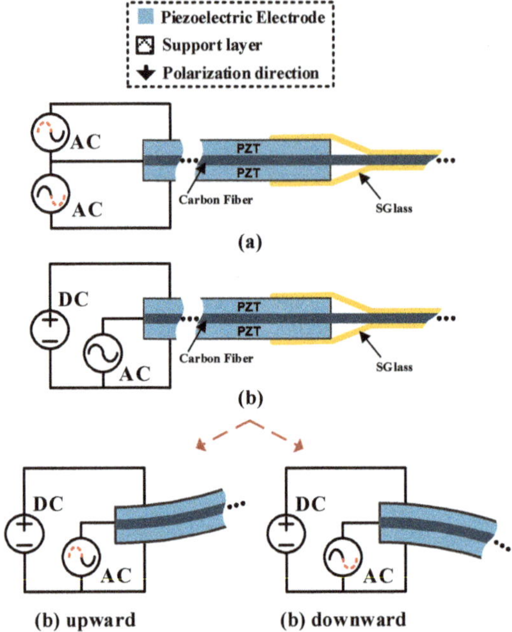

Figure 3. Driving methods for piezoelectric actuators: (a) Bimorph with alternating driving; (b) bimorph with simultaneous driving.

When the AC voltage is operating in the positive half cycle, the upper layer of the bimorph produces less contraction than the bottom layer, which causes the bending actuator to exhibit downward displacement. The effect of material contraction on the other half of the cycle is the opposite, which causes the piezoelectric bimorph to achieve an upward displacement. The simultaneous driving process is illustrated in Figure 3b.

3. Design and Control of the Drive Circuit

A dual-stage circuit is designed to meet the following requirements: The first stage should step up the low battery voltage to a high voltage, whereas the second stage should transform the DC voltage into a time-varying drive signal.

The drive circuit topology of the piezoelectric bimorph proposed in this study is depicted in Figure 4. It includes a flyback converter as the DC/DC stage and a bidirectional active half-bridge converter as the high-voltage DC/AC drive stage [28]. The flyback converter boosts the low voltage of the lithium battery to the DC high voltage required for the driving actuator. The power metal-oxide-semiconductor field-effect transistor (MOSFET) in both stages is controlled, via pulse width modulation (PWM), in which a novel control strategy is utilized in a half-bridge drive stage to output unipolar signals of any shape to satisfy the actuator drive requirements. The selected inductor with a high quality factor in the half-bridge can effectively achieve energy transfer. The half-bridge structure has less footprint area than other DC/AC converters [29], thereby increasing the energy density of the total system.

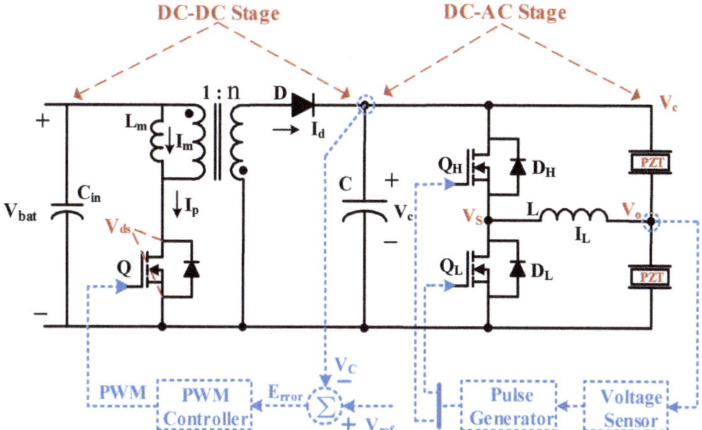

Figure 4. Bidirectional converter topology and control architecture.

3.1. DC/DC Voltage Conversion Stage

The boost stage of the circuit is implemented by the flyback converter, which evolves from the buck-boost circuit. Unlike a forward converter, the flyback converter has no secondary output inductance, thereby considerably improving its cost and volume advantages. The flyback converter is suitable for applications where low power and high boost ratio are required. In addition, the flyback converter minimizes reduction in energy efficiency and difficulty in manufacturability, because of the miniaturization of the circuit topology.

When the power MOSFET (Q) is turned on, the primary current I_p starts to build up under the action of the power supply and energy is stored in the magnetizing inductance. When Q is turned off, the energy stored in the transformer starts to transmit to the secondary side through magnetic coupling and the rectifier diodes (D), thereby outputting a high DC voltage.

The voltage and current waveforms of the flyback converter operating in a steady state are shown in Figure 5a. The converter operates in a discontinuous mode (DCM), in which the flyback transformer current must return to zero before a new switching cycle begins. This operation mode generally results in high efficiency at low output power levels [29]; it also simplifies the design of the control loop while achieving a high step-up ratio. The voltage step-up gain in DCM mode is given by

$$\frac{V_c}{V_{bat}} = D \cdot n \cdot \sqrt{\frac{R}{2L_m f_s}}, \tag{4}$$

where V_c and V_{bat} are the input and output voltages, respectively; D is the controllable on-time duty cycle; n is the transformer turn ratio; R is the equivalent load impedance, which represents the impedance of the half-bridge converter and the capacitive loads; L_m is the magnetizing inductance of the transformer; and f_s denotes switching frequency. As mentioned in [12], the voltage step-up gain of the flyback converter in DCM operation is considerably higher than that in continuous conduction mode operation.

The output voltage is stabilized by a resistive feedback divider and an analog comparator. When the output voltage is higher or lower than the reference voltage, the microcontroller unit adjusts the switching duty ratio accordingly to monitor it in real time, as shown in Figure 5b. In this manner, the flyback converter operates at the desired level over a range of load currents.

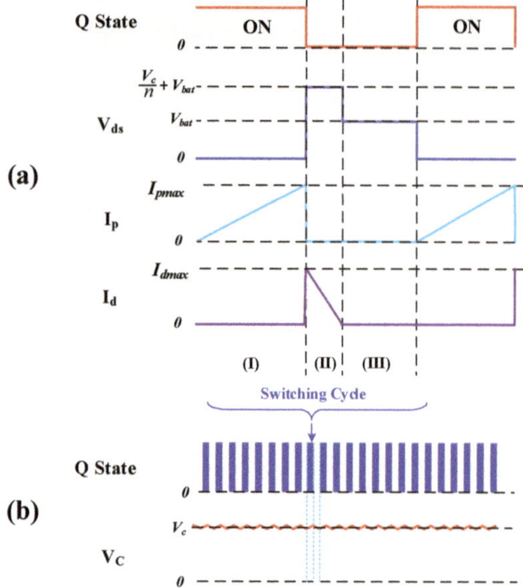

Figure 5. Switching waveforms of the flyback converter during (**a**) one switching period; (**b**) steady-state operation.

3.2. DC/AC High-Voltage Drive Stage

The DC/AC drive stage can convert DC voltage into high-voltage unipolar drive signals. Moreover, the energy stored in a piezoelectric actuator should be recovered to increase conversion efficiency. Many prior works have been conducted on the high-voltage driving of piezoelectric actuators. Steltz et al. proposed a drive stage without energy recovery [15]. Campolo et al. proposed a method for driving a piezoelectric actuator using a low-frequency square wave signal at the cost of increasing the complexity of a circuit [20].

Unipolar signals of any shape can be generated by utilizing a half-bridge inverter that is composed of a high-quality factor inductor L and two inherent capacitors of the electrostatic actuators. Different operating modes of the high-voltage half-bridge drive stage are illustrated in Figure 6.

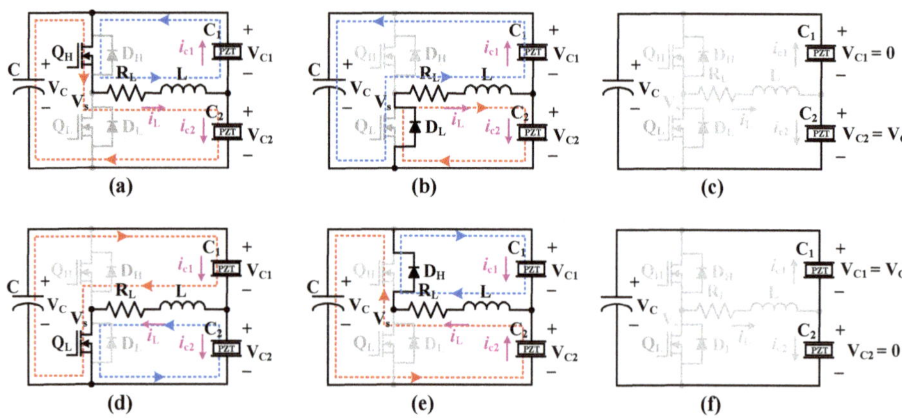

Figure 6. Six switching modes of the high-voltage half-bridge drive stage.

In mode (a), the inductor current i_L begins to rise once the switch Q_H is turned on. The high-side equivalent capacitor C_1 of the piezoelectric actuator starts to transfer the most unused energy to the low side, whereas the remaining energy is recovered in capacitor C. When the equivalent series resistance in a circuit is considered, the governing differential equations can be obtained as follows according to Kirchhoff's voltage law:

$$\begin{cases} L \cdot \frac{di_L}{dt} + R_L \cdot i_L + v_{c2} = V_C \\ C_2 \cdot \frac{dv_{c2}}{dt} = i_{c2} = i_L - i_{c1} = i_L - C_1 \cdot \frac{dv_{c1}}{dt} \end{cases} \cdot \tag{5}$$

When the two equations are combined, a second-order differential equation in this mode can be obtained as

$$\frac{\partial^2 v_{c2}}{\partial t^2} = 2\beta \cdot \frac{\partial v_{c2}}{\partial t} + \omega_0^2 \cdot v_{c2} = \omega_0^2 \cdot V_C, \tag{6}$$

where $\beta = R_L/2L$ is the neper frequency, and $\omega_0 = 1/(2LC_2)^{0.5}$ is the angular resonance frequency. Under an underdamped oscillation ($\beta^2 - \omega_0^2 < 0$),

$$v_{c2}(t) = V_C - V_C \cdot e^{-\beta t}\left(\cos \omega_d t + \frac{\beta}{\omega_d} \cdot \sin \omega_d t\right), \tag{7}$$

where the drive signal voltage v_{c2} is an underdamped voltage with a decaying oscillation at frequency $\omega_d = (\omega_0^2 - \beta^2)^{0.5}$. Mode (a) ends as soon as v_{c2} increases and becomes equal to V_C in the first oscillation cycle.

At the end of mode (a), Q_H is turned off and the inductor current i_L starts to decrease via the freewheeling diode D_L. A new oscillation occurs between inductor L and the capacitive loads. The circuit starts operating alternatively in modes (b) and (e). The initial voltage in mode (b) is V_C. The governing equations can be obtained as

$$\begin{cases} L \cdot \frac{di_L}{dt} + R_L \cdot i_L + v_{c2} = 0 \\ C_2 \cdot \frac{dv_{c2}}{dt} = i_{c2} = i_L - i_{c1} = i_L - C_1 \cdot \frac{dv_{c1}}{dt} \end{cases} \cdot \tag{8}$$

Under an underdamped oscillation ($\beta^2 - \omega_0^2 < 0$), the drive signal voltage becomes

$$v_{c2}(t) = v_{c2}(t_a) \cdot e^{-\beta(t-t_a)} \cdot \left[\cos \omega_d(t - t_a) + \frac{\beta}{\omega_d} \cdot \sin \omega_d(t - t_a)\right], \tag{9}$$

where ($v_{c2}(t_a) = V_C$) represents the initial voltage condition at the end of mode (a) at t_a. The drive signal voltage v_{c2} continues oscillating until it becomes stable and equal to V_C during the rest of the period.

In mode (c), the lower side electrostatic actuator is completely charged, whereas the upper side electrostatic actuator is empty. The drive signal voltage v_{c2} is the same as V_C during this period. All the unused energy is transferred from the upper side actuator to the lower side actuator, which enhances the overall efficiency of the system.

Mode (d) starts as soon as the transistor Q_L is turned on. The unused energy in the electrostatic actuator at the lower side begins to be transferred to the upper side. The inductor current i_L rises reversely, thereby causing the discharge of the low side actuator C_2. The underdamped drive signal v_{c2} descends with attenuated oscillation as follows:

$$v_{c2}(t) = V_C - V_C \cdot e^{-\beta t}\left(\cos \omega_d t + \frac{\beta}{\omega_d} \cdot \sin \omega_d t\right), \tag{10}$$

Mode (d) ends as soon as v_{c2} drops to zero in the first oscillation cycle. At the end of mode (d), Q_L is turned off, and the inductor current i_L increases through the freewheel diode D_H. The circuit operates alternatively in modes (e) and (b). The initial voltage condition in mode (e) is zero. Similar to mode (b), the underdamped drive signal v_{c2} in mode (e) can be expressed as

$$v_{c2}(t) = -v_{c2}(t_d) \cdot e^{-\beta(t-t_d)} \cdot \left[\cos \omega_d(t - t_d) + \frac{\beta}{\omega_d} \cdot \sin \omega_d(t - t_d) \right], \tag{11}$$

where $(v_{c2}(t_d) = 0)$ represents the initial voltage condition at the end of mode (d) at t_d. The high side actuator C_1 is being charged by the stored energy of inductor L. After the oscillation ends, the drive signal v_{c2} becomes zero as soon as the inductor current turns zero until the next operation cycle begins.

In mode (f), the drive signal voltage v_{c1} is charged up to V_C until the unused energy is fully transferred from the lower actuator to the upper one.

The inductor current i_L, the drain-to-source voltage v_{ds}, and the drive voltage v_{c2} waveforms during each switching cycle are illustrated in Figure 7, with an operation sequence of a-b-e-b-c-d-e-b-e-f. Modes (a) and (d) are the modes at the beginning of two half-operation cycles. Modes (b) and (e) are the oscillation modes, which occur alternatively in both half-operation cycles. Modes (c) and (f) are the stable modes at the end of two half-operation cycles. Energy conversion is replaced by energy transfer in a working cycle. With the exception of the loss of the equivalent resistance of inductor i_L, the half-bridge stage is a theoretically lossless structure that ensures high energy efficiency of the entire system.

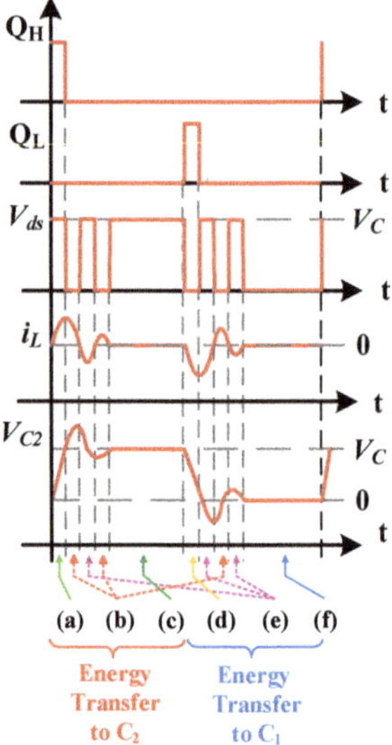

Figure 7. Switching pulse waveforms (Q_H and Q_L), drain-to-source voltage (v_{ds}), inductor current (i_L), and output voltage (v_{c2}) of the drive stage during one switching cycle.

3.3. Control Method of the Drive Stage

A traditional method used to generate a sinusoidal signal from a DC voltage is adopting an LC resonant network. For a sinusoidal PWM control technology, the inductor and two capacitors in a half-bridge inverter are generally required to have a large volume [30]. However, power density is particularly important for a miniaturized system in harsh applications, which is inconsistent with the ideal design for this study.

Janocha et al. considered the limitations of this method and proposed a control method for a reactive load (e.g., a piezoelectric actuator) to reduce the size of the inductor and meet the requirements for compactness [31]. In every switching cycle, the amplifier transfers the exact amount of energy that is necessary to achieve the desired output value at the load. This approach allows unnecessary switching cycles, thereby permitting higher system dynamics than those of converters using traditional (PWM or current mode) controllers.

However, this method is current-controlled. Therefore, the controller is required to calculate the switching time of the MOSFET. In such case, the parameters of the load and the components in the circuit should be determined in advance. If the parameters of the circuit are slightly changed, then the control method will become highly complicated.

A modified version of this method that can generate arbitrary waveform drive signals is embodied in this work through a four-phase control algorithm: Acquisition, Lookup, Charge, and Discharge.

During Acquisition, the output voltage is sampled using a resistive divider, which is regarded as the input of the analog comparator. In the Lookup phase, the sampled result and the desired value of V_o are used to generate a control pulse for switches Q_H and Q_L. The associated control architecture is shown in Figure 8.

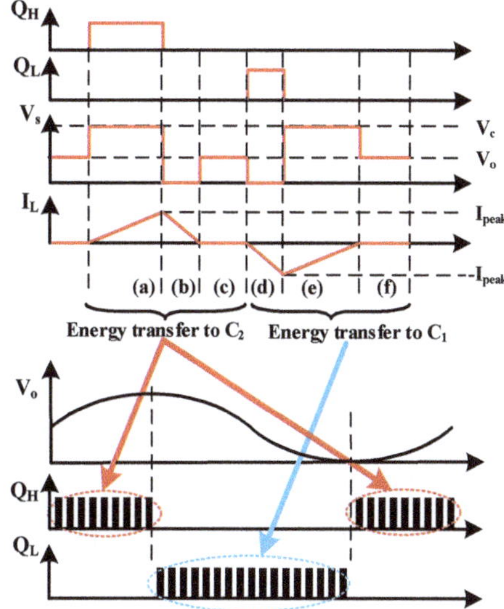

Figure 8. Inductor current, output voltage, and switching pulse waveforms during each half cycle.

When the output voltage is increased, the high-side power MOSFET is constantly turned on or off, which corresponds to the a–b cycle when energy is transferred to C_2 in Figure 8. Conversely, it corresponds to the d–e cycle in Figure 6 when voltage drops. Apart from the amplitude of the drive signal, each working interval is timed by a timer overflow interruption when the frequency of the output signal requires adjustment.

4. Simulation Verification

To verify the feasibility of the circuit, its topology was simulated in the MATLAB Simulink environment. The specifications of the components used in the simulation are provided in Table 2. Among these, magnetizing inductance, output filter capacitor, resonant inductor, and transformer turn ratio are the same as those of the transformer and other circuit parameters used in subsequent experiments.

Table 2. Specifications of the utilized components.

Components	Specifications
V_{bat}	3.7 V
f_s (switching frequency)	55.5 kHz
D (duty cycle)	50%
n (turn ratio)	1:10
L_m	2 μH
C	1 μF
L	15 μH
C_1, C_2	48 nF

For the DC/DC stage, Figure 9 shows the simulated waveforms of the control signal of the power MOSFET Q, the input voltage V_{bat}, the currents I_p and I_d through the primary and secondary sides of the transformer, and the output voltage V_c. Before the switch is turned on, the current through the rectifier diode drops to zero. Therefore, the circuit operates in DCM, which is consistent with the theoretical analysis.

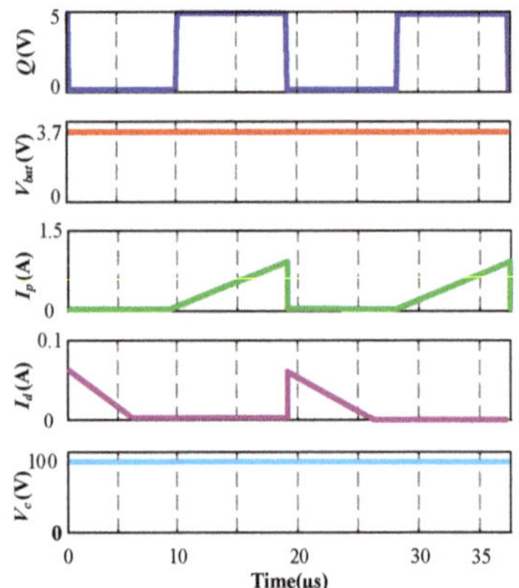

Figure 9. Simulation waveforms of each node in flyback DC/DC stage.

The switching frequency of the MOSFET Q is 55.5 kHz. When Q is turned on, the primary inductance is linearly charged by the battery's DC voltage of 3.7 V. During this phase, the current I_d and the drain–source voltage v_{ds} of Q are both zero. When Q is turned off, the current I_p becomes zero, and the current of the rectifier diode I_d starts to decrease from the maximum value I_d/n. When the current I_d drops to zero, the load is supplied by the output filter capacitor, and the primary and secondary sides of the transformer have zero current. When Q is turned on again, the new switching cycle of the operation is implemented.

The voltage signals of V_{c1} and V_{c2} generated by the DC/AC half-bridge drive circuit should meet the requirements of drive signals for piezoelectric bimorph actuators. The simulation results in the DC/AC stage are presented in Figure 10. The voltages of V_{c1} and V_{c2} are 100 V sinusoidal signals with 100 Hz resonant frequency and 180° phase delay. In addition, various types of signals can be generated based on the design of the DC/AC circuit, including square, triangle, and sinusoidal signals.

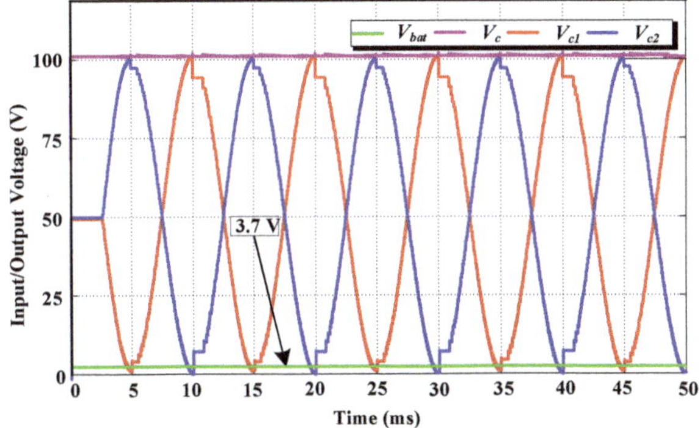

Figure 10. Simulation waveforms of the drive stage.

5. Experiment Realization

5.1. Fabrication Specifications

The circuit topology described in Figure 4 is soldered onto a well-designed printed circuit board using discrete components. Table 3 lists the specifications of the discrete components, which include a power MOSFET, a rectifier diode, a flyback transformer, and a filtering capacitor. The flyback step-up stage is responsible for increasing the input DC voltage. The basis of the selection is to ensure that the components can handle the predicted voltage and current without breakdown and failure. As mentioned earlier, if the input voltage varies within the range of 3.0–4.5 V, then the maximum average current can reach 200 mA. Therefore, when the output is approximately 100 V, the average output current should be less than 10 mA. In accordance with package specifications, the N-channel enhancement MOSFET (SI2304, Vishay Intertechnology Inc., Shelton, CT, USA) with the small outline transistor (SOT) package is used as the primary side switch, because of its high voltage stress, low on-state resistance, and sufficient compactness.

Table 3. Specifications of the discrete components.

Device Name	Chip Number	Weight (mg)	Size (mm)
Flyback transformer	lpr4012 (L_m = 2 µH)	64	4 × 4
Power MOSFET (power stage)	SI2304	8	2.8 × 2.1
Rectifier diode	SMD1200PL	8	2.8 × 2.1
Capacitor (power stage)	1 µF/200 V	1.2	1.6 × 0.8
Power MOSFET (drive stage)	TN2404K	8	2.8 × 2.1
Inductor (drive stage)	15 µH/40 mA	1.2	1.6 × 0.8

The flyback secondary diode is an SOD-123 packaged Schottky barrier rectifier diode, i.e., SMD1200PL (Micro Commercial Components Inc., Simi Valley, CA, USA), which can withstand more than 1 A average forward current and up to 200 V reverse voltage. Compared with a conventional diode with a turn-on voltage of 0.7 V, SMD1200PL has a voltage of only 0.4 V at a forward current of 20 mA. Therefore, conduction loss is low, which is a key factor in selecting this diode.

The flyback transformer LPR4012 (Coilcraft Inc., Cary, IL, USA) with a weight of 64 mg and made of ferrite material produced by Coilcraft has an excellent coupling coefficient (>0.95) and good electromagnetic interference (EMI) performance. Its primary inductance is 2 µH and the turn ratio is 1:10, which can withstand a peak current of 1.7 A under an unsaturated condition. The 1 µF/200 V output filter capacitor (Vishay Intertechnology Inc., Shelton, CT, USA) with a 0603 package is selected to diminish voltage ripple on the actuator outputs.

For the half-bridge drive circuit, two N-channel enhancement MOSFETs TN2404K (Vishay Intertechnology Inc., Shelton, CT, USA) with an SOT-23 package are adopted as high-side and low-side switches, because of their high capability to withstand voltage and low on-state resistance. A 0603 package passive inductor (Vishay Intertechnology Inc., Shelton, CT, USA) with a high quality factor and good EMI characteristic is used to obtain excellent output signals. The image of the fabricated converter is shown in Figure 11.

Figure 11. Image of the fabricated converter.

Using the above components, a prototype weighing 108 mg was fabricated. The length, width, and thickness of the circuit are 13, 13, and 4 mm, respectively. The converter realizes miniaturization at the gram level, thereby enabling it to meet the requirement of compactness for MMRs in harsh applications.

5.2. Circuit Experimental Analyses

A low-voltage source lithium battery is utilized to generate high drive voltage in the developed topology. The input voltage V_{bat} of the dual-stage drive circuit is 3.7 V, and the output voltage V_o is regulated at 100 V. One layer of the bimorph actuator can be modeled as a 48 nF capacitor in parallel with a few megaohm resistance. Multiple actuators are cascaded to increase load condition in circuit validation. The dual-stage converter is capable of driving capacitive loads at nanofarad level, with up to 0.5 W (full load) unipolar actuation.

In the DC/DC stage, the node waveforms of V_{gs} (driven gate voltage), V_{ds}, and I_p were obtained, as shown in Figure 12. The I_p current is fully provided by the lithium battery. The waveforms are consistent with the theoretical analysis. The maximum instantaneous output power of the lithium battery can reach four watts. The zero current shows that the flyback circuit worked in DCM mode.

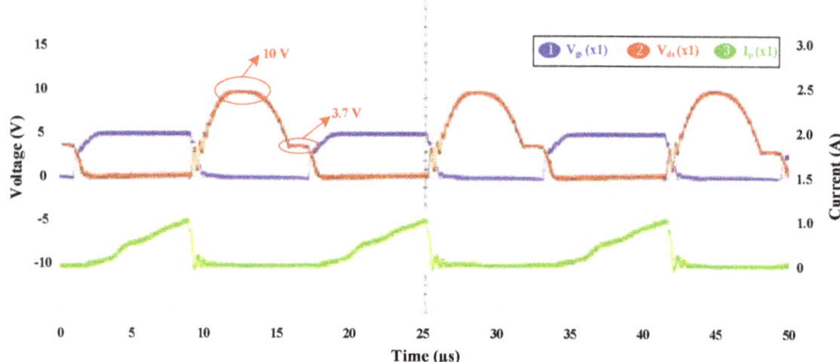

Figure 12. Node waveforms of the flyback converter.

This circuit topology is capable of generating three waveforms (i.e., square, triangle, and sinusoidal waveforms; Figure 13). An oscilloscope probe is used five and ten times to measure a voltage of 100 V.

Figure 14 shows the output waveforms of the three signals and the corresponding control pulses of the two power MOSFETs. When the sinusoidal signal is used as an example, the timing pulses are turned on and the high-side transistor Q_H is constantly switched during the Charge phase, thereby initiating the transfer of energy to node V_o. During the Discharge phase, the low-side transistor Q_L is turned on and the high-side transistor Q_H is turned off, thereby lowering voltage V_o. As shown in the enlarged view, the switch is only turned on when the capacitor is required to reach the ideal voltage, thereby avoiding unnecessary switching actions and significantly reducing switching loss.

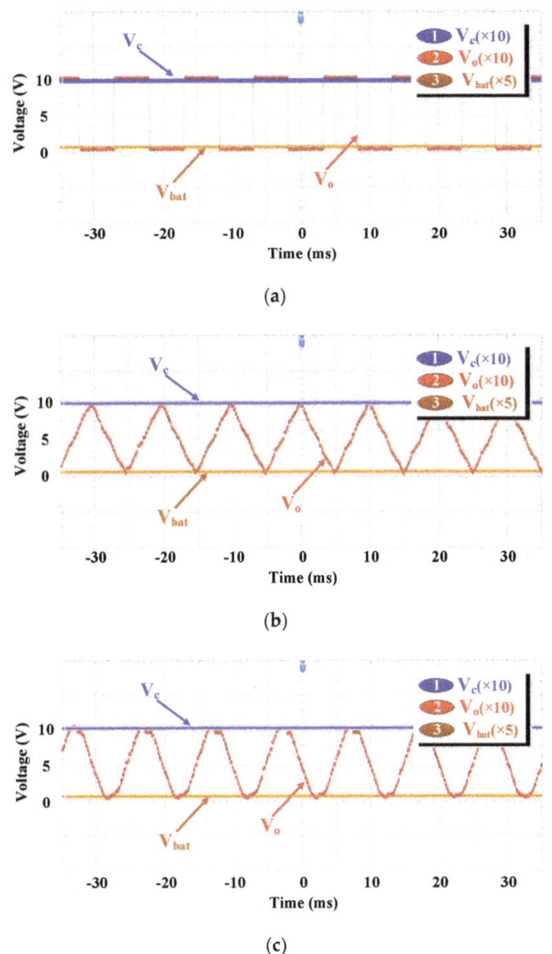

Figure 13. Output voltages of each node (i.e., V_{bat}, V_c and V_o) in (**a**) square driving; (**b**) triangle driving; and (**c**) sinusoidal driving modes.

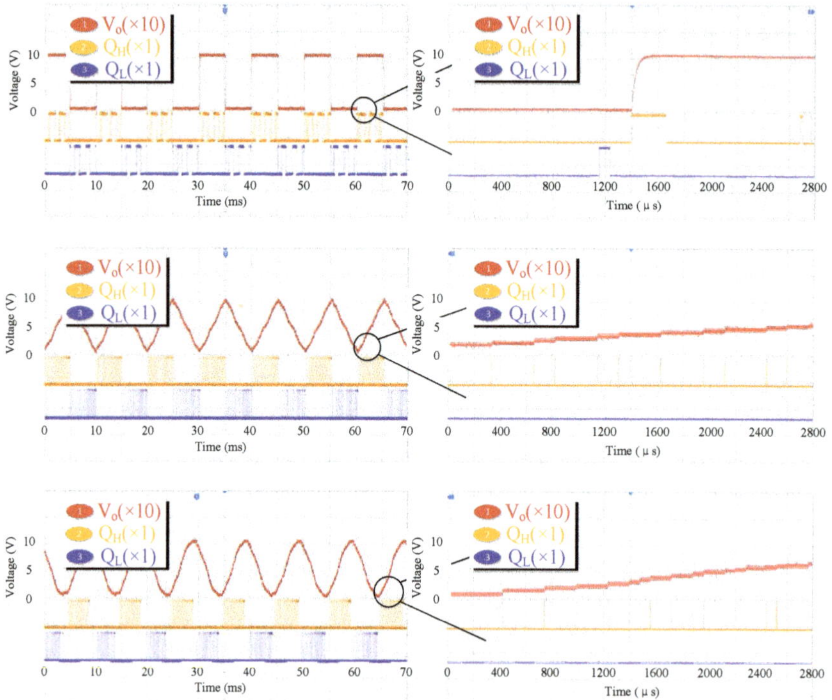

Figure 14. Images of the control pulses and output waveforms.

The proposed converter's conversion efficiency can be calculated as follows:

$$\eta = \frac{P_{load}}{P_{in}} \times 100\%,\tag{12}$$

where P_{load} is the average power of the PZT load, and P_{in} is the average input power of the developed converter.

Figure 15 demonstrates the efficiency of the proposed interface in terms of various loads and drive signal frequencies.

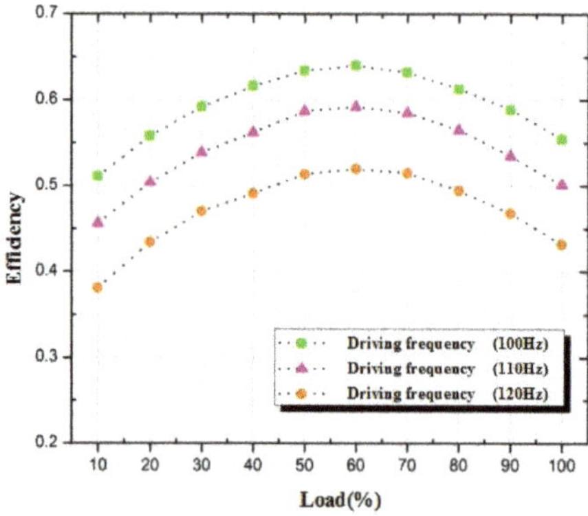

Figure 15. Efficiency vs. load for the three different driving frequencies.

The loads continuously increase from 10% to 100% with 10% intervals. The three driving frequencies are 100, 110, and 120 Hz. The peak efficiency (64.5%) can be reached at 60% load and 100 Hz driving frequency (Figure 14), which is higher than the maximum conversion efficiency of 42.4% obtained in [9]. Table 4 outlines the electrical characteristics of the proposed drive circuit.

Table 4. Specific characteristics of the dual-stage circuit.

Output Power (maximum)	500 mW
Efficiency (maximum)	64.5%

5.3. Performance Optimization

To optimize the parameters and improve circuit performance, energy efficiency optimization experiments of the DC/DC and DC/AC stages are conducted to understand the design tradeoffs. For the DC/DC step-up stage, four parameters, namely, magnetizing inductance, switching frequency, drive voltage, and output power, are considered as critical parameters that affect the efficiency of the circuit. Six flyback transformers with different magnetizing inductances are selected for the optimization test. In addition, switching frequency varies from 40 kHz to 200 kHz with 20 kHz steps. Within a drive voltage range of up to 160 V, the circuit changes from 20% load (0.1 W) to full load (0.5 W).

Figure 16 depicts the efficiency variation of the flyback stage with different magnetizing inductances and switching frequencies at 60% load and 90 V drive voltage. When the switching frequency is constant, a large magnetizing inductance results in high conversion efficiency, because a large magnetizing inductance can reduce current swing, thereby decreasing hysteresis eddy losses.

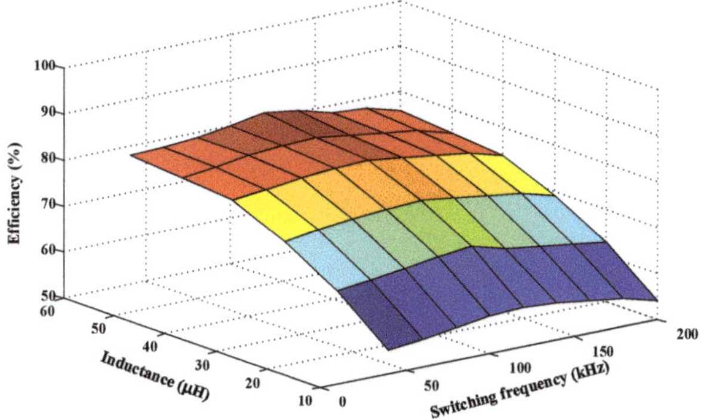

Figure 16. Conversion efficiency of flyback stage vs. switching frequency and magnetizing inductance.

When magnetizing inductance is constant, switching frequency significantly affects conversion efficiency, as reflected in switching and iron losses. At low switching frequencies (<80 kHz), iron losses become dominant, due to the high magnetizing forces caused by a large current swing of the coupled inductors. A low switching frequency results in a small current swing, which reduces iron losses and improves system efficiency. However, MOSFET switching losses become dominant when operating at high switching frequencies (>80 kHz).

Figure 17 shows that at 60% load operation and 90 V drive voltage, a maximum efficiency of 81.8% is detected with 60 μH magnetizing inductance at 100 kHz switching frequency.

This figure illustrates how conversion efficiency varies with output power and drive voltage when switching frequency is 100 kHz and magnetizing inductance is 60 μH. Efficiency is measured under the condition that the output power range is 0.1–0.5 W and the drive voltage range is 40–160 V. At a low output power, parasitic MOSFET losses are dominant. When output power increases to a certain value, the conduction loss of a MOSFET and the copper loss of the transformer become dominant.

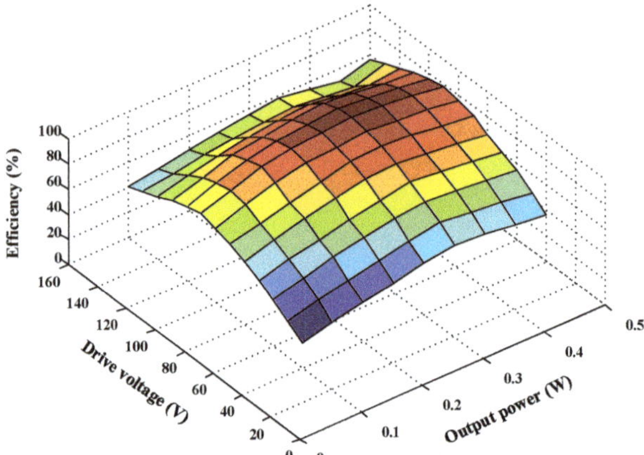

Figure 17. Conversion efficiency of the flyback stage vs. drive voltage and output power.

Under the condition of a constant load, low drive voltages lead to large currents, with conduction losses being dominant. At high drive voltages, conduction losses will be reduced in correspondence to the smaller currents, thereby increasing efficiency. When drive voltage exceeds a certain value, iron losses will become dominant, because of the large current swing caused by large duty cycles. Maximum efficiency occurs at nearly 0.3 W output power and 90 V driving voltage with a switching frequency of 100 kHz and a magnetizing inductance of 60 μH.

In the half-bridge stage, a high-quality factor inductor can effectively reduce conduction loss. Simultaneously, an inductance with a small value is selected to increase resonance frequency. Similar to the flyback converter, the experimental results exhibit high efficiency (ranging from 70.1% to 80.3%) under the desired operation as illustrated in Figure 18.

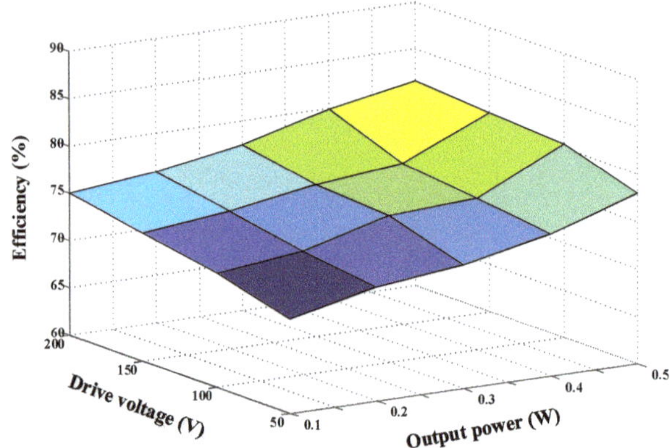

Figure 18. Efficiency of the half-bridge stage vs. drive voltage and output power.

5.4. PZT Driving Validation

The dynamic performance of the fabricated converter was tested by observing the displacement of the piezoelectric bimorph (QDTE52-7.0-0.82-4, PANT, Suzhou, Jiangsu, China). Figure 19 shows the image of the entire system, which comprises a battery box, the proposed drive circuit, and a manual probe station that is used in conjunction with a B1505A I/V Agilent power device analyzer (Keysight Technologies, Inc., Santa Rosa, CA, USA). The output of the drive circuit is connected to the terminal of the probe, and the piezoelectric actuator is physically anchored by the terminal under the electron microscope.

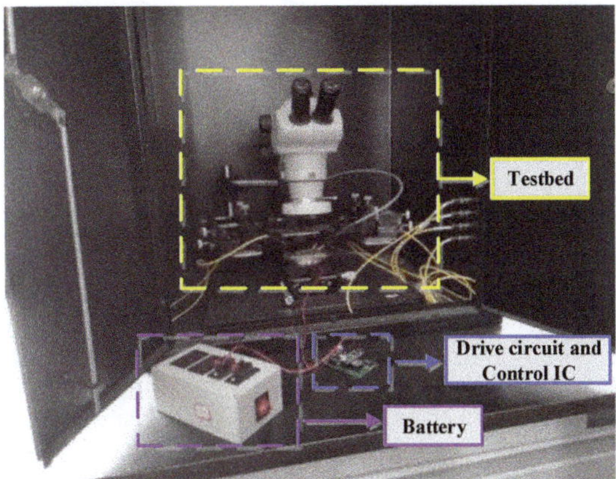

Figure 19. Image of the entire system.

The parameters of the piezoelectric bimorph used in the experiment are 52 × 7.0 × 0.82 mm (size), 1.2 mm (bidirectional displacement), 100 Hz (resonant frequency), and 200 V (maximum driving voltage). The displacement of the front end of the actuator at 0, 60, 80, and 100 V drive circuit was photographed using a high-speed camera and the images in (a)–(d) of Figure 20 present the corresponding observations. The figure shows that the displacement degree of the actuator increases as driving voltage increases. The arrow indicates the direction, in which 0.5 mm unidirectional displacement can be obtained when the actuator is applied at a drive voltage of 100 V.

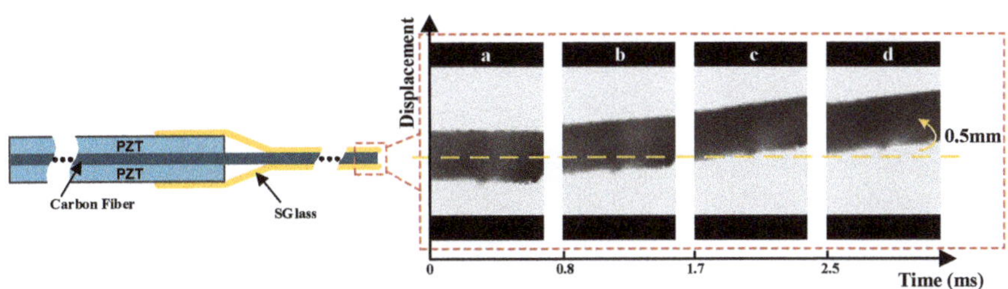

Figure 20. Displacement images of the piezoelectric actuator driven at four different voltages: (**a**) 0 V; (**b**) 60 V; (**c**) 80 V; and (**d**) 100 V.

Similarly, the bidirectional displacements, shown in Figure 21, are achieved with −100, 0, and 100 V excitation generated by the proposed circuit, which illustrate the piezoelectric bimorph at (a) down, (b) equilibrium, and (c) up, respectively. The experimental results show that the mechanical displacement is proportional to the square of the driving voltage. The displacement of the up–down motion is close to 1 mm, which can be further expanded by mechanically magnifying the structure to obtain potential applications in MMR.

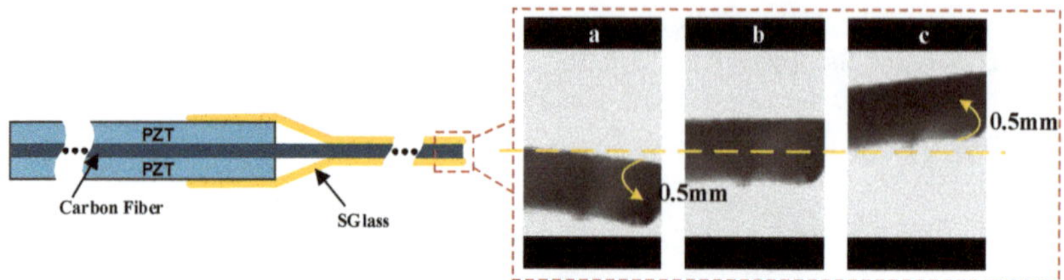

Figure 21. Image of the piezoelectric actuator displacement: (**a**) Down; (**b**) at equilibrium; and (**c**) up.

6. Conclusion

In summary, this paper discusses a dual-stage converter that exhibits high performance at low mass and microsize for driving piezoelectric bimorphs at system level in MMR applications. The operations and control strategies of high step-up DC/DC stage and DC/AC driving stage are presented in detail. Combined with mechanical loads, the dual stage converter can fully utilize its many advantages to accomplish specific tasks, including, but not limited to, military reconnaissance and environmental monitoring. Future research will focus on reductions in circuit losses to improve efficiency and the miniaturization of prototypes. Simultaneously, a study on coupling theory of high-performance composites and piezoelectric ceramics will also be conducted to produce a piezoelectric composite actuator with lighter mass and higher strength energy density. With advanced drive technology, greater mechanical displacement is generated for MMR. For the drive circuit, custom bare die package components have potential uses when fabricated. On the basis of this study, the suppression of switching losses can be realized by using soft switching technology to improve the efficiency of a system.

Author Contributions: Y.W. conceived and designed the research. C.C. and M.L. performed the research. C.C. and M.L. wrote the paper.

Funding: This research was funded by the National Key R&D Program of China grant number: 2016YFC0303902, the Science and Technology Department of Jilin Province of China grant number: 20180201022GX, SXGJQY201711, and the Education Department of Jilin Province of China grant number: JJKH20180154KJ.

Acknowledgments: This research was supported by the National Key R&D Program of China (No. 2016YFC0303902), the Science and Technology Department of Jilin Province of China (Nos. 20180201022GX, SXGJQY201711), and the Education Department of Jilin Province of China (No. JJKH20180154KJ).

Conflicts of Interest: The authors declare no conflict of interest.

References

1. Ceylan, H.; Giltinan, J.; Kozielski, K.; Sitti, M. Mobile microrobots for bioengineering applications. *Lab Chip* **2017**, *17*, 1705. [CrossRef] [PubMed]
2. Gregory, J.; Fink, J.; Stump, E.; Twigg, J.; Rogers, J.; Baran, D.; Fung, N.; Young, S. Application of Multi-Robot Systems to Disaster-Relief Scenarios with Limited Communication. In *Field and Service Robotics*; David, W., Timothy, D.B., Eds.; Springer International Publishing: New York, NY, USA, 2016; pp. 639–653, 1610–7438.
3. Cai, H.M.; Ang, H.S.; Zhang, H.L.; Duan, W.B. Experiment on aerodynamic characteristics of a novel micro tilt ducted fan aircraft with wings. *Acta Autom. Sin.* **2012**, *30*, 777–781.
4. Rios, S.A.; Fleming, A.J.; Yong, Y.K. Design of a two degree of freedom resonant miniature robotic leg. In Proceedings of the IEEE International Conference on Advanced Intelligent Mechatronics, Busan, Korea, 7–11 July 2015; pp. 318–323.
5. Tang, Y.; Chen, C.; Khaligh, A.; Penskiy, I.; Bergbreiter, S. An ultracompact dual-stage converter for driving electrostatic actuators in mobile microrobots. *IEEE Trans. Power Electron.* **2014**, *29*, 2991–3000. [CrossRef]
6. Suzumori, K.; Endo, S.; Kanda, T.; Kato, N.; Suzuki, H. A Bending Pneumatic Rubber Actuator Realizing Soft-bodied Manta Swimming Robot. In Proceedings of the IEEE International Conference on Robotics and Automation, Roma, Italy, 21 May 2007; pp. 4975–4980.

7. Lv, X.; Wei, W.; Mao, X.; Chen, Y.; Yang, J.; Yang, F. A novel MEMS electromagnetic actuator with large displacement. *Sens. Actuators A* **2015**, *221*, 22–28. [CrossRef]

8. Mineta, T.; Mitsui, T.; Watanabe, Y.; Kobayashi, S.; Haga, Y.; Esashi, M. An active guide wire with shape memory alloy bending actuator fabricated by room temperature process. *Sens. Actuators A* **2002**, *97–98*, 632–637. [CrossRef]

9. Towfighian, S.; Seleim, A.; Abdel-Rahman, E.M.; Heppler, G.R. A large-stroke electrostatic micro-actuator. *J. Micromech. Microeng.* **2011**, *21*, 075023. [CrossRef]

10. Kovacs, G.; Düring, L.; Michel, S.; Terrasi, G. Stacked dielectric elastomer actuator for tensile force transmission. *Sens. Actuators A* **2009**, *155*, 299–307. [CrossRef]

11. Wood, R.J.; Avadhanula, S.; Sahai, R.; Steltz, E.; Fearing, R.S. Micro robot Design Using Fiber Reinforced Composites. *J. Mech. Des.* **2008**, *130*, 680–682. [CrossRef]

12. Karpelson, M.; Wei, G.Y.; Wood, R.J. Driving high voltage piezoelectric actuators in microrobotic applications. *Sens. Actuators A* **2012**, *176*, 78–89. [CrossRef]

13. Nabawy, M.R.A.; Parslew, B.; Crowther, W.J. Dynamic performance of unimorph piezoelectric bending actuators. *Proc. Inst. Mech. Eng. Part I J. Syst. Control Eng.* **2015**, *229*, 118–129. [CrossRef]

14. Nabawy, M.R.A.; Crowther, W.J. Dynamic electromechanical coupling of piezoelectric bending actuators. *Micromachines* **2016**, *7*, 12. [CrossRef]

15. Steltz, E.; Seeman, M.; Avadhanula, S.; Fearing, R.S. Power electronics design choice for piezoelectric micro robots. In Proceedings of the IEEE/RSJ International Conference on Intelligent Robots and Systems, Beijing, China, 9–15 October 2006; pp. 1322–1328.

16. Chen, C.; Liu, M.; Lin, J.; Wang, B.; Wang, Y. Piezoelectric transformer-based high conversion ratio interface for driving dielectric actuator in micro robotic applications. *Int. J. Adv. Robot. Syst.* **2016**, *13*. [CrossRef]

17. Luo, F.L.; Ye, H. Positive output cascade boost converters. *IEE Proc. Electr. Power Appl.* **2004**, *151*, 590–606. [CrossRef]

18. Liang, T.J.; Chen, S.M.; Yang, S.L.; Chen, J.F.; Ioinovici, A. A single switch boost-flyback dc-dc converter integrated with switched-capacitor cell. In Proceedings of the IEEE 8th International Conference on Power Electronics and Ecce Asia, Jeju, Korea, 30 May–3 June 2011; pp. 2782–2787.

19. Liao, W.C.; Liang, T.J.; Liang, H.H.; Liao, H.K.; Juang, K.C.; Chen, J.F. Study and implementation of a novel bidirectional dc-dc converter with high conversion ratio. In Proceedings of the IEEE Energy Conversion Congress and Exposition, Phoenix, AZ, USA, 17–22 September 2011.

20. Campolo, D.; Sitti, M.; Fearing, R.S. Efficient charge recovery method for driving piezoelectric actuators with quasi-square waves. *IEEE Trans. Ultrason. Ferroelectr. Freq. Control* **2003**, *50*, 237–244. [CrossRef] [PubMed]

21. Edamana, B.; Oldham, K. Optimal on-off controller with charge recovery for thin-film piezoelectric actuators for an autonomous mobile micro-robot. In Proceedings of the American Controls Conference, San Francisco, CA, USA, 29 June–1 July 2011.

22. Zsurzsan, G.T.; Zhang, Z.; Andersen, M.M.A.; Andersen, N.A. Class-D amplifier design and performance for driving a piezo actuator drive servomotor. In Proceedings of the IEEE International Conference on Industrial Technology (ICIT), Taipei, Taiwan, 14–17 March 2016.

23. Clingman, J.C.; Gamble, M. High-voltage switching piezo drive amplifier. In Proceedings of the 5th Annual International Symposium on Smart Structures and Materials, San Diego, CA, USA, 1–5 March 1998.

24. Luan, J. Design and Development of High-Frequency Switching Amplifiers Used for Smart Material Actuators with Current-Mode Control. Master's Thesis, Virginia State University, Blacksburg, VA, USA, 29 July 1998.

25. Gnad, G.; Kasper, R. Power Drive Circuits for Piezo-Electric Actuators in Automotive Applications. In Proceedings of the IEEE International Conference on Industrial Technology, Mumbai, India, 15–17 December 2006; pp. 1597–1600.

26. Guan, M.; Liao, W.H. Studies on the circuit models of piezoelectric ceramics. In Proceedings of the International Conference on Information Acquisition, 2004, Proceedings, Hefei, China, 21–25 June 2004; pp. 26–31.

27. Rios, S.A.; Fleming, A.J. A new electrical configuration for improving the range of piezoelectric bimorph benders. *Sens. Actuators A* **2015**, *224*, 106–110. [CrossRef]

28. Small, high-voltage boost converters. Maxim Semiconductor Application Note 1109. 2002. Available online: https://www.maximintegrated.com/en/app-notes/index.mvp/id/1109 (accessed on 12 September 2018).

29. Karpelson, M.; Whitney, J.P.; Wei, G.Y.; Wood, R.J. Design and Fabrication of Ultralight High-Voltage Power Circuits for Flapping-Wing Robotic Insects. In Proceedings of the Twenty-Sixth Annual IEEE Applied Power Electronics Conference and Exposition (APEC), Fort Worth, TX, USA, 6–11 March 2011; pp. 2070–2077.
30. Honda, J.; Adams, J. Class D Audio Amplifier Basics. *Amp Nch.* 2005. International Rectifier Application Note AN-1071. Available online: https://www.infineon.com/dgdl/an-1071.pdf?fileId=5546d462533600a40153559538eb0ff1 (accessed on 12 September 2018).
31. Janocha, H.; Stiebel, C. New approach to a switching amplifier for piezoelectric actuators. *Energy* **1998**, *2*, 2.

Review

A State-of-the-Art Review on Robots and Medical Devices Using Smart Fluids and Shape Memory Alloys

Jung Woo Sohn [1], Gi-Woo Kim [2] and Seung-Bok Choi [2,*]

[1] Department of Mechanical Design Engineering, Kumoh National Institute of Technology, Gumi 39177, Korea; jwsohn@kumoh.ac.kr
[2] Department of Mechanical Engineering, Inha University, Incheon 22212, Korea; gwkim@inha.ac.kr
*   Correspondence: seungbok@inha.ac.kr

Received: 21 September 2018; Accepted: 14 October 2018; Published: 15 October 2018

Abstract: Over the last two decades, smart materials have received significant attention over a broad range of engineering applications because of their unique and inherent characteristics for actuating and sensing aspects. In this review article, recent research works on various robots, medical devices and rehabilitation mechanisms whose main functions are activated by smart materials are introduced and discussed. Among many smart materials, electro-rheological fluids, magneto-rheological fluids, and shape memory alloys are considered since there are mostly appropriate application candidates for the robot and medical devices. Many different types of robots proposed to date, such as parallel planar robots, are investigated focusing on design configuration and operating principles. In addition, specific mechanism and operating principles of medical devices and rehabilitation systems are introduced and commented in terms of practical realization.

Keywords: smart materials; actuators; robots; electro-rheological fluids; magneto-rheological fluids; shape memory alloys; medical devices; rehabilitation system

---

1. Introduction

Historically, material technologies have had a profound influence on human civilization, and hence historians have defined distinct time periods by the dominant materials used during those eras. The term 'smart materials' first appeared in the late 1980s, and since then, a multitude of research activities on smart materials has been, or is being, conducted in diverse industry areas. Among the many inherent characteristics of smart materials, the crucial common denominators to achieve high performances in application systems are their capabilities of sensing, actuating, and controlling under external stimuli that govern the responses of the systems. To date, more than 100 types of smart materials with these capabilities have been proposed, and their adaptability as 'smart' has been validated. These include the identification of material characteristics showing sensitivity as sensors, generating forces as actuators, and performing control responses as a controller. In the last two decades, several classes of smart materials have received significant attention, over a broad range of engineering applications, because of their unique and useful actuator properties. Potential smart materials for robotic and medical applications include electro-rheological (ER) fluids (ERFs), magneto-rheological (MR) fluids (MRFs), and shape memory alloys (SMAs). It is noted here that ERFs also include homogeneous types such as liquid crystal in the sense of the electric field response smart material. However, in this review article, ERFs which can be expressed by Bingham model, in which the field-dependent yield stress is the dominant behavior, are considered only.

In the robotic research society, many researchers have developed various types of robots using traditional actuators (e.g., electric motors, hydraulic fluidic actuators) to achieve some specific requirements for desired movements. Although some robots with conventional actuators have demonstrated excellent performance, alternative actuators need to be explored when the capability to perform flexible and complex movements is required. Smart material-based actuators are one of these alternative solutions, because of their relatively small weight and volume, compared to those fabricated using conventional actuating methods. For example, SMAs have been widely used in a diverse range of humanoid-robotic applications since the 1980s, especially as artificial muscles, because SMA wire actuators can mimic the human muscle. The use of smart materials as actuators can be categorized in terms of their application areas: automotive, aerospace, robotics, and medical, amongst others. Despite numerous research works on smart materials applications, a comprehensive overview on the specific application areas of diverse robots, medical devices and rehabilitation systems is not readily available. Therefore, it is worthwhile to timely review one of attractive applications of smart materials, namely robot and medical applications.

Consequently, the main objective of this review article is to present a broad perspective of the research efforts during the last two decades in relation to robot and medical devices, and their prototypes, based on ERFs, MRFs, and SMAs. This review describes the attributes of specific smart materials that make them ideal for actuating robotic and medical applications, and discusses their associated technical capabilities and limitations in order to emphasize the design challenges. Therefore, this article provides an updated review of recent smart material research of robot and medical applications, with 100 state-of-the-art references categorized into three types of smart materials. It is remarked here that this review article is focused on the operating principle and design concept of each application for easy understanding of the potential readers of this attractive research field, instead of the specific dynamics and control strategies of the adopted devices and systems.

2. Robots Using ERF and MRF

2.1. Material Characteristics

ERFs are a class of colloidal dispersions that exhibit a large reversible change in their rheological behavior when subjected to external electric fields. These changes in rheological behavior are manifested by a dramatic increase in flow resistance, which depends upon the flow regime and the composition of the ERF. The flow resistance can be considered an actuating force controlling dynamic motions, such as vibration and position. In order to achieve flow resistance, ERFs are frequently modeled by a simple Bingham plastic model, in which the field-dependent yield stress is expressed as a controllable resistance. Among the many inherent characteristics of ERFs, the most salient property is the fast response of the actuating force (or torque) caused by application of an external electric field. This important property has triggered numerous application research works on ERF, including automotive shock absorbers, brakes, clutches, and smart structures. In addition, owing to their fast and easy controllability, ERFs are applied to robot systems, where they are embedded into the flexible arms of manipulators. Some other applications of ERF in robotic areas include tactile sensors, and human-friendly soft hands. It is also known that, when subjected to an external magnetic field, MRFs have exactly the same characteristics as ERFs. The rheological properties of MRFs, such as complex moduli, can be tuned or controlled by changing the intensity of the magnetic field associated with appropriate control schemes. Thanks to the robustness and higher material performance of MRFs compared to ERFs in a practical environment, numerous studies on the applications of MRFs are being actively undertaken in many industrial fields such as automotive, aerospace, and civil engineering. MRFs have a much higher field-dependent yield stress than ERFs, and hence, they satisfy the force or torque requirements of many practical systems or devices over a wide temperature range. However, MRF has slower response time compared to ERF. Moreover, the weight of MRF-based devices is much heavier than ERF-based devices because of high density of ferromagnetic metals.

2.2. ERF-Based Robots

As far as is known, the first application of ERFs to robotic systems was reported by Gandhi et al. [1]. In this work, the forearm of a commercial parallel robot was retrofitted by inserting ERF. The deflection of the forearm was successfully reduced during dynamic maneuvering and transient motions, owing to the increment of the field-dependent damping property of the ERF domain. Choi et al. [2,3] performed a study similar to Gandhi et al. [1], with primary focusing on the vibration control of a flexible robot arm incorporating ERF. In this work, a single-link flexible robot arm system was made by inserting ERF into the box-type aluminum arm. The simulation and experiment were both conducted to evaluate effective control of the transient vibration during a rotational maneuver. A number of other studies were conducted using the same concept for vibration control of a flexible gantry robot [4,5]. In the gantry robot system, the translational motion was controlled by a bi-directional ER clutch, and the vibration due to the flexibility of the link was controlled by a piezoelectric actuator bonded on the surface of the link. As for the controller, a robust, H-infinity controller was designed and built, using a microprocessor. Excellent positional and vibrational control responses were achieved in the presence of uncertain system parameters, such as the variation of natural frequencies. Monkman [6] proposed a high-resolution tactile display using ERF, applicable to the virtual reality (VR) environments encountered underwater or in space exploration. Figure 1 shows a compressive-type tactile display whose hardness can be tuned by the electric field and the changed hardness can be felt by fingers exploring the surface. The movement of the fingers over the surface allows the shape of an object to be detected by touch, with the regions that correspond to the position of the object being harder than those outside the image. In Figure 1, the white circles represent the dielectric separating mesh. The dielectric separating mesh is essential to prevent the earthed compliant surface from coming into direct contact with the uncharged tactor elements resulting in a short circuit when they become activated. More recently, a study on the feasibility and suitability of an ERF-based actuator in human-friendly manipulators was reported in [7]. In this work, the actuator criteria were checked, based on ERF properties and the application of an ER clutch. It has been shown that ERFs exhibit promising characteristics for use as robotic actuators. Specifically, they are well suited for actuation systems that interact physically with humans, because of their inherent characteristics such as intrinsic back-drivability, low-output inertia, superior performance and bandwidth, and precision controllability of output torque. The state-of-the-art robotic applications using ERF actuators are summarized in Table 1.

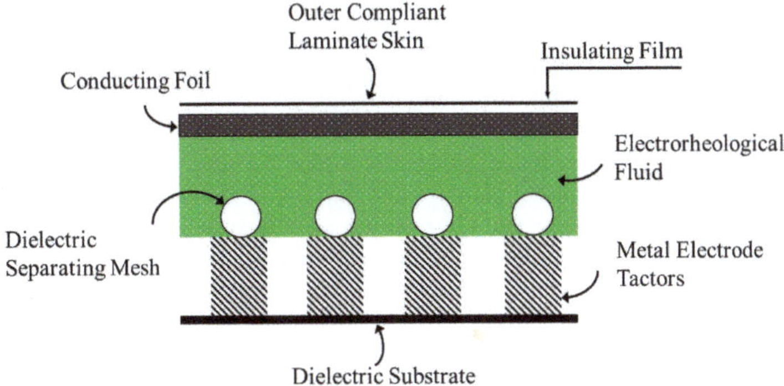

Figure 1. Tactile display featuring ERF [6].

Table 1. Summary of state-of-the-art robotic applications using ERF.

Robot	References
Flexible robot arm	[1–3]
Flexible gantry robot	[4,5]
Tactile display	[6]
Manipulator	[7]

2.3. MRF-Based Robots

MRFs have been for the first time applied to vehicle shock absorbers, and some of the passenger vehicles integrated with MR dampers are currently available on the market [8,9]. The application of MRFs to robotics has been actively researched since 2000. Initially, the feasibility of MRFs as robotic actuators was both conceptually and experimentally studied [10]. There are two possible approaches when investigating the actuating characteristics of MRFs. One is to make valve systems and produce an active control force in closed-loop systems. The other is to directly use MRFs as actuators, by generating a flow motion in three different flow modes (flow, shear, and squeeze), in which a semi-active damping force is normally produced in open- or closed-loop control systems. Recently, Yadmellant and Kermani [11] proposed a new adaptive control method to compensate for the inherent hysteretic phenomenon of MRFs, and to develop high-performance robot actuators. In order to validate the proposed control scheme, a small-sized two degrees-of-freedom (2-DOF) MR-actuator robot manipulator was manufactured, and the higher accuracy of its torque-control results, compared to the case without the hysteretic compensator, was presented with various frequency torque trajectories.

The state-of-the-art robotic applications using MRF actuators are summarized in Table 2. A robot application device with MRFs, a deformable gripper that handled implements such as a writing pen, was invented in [12]. This gripper could be deformed in conformance with the user's hands and fingers, with the gripper comprising a tubular sleeve in which the MRF flowed radially. The deformability of the gripper was controlled by a sliding mechanism activated by an electromagnet. A robot gripper for handling delicate food products using MRFs was also developed in [13]. In this work, a two-finger gripper was manufactured. When the MRF, in pouches, reached the sides of the object, the pouches started to deform without stretching the pouch material. One of the gripper arms was fitted with a strain-gauge force sensor, which allowed for multiple gripping modes. The gripping force when handling various fruits and vegetables, such as carrots and strawberries, was measured through both open-loop setting speeds and closed-loop control methods, associated with a proportional–integral–derivative (PID) controller. It was concluded that the gripping force was evenly spread over a large surface, reducing the risk of bruising, and that the geometric dimensions of the gripper were crucial in achieving a universal gripper, able to handle any size or shape of food products. More recently, Nguyen et al. [14] proposed a new type of 3D-haptic gripper for tele-manipulation, using an MR brake system. This haptic gripper transfers the rolling torque, grasping force, and approach force from the slave manipulator to the master operator. Figure 2 shows a schematic configuration of the 3D-haptic gripper which can be applied in surgical robots, where the gripper is attached to a stationary base, and the master manipulator and the haptic gripper are separated. The haptic gripper consists of two rotary MR brakes. These are the grasping and rolling MRBs, and two identical linear MRBs to reflect the approach force. The shaft of the rolling MRB is fixed to the body of the haptic griper while its housing is attached to the stationary base. The shaft of the grasping MRB is connected to gear 2 while its housing is fixed to the gripper body. The gear 2 externally mates with gear 1. On gears 1 and 2, the two shafts of the linear MRBs are attached. The housings of the linear MRBs, on which the handles of the gripper are attached, can move relative to their shafts. By controlling the applied currents of the MRBs according to signals received from the slave sensors, the grasping force, rolling torque and approach force from the slave can be reflected to the human operator [14]. In this work, the principal design parameters, such as the magnetic pole, are determined by an objective function considering both the off-state braking torque/force, and the mass of the brake itself.

Table 2. Summary of state-of-the-art robotic applications using MRF.

Robot	References
Manipulator	[11]
Deformable gripper	[12,13]
Haptic robot	[14–18]
Rehabilitation	[19–22]
Collaborative robot	[23–28]
Planar robot	[29]
Climbing robot	[30]
Spherical robot	[31]

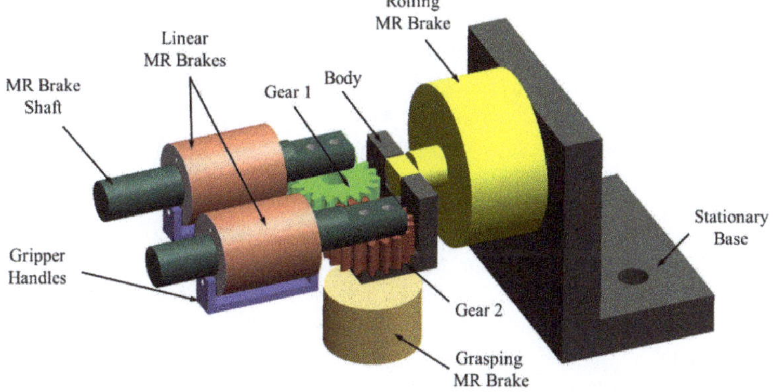

Figure 2. 3-D haptic MRF gripper for surgical robot applications [14].

An innovative haptic display for whole-hand immersive exploration was proposed, and its practical feasibility was demonstrated via a conceptual experiment proposed by Scilingo et al. [15]. In this work, a haptic black box, which can be imagined as a box into which the operator can poke his or her bare hand, was devised using MRFs and its effectiveness was tested through psychophysical experiments to assess performances of softness and shape. A compact haptic glove using MR brakes was developed and its time required to grasp a virtual object was investigated with a force feedback controller by Blake and Gurocak [16]. The proposed glove was made using six MR brakes to generate the movement three fingers, and four-bar mechanisms, to connect the brakes to the digits of the fingers. In the design process of the glove, easy controllability of the fingers, minimization of user fatigue, user's safety, and freedom of motion were considered for satisfying the general requirements of the force feedback gloves. The thumb, index finger, and middle finger are attached to two MR brakes in which the motion of the thumb is constrained to 2-DOF. The braking torque is then transmitted to each finger joint by using the four-bar linkage, which allows the MR glove to be adjusted to different hand sizes. The effectiveness of the MR glove, which weighs 640 g, was evaluated by two different methods: virtual object manipulation, and by reflecting the stiffness of a virtual object held between two fingers. It was shown that the task completion times were reduced by 79% by activating the MR glove, and the difference in stiffness of the springs was identified with a higher percentage of success. Several works on robotic systems closely related to medical surgical applications have been performed utilizing MRF technology. For applications in robotic surgery, a new type of 5-DOF MRF-based tele-robotic haptic was proposed and applied to remotely control a slave robot by Ahmadkhanlou et al. [17]. In this work, after analyzing the MR structure, a force feedback control was employed to replicate in the master, those forces encountered in the slave. Recently, using the salient property of the variable impedance of MRFs as a function of a magnetic field, a series-clutch element was devised for autonomous and tele-robotic operation by Walker et al. [18]. In this work, a series-clutch actuator, which transfers torque up to some saturation level, was made by utilizing a clutch between the DC motor and output link. After experimentally identifying its step response and hysteretic behavior, the clutch was modeled as a first-order system, and a combined tele-robotic control architecture, to reduce the impact force, was formulated and implemented. It was shown that unwanted impact, or collision forces, at the output, are significantly reduced by activating the proposed system, which in turn indicated a great improvement of safety in uncertain environments, or with human contact.

One potential application of MRF technology in the robotic area is in rehabilitation elements or components. Over the last decade, many studies on lower-limb rehabilitation, utilizing MR mechanisms such as clutches or brakes, were undertaken. Specifically, research work on lower-limb robotic rehabilitation related to various gait motions was performed by Diaz et al. [19]. It is known that many patients with cerebral lesions have a spastic movement disorder, which slows voluntary limb movement. This has led to much research on lower-limb rehabilitation using servo-motors. However, it is challenging to guarantee both stability, and powerful motion, by employing a servo-motor-based leg system. Kikuchi et al. [20,21] proposed a haptic controllable leg-shaped robot (Leg-Robot) to be used by patients with spastic ankle joints. The Leg-Robot was made with an MR clutch, which provided a safety mechanism, and a haptic generator of the system. In the design process of the Leg-Robot, they considered mass characteristics at a level approximating that of elderly people (over 75-year-old males). Figure 3a shows the basic structure of the Leg-Robot, in which the MR clutch is built in the knee joint. It was demonstrated that the desired torque levels, corresponding to the spasticity mode and the ankle-clone mode, could be successfully achieved by controlling the MR brake. Jiang et al. [22] proposed a robot leg comprising a rotational MR damper, a torsional spring, a steel plate, and a curved part of the foot that makes contact with the ground, as shown in Figure 3b. As it walks, the foot rotates to form an angle between the core and outer cylinder of the MR damper, while the forces acting on the foot vary according to the synthesized effect of the driving torque of the motor on the axle, and to the dynamic force of the terrain acting on the foot. It was demonstrated that the vertical forces at different walking speeds can be controlled in real time to adapt to the surroundings by controlling the magnetic field applied to the MR damper. In this case, a higher angular speed than other tunable legs were achieved, due to the guaranteed stability associated with the semi-active control of the MR damper.

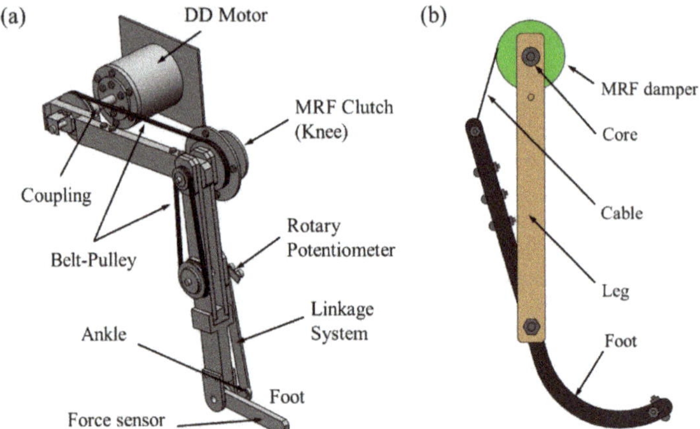

Figure 3. Leg-Robot using MR devices; (**a**) MR fluid clutch [20,21]; (**b**) MR fluid damper [22].

There are several types of collaborative robot systems for robot-to-robot or robot-to-human interactions. Specifically, a robot sharing its working space with humans has been actively researched for many purposes such as carrying heavy objects in factories. Saito and Ikeda [23] proposed a

MR-clutch type safety device for human-collaborative robots. In this work, an MR clutch comprising magnetic circuits associated with the permanent magnet and electromagnetic coil, yoke, rotor, and driving shaft was developed to achieve high levels of output torque. After verifying the safety function of the MR clutch, the transmitted torque of two different types, single- and triple-rotor, were evaluated as a function of the input current (magnetic field). They concluded that, in order to build a safety device capable of controlling the torque output of the robot joint axis and securing the holding torque in an emergency case, the proposed MR clutch was a promising candidate for practical implementation. Ahmed et al. [24,25] developed a robot-arm system featuring an MR clutch/brake to accomplish a safe human–robot interaction, which is a type of compliance control. Fauteux et al. [26] proposed a dual-differential rheological actuator (DDRA) to investigate the feasibility of high-performance robotic interactions, in which safety, robustness, and versatility are paramount. The DDRA was manufactured based on the synergistic use of two differentially coupled MR brakes, and an electromagnetic (EM) motor. After analyzing the DDRA in terms of output friction, torque level, backlash, and interaction force, a one-link manipulator was integrated with the DDRA, as shown in Figure 4, in order to control the joint torque, and hence the position of the manipulator. It has shown from the experiment that more accurate position control, higher torque bandwidth, and smaller inertia of the DDRA compared to conventional EM actuators can be achieved utilizing the DDRA. Shafer and Kermani [27,28] undertook a research similar to Fauteux et al. [26]. In these works, MR clutch actuators were proposed as promising candidates for human-friendly manipulators because of the clear benefits of MRFs. After surveying the actuator requirements for human-friendly manipulators, such as torque levels and control bandwidth, a distributed active semi-active (DASA) actuation mechanism, comprising a driving motor and semi-active MR clutch, was designed and manufactured. In the design process, the actuator inertia, mass of MR clutch, and output impedance were crucially considered in order to achieve maximum efficiency as a robot actuator. It was demonstrated, via experimental work, that accurate torque control with high control bandwidth could be achieved, thus validating an excellent actuator for human-friendly manipulators, which require intrinsic back drivability, low output inertia, and accurate control performance with high bandwidth.

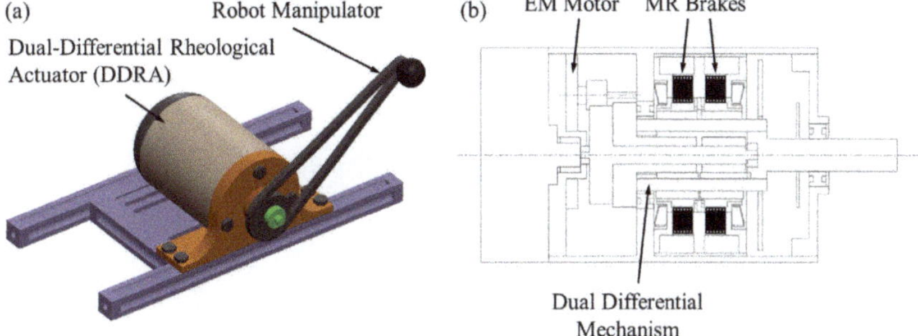

Figure 4. Robot arm featuring dual-differential MR actuator; (**a**) configuration (**b**) section view of DDRA [26].

A novel parallel-planar robot using an MR damper was developed, and its effectiveness was validated by demonstrating tracking control performance in the x–y plane by Hoyle et al. [29]. In this work, a linear motor and two MR dampers were used as actuators. The linear motor is responsible for moving the platform, while the MR dampers guide it through predefined paths by creating adjustable and controllable resistive forces. The platform then follows the path, based on the concept that moving objects intuitively tend to follow the direction that imposes minimum resistance against motion. Figure 5 shows the planar robot manufactured in this work. In order to demonstrate control accuracy of the robot, three different trajectories, straight, inclined, and sinusoidal, were adopted as those to be followed by the robot, and excellent tracking control performance was achieved by implementing a PD controller. This work noted one salient property of the planar robot using an MR damper. If an excessive load is applied to the joints by an incorrect controller command, the joints and links will not be easily damaged because the MR dampers can easily absorb the majority of such loads. MR fluid technology has been extended to a controllable climbing robot, using the adhesive effects of MR fluids. Wiltsie et al. [30] developed a novel type of climbing robot using the field-dependent adhesive effect of an MR fluid. Before building the robot, they investigated the adhesion property of MR fluids with, and without, a magnetic field. From this test, the fluid thickness between two parallel plates was found to have little effect on the adhesive failure strength, and a positive effect on time to failure, while the target surface roughness and orientation were found to have a significant effect on pull-off adhesion, and a positive effect on shearing loads. In addition, in order to investigate the holding time of the MR fluid adhesive, a pseudo-creep test was undertaken, and the maximum allowed holding time until failure was identified as 5 min. They manufactured four feet using MRF, by considering the holding time, failure time, and surface conditions of climbing materials, and tested them on a vertical board covered with 150-grit sanding cloth. It was found in the test that the proposed robot could climb the vertical board with a shear of approximately 7.3 kPa, which was generated by activating the magnets. On the other hand, Yue and Liu [31] applied MRF technology to control unwanted oscillations of a spherical robot. In general, the inner suspension platform of a spherical robot undergoes a severe oscillation when the robot is rolling forwards, which could cause instability of the robot system. Thus, in order to reduce oscillations during the rolling motion, they used an MR damper integrated with a robust sliding-mode controller, which was robust against disturbances and parameter insensibility.

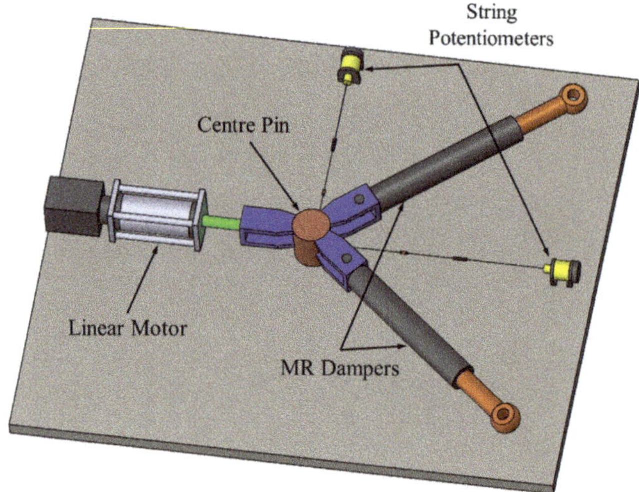

Figure 5. Schematic of parallel planar robot using MR damper [29].

One potential smart material, with material characteristics controllable by external magnetic fields, is the magneto-rheological elastomer (MRE). A controllable MRE, which belongs to the MR-material family, is a composite material with magnetic sensitive particles suspended, or arranged, within a non-magnetic elastomer matrix [32,33]. The operational modes of an MRE are quite different to those of MRFs in that the iron particles are locked within the polymeric matrix, and under external excitations. This restricts the particle movement around their original locations, and thus the direction of the chain structures no longer coincides with the magnetic field. The three operational modes of an MRE are classified as shear, squeeze/elongation, and field-active modes. Depending upon the dynamic motion of MRE applications, an appropriate operational mode needs to be selected to maximize actuating performance. Recently, a new concept for locomotion of miniature robots, based on the periodic electromagnetic actuation of the MRE body structure, was introduced by Zimmermann et al. [34]. Figure 6 presents a basic configuration of two different micro-robots with MRE body structures. The left one incorporates an inelastic-polymeric frame with an integrated micro-coil, and the movement is bidirectional, controlled by the driven frequency. The right one consists of only a symmetric MRE body with six embedded micro-coils, and thus, is applicable to compliant planar-locomotion systems. In this work, prototypes of the two MRE activated micro-robots were manufactured, and their dynamic motions were investigated by observing the driven frequency-dependent movement for two cases, uniaxial- and the planar-locomotion systems. It was seen from the experiments that the maximum average speed of the uniaxial locomotion system was 4.1 mm/s at a driving frequency of 20 Hz, while that of the planar locomotion system was 5 mm/s, at the same driving frequency.

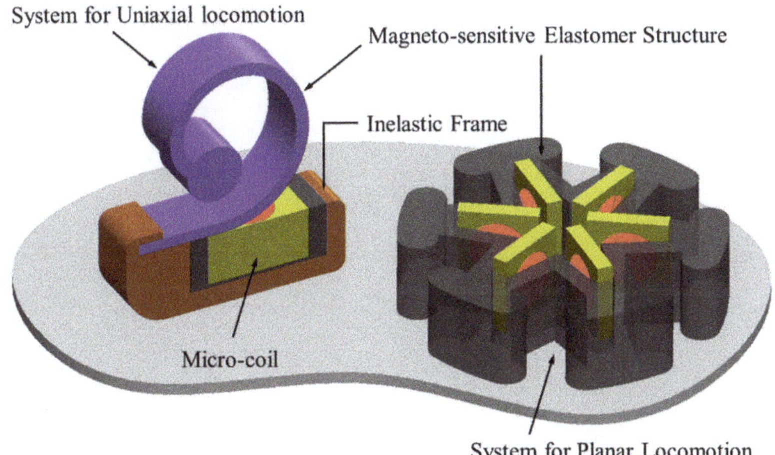

Figure 6. Two different micro-robots featuring MRE body structures [34].

3. Robots Using SMA

3.1. Material Characteristics

Shape memory alloys (SMA) are a class of metallic alloys that can return to their original shape because of the phase-transformation phenomenon, known as the shape memory effect (SME), when subjected to changes in temperature or magnetic field. The primary application of these materials is therefore in actuation systems, where the SMAs can be readily contracted or recovered to their original form. The actuating motions could be achieved by controlling temperature beyond a certain threshold temperature, by internal Joule heating. SMAs possess two different phases with three different crystal structures: twinned martensite, detwinned martensite, and austenite. Typically, the austenite phase is stable at high temperatures, and the martensite phase is stable at lower temperatures. Upon heating an SMA, the initial martensite phase begins to transform into the austenite phase, as shown in Figure 7. The onset temperature is A_s, where the austenite transformation starts, and A_f is the terminal temperature, where this transformation is complete. Once an SMA is heated beyond A_s, it begins to transform into the austenite structure, resulting in recovering (i.e., contracting) to its original shape. During cooling, the transformation starts to reverse to the martensite at M_s, and is completed when it reaches M_f. Above the highest temperature (i.e., A_f), SMA is permanently deformed, similar to ordinary metals and alloys. In addition, SMAs exhibit other shape-memory characteristics, such as pseudoelasticity or superelasticity. The SMAs can return to their original shapes after applying mechanical loading at temperatures between A_f and M_d, without any thermal activation. This property is currently used for passive vibration-damping applications. Hysteresis is a measure of the difference in transition temperatures between heating and cooling, and is generally defined between the temperatures at which 50% of the material is transformed to austenite upon heating, and 50% is transformed to martensite upon cooling. This hysteresis has significant design considerations for SMA applications. For example, a small hysteresis is typically required for fast actuation applications, such as in robotics, while larger hysteresis may be required to retain the predefined shape, such as in deployable structures within a large temperature range. The hysteresis loop associated with different transition temperatures is known to be affected by the composition of the SMA material, and the thermomechanical processing. In addition, some of the SMA's properties also vary between these two phases, including the Young's modulus, electrical resistivity, thermal conductivity, and thermal expansion coefficient. Detailed information on the mechanical properties of SMAs is well summarized in relevant references [35–37].

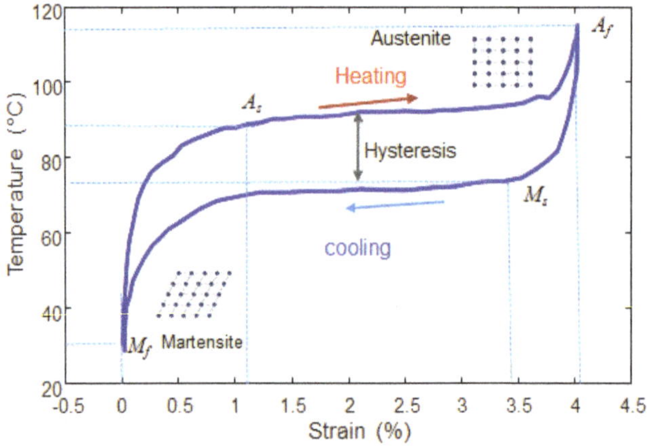

Figure 7. Typical phase transformation curve of SMA wire [35].

SMAs are particularly attractive for actuator applications because their strain (length change or stroke) during the contraction phase is relatively larger than other smart materials, such as piezoelectric transducers, and it is typically 4–5% of their initial length. If the contraction is constrained, large block forces can be generated, and stresses may reach the order of magnitude of 1000 MPa. However, for actuator applications with a large number of load cycles, up to 10^5, the stress level should be alleviated, up to 200 MPa, to prevent fatigue damage. Compared to other actuation systems, high work output along with lower weight (i.e., high power density) can be achieved with SMAs, which is important for robot applications. While the temperature during the contraction process is controlled electrically through Joule heating, cooling is simply controlled by convection in surrounding air, which results in a much slower recovery response. This asymmetric actuation property limits SMA actuators to nearly quasi-static applications, with an actuation frequency of approximately 1 Hz. A few feasible solutions have been suggested to resolve this narrow bandwidth issue, including the use of heat sinks, as in Peltier cooling elements, heat conductive compounds, oil, or water with glycol [35]. However, the cooling rates are not very promising, and special additional devices are required. In addition, SMAs have another technical challenge associated with their lower energy efficiency. Theoretically, the maximum energy efficiency of SMAs is in the range of 10–15%, depending on actuation modes (i.e., loading type). As a result, SMA wire-actuator applications are limited to special areas where energy efficiency is not a significant issue.

3.2. SMA-Based Robots

SMA actuators for robotic applications can be categorized according to various actuation configurations in terms of loading types. Most of the actuator designs are based on an SMA spring-wire type, as a large stroke change can be generated with a relatively small microscopic strain, and because of its simplicity of fabrication, by annealing an SMA wire wound on a rod [38]. Most key parameters—such as the wire diameter, rod diameter, pitch angle, and number of active coils—can be successfully designed, based on reliable SMA models [39]. However, the stress distribution over the cross-section of an SMA spring is no longer constant, and therefore it requires greater material volume to generate the same force, which will have an effect on the efficiency and the bandwidth of the actuator. Therefore, a straight, linear wire type for tension loading is more advantageous because of its significantly higher efficiency (i.e., more work generated from a minimal amount of SMA material). Various versions of linear SMA wire actuators can be found in recent literature. For example, Flemming et al. [40] suggested a wet SMA actuator embedded within a compliant fluid-filled tube to produce a linear contraction and extension of the SMA wire. Zhang et al. [41] proposed a hybrid linear actuator by using an SMA wire and DC motor to resolve the low driving frequency issue. A new flexible SMA actuator, composed mainly of an SMA wire and a multilayer tubular structure, was proposed to make the conventional linear SMA wire actuator flexible, with high axial stiffness [42]. As robots have become a focus of interest, they are now required to complete various diverse tasks in a variety of environments, using factors such as a torsional motion. In order to produce a torsional (or twist) deformation, a smart soft-composite torsional actuator using SMA wires has been proposed by Shim et al. [43]. The proposed twisting actuator is composed of a torsional-restrained SMA wire embedded in the center of a polydimethylsiloxane matrix. Sheng et al. [44] proposed a torsion actuator composed of a pair of antagonistic SMA torsion springs, capable of bi-directional actuation. Rodrigue at al. [45] manufactured a hollow-tube-shaped actuator, with multiple curved SMA wires that follow the curvature of the tube, and which is capable of pure-twisting deformations, while sustaining a cantilever load. This wrist actuator showed twisting deformation up to 25° while holding objects weighing 100 g, and could sustain loads above 2 N without buckling. Recently, different loading types of SMA wire actuation have been explored for robotic origami applications, which require multiple sequential folding of tiled sheets for recovering the original shape. Paik et al. [46,47] presented a preliminary research on a low-profile bidirectional folding actuator based on annealed thin NiTi sheets (thickness of 0.1 mm), for meso- and micro-scale robotic applications. A torsional actuator composed of an SMA wire, with a diameter of 0.25 mm in the array form, and two supporting elements, was developed to provide large rotational motion by connecting single actuator modules in series [48]. A new soft SMA actuator was developed, capable of fast bending actuation with large deformations. Multiple thin SMA wires were used to increase heat dissipation for faster cooling, and the bending driving frequency of an SMA wire was increased up to 35 Hz [49].

SMAs have been widely used in a diverse range of humanoid-robotic applications since the 1980s, especially in artificial muscles, because SMA wire actuators can mimic human muscle. For example, many researchers created a dexterous robot manipulator that can mimic the human hand using SMAs. A recent study by Thayer and Priya [50] divided the robot hand into several categories based on locomotion styles and multi-DOF. For the design of a facial-expressive baby robot, SMAs were embedded in the skull of a humanoid head and connected to the elastomeric skin at control points. The SMA wires, with 35 routine pulleys were used as the skull actuators [51]. Recently, the majority of SMA robotic researchers have been more interested in biologically inspired (or biometric) robotics, as these robots are useful in solving problems that are challenging for humans by providing pertinent information from underwater, space, air, and land. The state-of-the-art robotic applications using SMA wire actuators are summarized in Table 3, while emphasizing biomimetic robot applications.

Over the last decades, SMAs have been utilized as alternative actuators for underwater robots due to their ability to perform flexible and complex movements, inspired by biological mechanisms [52]. Wang et al. [53,54] designed a micro-robot fish driven by a flexible actuated biomimetic fin that simulates the musculature, and flexible bending, of a squid/cuttlefish. Rossi et al. [55] suggested a new concept, using V-shape configured SMA actuators to bend a continuous flexible structure, representing the backbone of a robot fish. A small crawling robot has been developed by mimicking the model organism of *Caenorhabditis elegans* [56]. A thermal SMA was selected as an actuator because of having properties similar to those of *C. elegans* muscles. The starfish robot has a number of tentacles or arms extending from its central body in the form of a disk, like the topology of a real starfish. The arm is a soft and composite structure, generating a planar reciprocal motion with a fast response speed upon actuation provided by the SMA wires [57,58]. An SMA composite-based soft actuator was designed to provide a large deformation profile inspired by the contraction of a jellyfish bell, utilizing the rowing mechanism for locomotion (e.g., buoyance) [59]. This actuator was found to achieve 80% of maximum deformation, consuming 7.9 J/cycle when driven at 16.2 V/0.98 A and a frequency of 0.25 Hz [59]. Marut et al. [60] designed another type of jellyfish-inspired jet propulsion robot (JetPRo) mimicking the proficient jetting propulsion mechanism (iris mechanism) used by the hydromedusa *Sarsia tubulosa*. A biomimetic swimming robot, based on the locomotion of a marine turtle and SMA wire, was developed by Kim et al. to perform the smooth, soft flapping motions of this type of turtle [61]. The motion of such a structure can be designed by specifying the angle between a filament of the scaffold structure, and a shape-memory alloy (SMA) wire. Based on the analysis of the *Chelonia mydas* turtle, using two swimming gaits (routine and vigorous swimming gaits), the flipper actuator was designed to be divided into three segments, containing a scaffold structure fabricated with a 3-D printer [62]. A soft-bodied robot was proposed using an SMA-based soft composite with inchworm-inspired locomotion, capable of both two-way linear and turning movement [63]. The centimeter-scale work-like robot, inspired by a helical kirigami-enabled parallel structure, was designed using an SMA coil spring actuator [64]. Unlike the inchworm-inspired locomotion, a quick actuation mechanism for a flytrap-inspired robot has been studied by using SMAs [65], which actuates artificial leaves made from asymmetrically laminated carbon fiber. Similar to the flytrap-inspired robot, a novel direction-changing concept for miniature jumping robots, inspired by Froghopper, has been developed by using a coil spring actuator for triggering the jumping [66]. SMAs acting as artificial biceps and triceps are used for mimicking the morphing wing mechanism of the bat [67], or dragonfly [68].

Table 3. Summary of state-of-the-art bio-inspired robotic applications using SMA actuators.

Robots	Biomimetic	References
Micro-fish	Fish fin	[53–55]
Crawling	*C. elegans*	[56]
Tentacle	Starfish	[57,58]
Buoyance	Jellyfish	[59,60]
Swimming	Turtle	[61,62]
Linear	Inchworm	[63,64]
Flytrap	Venus flytrap	[65]
Jumping	Froghopper	[66]
Flying	Bat, Dragonfly	[67,68]

Although SMA actuators provide several attractive advantages over traditional actuators, including silent and smooth operation, direct actuation, simple driving circuitry, and high-power density, it is well-known that there are some technical limitations, such as low energy efficiency, associated with the energy conversion of heat to mechanical work, slow response, and difficulties in positional control [69]. In particular, a complex thermal–electrical–mechanical model is required to correctly represent the behavior of the SMA described in Section 3.1. However, this is challenging because of its strong temperature dependence. The property changes will cause a significant backlash-like hysteresis loop in a highly nonlinear manner in open-loop control responses, which results in response delays and steady-state errors, and limit-cycle issues in the position control of SMA actuators with a conventional feedback controller, such as a PID controller [70]. The robust position control of SMA actuators for compensating hysteresis is therefore one of the interesting research areas in the smart-material community. For example, the sliding mode control (SMC) was applied to control the arm positions of a flexible robot [71,72]. Advanced SMC schemes, such as an adaptive sliding-mode control with a PID tuning method [73], and an intelligent sliding mode control [74], are proposed to achieve robustness against the SMA hysteresis phenomenon. Lee et al. [75] experimentally demonstrated that a simple time delay control (SDC), without a precise actuator model, can be effective for actuating an SMA wire actuator. The open-loop pulse width modulation (PWM) control was also developed to reduce the SMA actuator energy consumption [76]. This control method with a metal-oxide-semiconductor field-effect transistor (MOSFET) switching circuit, can also be used for safely powering the SMA wire actuators across a wide range of speeds or input voltages, which will in turn ensure a better durability, by preventing overheating of the SMA wires [77]. Numerous new approaches for the design and control of SMA actuators have been explored. For example, Selden et al. [78] proposed a segmented SMA wire actuator, which is divided into many segments, and their thermal states are controlled individually as a group of finite-state machines. Then, instead of driving a current to the entire SMA wire and controlling the wire length based on the analog strain–temperature characteristics, this approach controls the binary state (hot or cold) of individual segments, which improves efficiency. The SMA wire itself can be the sensing element for the control of the SMA actuator. The classic example of this is to take advantage of the electrical resistance feedback, which eliminates the need for a position sensor, in this case by utilizing the SMA's electrical resistance feedback. The position control system for the SMA wire actuator with electrical-resistance feedback was proposed by Ma et al. [79], and has been recently advanced by Lynch et al. [80]. We use the embedded SMA wire in composites, both as an actuator and as a strain sensor. Nagai et al. [81] proposed using the SMA wire as a strain sensor to obtain an actuation trigger signal, based on the variation of the SMA electric resistance correlated with the strain.

4. Medical and Rehabilitation Applications

In recent years, robots are being actively studied for utilization in biomedical fields. Especially, various efforts have been made to apply smart fluid to the rehabilitation robot based on the characteristic that the torque can be controlled continuously with a simple system. In 2007, Weinberg et al. [82,83] reported for design and test of ER brake-based active knee rehabilitation orthotic device. The knee brace provides variable and controllable damping that foster motor recovery in stroke patients. An electrorheological fluid-based brake component is used to facilitate knee flexion during stance by providing resistance to knee buckling. The schematic diagrams of the proposed orthotic device and ER fluid-based brake are shown in Figure 8. In the early 2000s, Kim and Oh [84] developed an above knee prosthesis using a rotary MR damper and its effectiveness was evaluated by using leg simulator. It was reported that by controlling the damping force of MR damper, the desired knee joint angle can be accurately achieved. Park et al. [85] designed and manufactured a prosthetic leg for above-knee amputees using MR damper and control performance was evaluated experimentally. The proposed device includes the wearable connector, encoder, flat motor, planetary gear head, gyro sensor, hinge, and MR damper. The MR damper generates reaction force, while the electronically commutated motor actively controls the knee joint angle during gait cycle. The configuration of the proposed above knee prosthesis and photograph of the fabricated above knee prosthesis are presented in Figure 9. Furusho et al. [86] proposed an ankle–foot orthosis using controllable MR brake in 2007. They fabricated two different prototypes by applying shear mode MR brakes with maximum torque of 0.71 Nm and 11.8 Nm, respectively. It is observed that the subject could maintain the dorsal flexion and prevent the drop foot in swing phase that by controlling the ankle torque using MR brake. Additionally, it is demonstrated that the higher bending moment and shorter walking cycle could be achieved by activating MR brake that in turn can prevent drop foot in swing phase and slap foot at heel strike. Kikuchi et al. [87] improved the model developed by Fursusho et al. by adopting a newly proposed compact size MR brake with enhanced maximum torque of 10 Nm. Avraam et al. [88] proposed portable smart wrist rehabilitation device with rotational MR fluid brake actuator for telemedicine application as shown in Figure 10. They designed a novel T-shaped MR fluid brake that can generate output torque of 22.5 Nm and adopted to the proposed portable wrist rehabilitation system. Egawa et al. [89] recently developed and tested a wearable haptic device with pneumatic artificial muscles and MR brake for the application of upper limb rehabilitation. It is observed that this haptic device can render various force senses such as elasticity, friction, and viscosity.

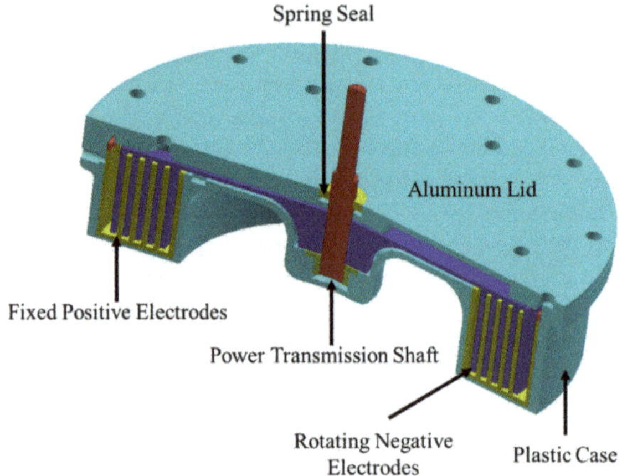

Figure 8. ER brake for active knee rehabilitation orthotic device [83].

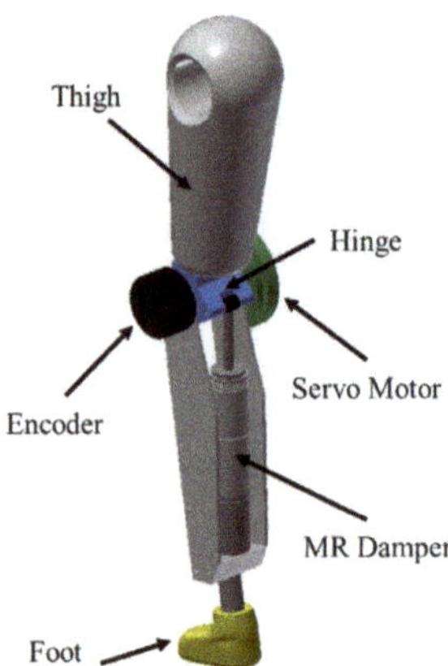

Figure 9. The configuration of above knee prosthesis using MR damper [85].

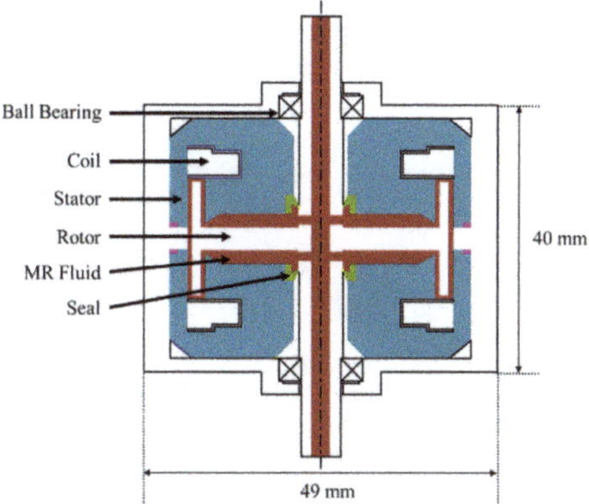

Figure 10. T-shaped MR brake for portable wrist rehabilitation device [88].

Research on the application of biomedical field using shape memory alloys is also proceeding variously. To take advantage of the shape memory effect that can generate a motion similar to human muscle behavior, many studies have been reported for SMA actuation mechanism in the prosthetic hand application. Matsubara et al. [90] reported a new prosthetic hand using SMA artificial muscles. It is demonstrated that the proposed device could grasp an object with the motion of the prosthetic finger. The photographs of the fabricated prosthetic hand and test result are presented in Figure 11. Kaplanoglu [91] proposed a tendon-driven SMA actuated finger for the prosthetic hand development that activates SMA actuation using an electro-myographic signal. In order to satisfy the requirement that a minimally-invasive surgical robot should have a large operating force in a small volume, research has been conducted to apply an SMA wire actuator to the robot end effector. Kode et al. [92] proposed a novel hybrid actuator to increase the number of degrees of freedom in minimally-invasive surgery (MIS) robots through local actuation of the end effector. The proposed system consists of a laparoscopic needle driver, SMA wire, and DC motor. The photograph of the proposed prototype is shown in Figure 12. The designed actuator is 5 mm in diameter and 40 mm in length and is used to actuate 10 mm long needle driver jaws, while generating a force of 15 N and a gripping force of 5.5 N. Giataganas et al. [93] reported prototype of minimally invasive surgery robotic tool using SMA wires in an antagonistic tendon configuration to obtain low stiffness. It is observed that the low weight (150 g) of the proposed tool could make it suitable for most surgical operations.

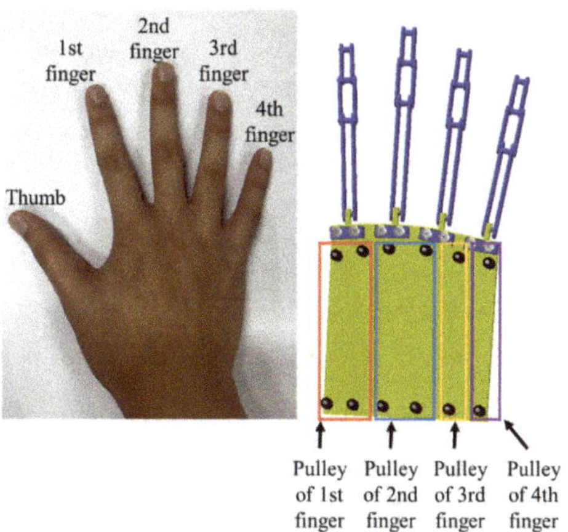

Figure 11. Prosthetic hand with SMA actuator [90].

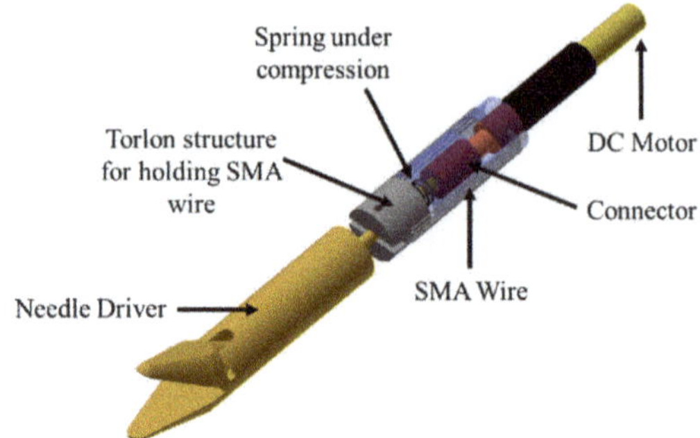

Figure 12. SMA wire-based end-effector for minimally invasive surgery [91].

5. Conclusions and Future Direction

This topical review presents state-of-the-art developments of robotic applications using smart materials. Smart materials used for robotic applications include electro-rheological (ER) fluids, magneto-rheological (MR) fluids, and shape memory alloys (SMAs). As reviewed, these smart materials can be effectively used as actuators for various types of robots, but there exist still many material and technical limitations for practical implementation. Both ERF and MRF are very effective to devise special robots such as leg robot in which unwanted vibrations need to be controlled in real time. To successfully realize those special robots in practical environment, the higher level of the field-dependent yield stress, the higher robustness to impurities, the wider temperature range, the less abrasiveness and wear properties, and the prevention of particle sedimentation are essentially required to be resolved or improved. Bio-inspired small-scale soft robotics using SMAs are very promising candidates for practical use because of their artificial muscle properties with large strain variation. However, in order to apply SMAs to robots, the narrow bandwidth issue associated with the asymmetric actuation property should be first resolved. Moreover, continuous improvement on their controllability, stability, and large actuation force are still essential to be ensured for future-oriented robot applications associated with SMA actuator. This review article has covered recent advances and trends on robotic applications using diverse smart materials.

The words representing future robots will be 'artificial intelligence' and 'natural mimicking', and the implementation of natural mimicking is proceeding in the form of soft robotics. Soft robotics is a newly emerging field of robot research. It is expected that friendly human–robot interaction will be obtained based on inherent soft characteristics. The soft robots can be adopted to their surroundings actively and can mimic the function of biological systems such as octopus leg and human muscles. The final goal of soft robotics is the integration of soft actuators, sensors, controls and power source without any rigid part. Soft materials, actuating mechanisms using the materials, and fabrication methods such as 3D printing have been extensively studied. However, there are still many challenges in the field of soft robotics. From the material aspect of the sensor and the actuator, soft and smart materials, such as shape memory alloys, shape memory polymers, electro-rheological fluids, electro-rheological elastomers, magneto-rheological fluids, magneto-rheological elastomers, and electro-active polymers (EAPs) are very important part for the successful development of soft robots. Smart materials will change the definition of a robot and the relationship between robot and human.

However, it should be noted here that the development of a new class of smart materials, which can be optimally and adaptively applicable to various robots, is still an on-going research issue. Continuous advances in the synthesis of smart materials are likely to motivate the creativity of researchers seeking to harness these materials for robotic applications. Finally, this review article provides very useful information on the potential research opportunities and emerging technologies using smart materials for the innovative robots of the future.

Appl. Sci. **2018**, *8*, 1928

Author Contributions: S.-B.C. as a corresponding author takes the primary responsibility for this article. J.-W.S. and G.-W.K. drafted the manuscript, and all co-authors reviewed.

Funding: This research was funded by INHA IST-NASA Deep Space Exploration Joint Research Center (NRF-2017K1A4A3013662).

Conflicts of Interest: The authors declare no conflicts of interest.

References

1. Gandhi, M.V.; Thompson, B.S.; Choi, S.B.; Shakir, S. Electro-rheological-fluid-based articulating robotic systems Transactions of the ASME-Mechanisms. *Transm. Autom. Des.* **1989**, *111*, 328–336. [CrossRef]
2. Choi, S.B. Control of Single-Link Flexible Manipulators Fabricated from Advanced Composite Laminates and Smart Materials Incorporating Electro-Rheological Fluids. Ph.D. Thesis, Department of Mechanical Engineering, Michigan State University, East Lansing, MI, USA, 1990.
3. Choi, S.B.; Thompson, B.S.; Gandhi, M.V. Experimental control of a single-link flexible arm incorporating electrorheological fluids. *J. Guid. Control Dyn.* **1995**, *18*, 916–919. [CrossRef]
4. Choi, S.B.; Han, S.; Kim, H.K.; Cheong, C.C. H-infinity control of a flexible gantry robot arm using smart actuators. *Mechatronics* **1999**, *9*, 271–286. [CrossRef]
5. Han, S.S.; Choi, S.B.; Kim, J.H. Position control of a flexible gantry robot arm using smart material actuators. *J. Robot. Syst.* **1999**, *16*, 581–595. [CrossRef]
6. Monkman, G.J. An electrorheological tactile display. *Presence Mass. Inst. Technol.* **1992**, *1*, 219–228. [CrossRef]
7. Shafer, A.S.; Kermani, M.R. On the feasibility and suitability of MR and ER based actuators in human friendly manipulators. In Proceedings of the 2009 IEEE/RSJ Ternational Conference on Intelligent Robots and Systems, St. Louis, IL, USA, 11–15 October 2009.
8. LORD Company. Available online: http://www.lord.com/ (accessed on 1 July 2018).
9. Choi, S.B.; Han, Y.M. *Magnetorheological Fluid Technology: Applications in Vehicle Systems Taylor & Francis Group*; CRC Press: Boca Raton, FL, USA, 2012.
10. Kordonsky, W.I.; Gorodkin, S.R.; Kolomentsev, A.V.; Kuzmin, V.A.; Luk'ianovich, A.V.; Protasevich, N.A.; Prokhorov, I.V.; Shulman, Z.P.; Byelocorp Scientific Inc. Magnetorheological Valve and Devices Incorporating Magnetorheological Elements. U.S. Patent No. 5353839, 11 October 1994.
11. Yadmellat, P.; Kermani, M.R. Adaptive Control of a Hysteretic Magneto-rheological robot actuator. *IEEE/ASME Trans. Mechatron.* **2016**, *21*, 1336–1344. [CrossRef]
12. Jolly, M.R.; Bryan, S.R. Magnetorheological Grip for Handheld Implements. U.S. Patent No. 6158910, 12 December 2000.
13. Pattersson, A.; Davis, S.; Gray, J.O.; Dodd, T.J.; Ohlsson, T. Design of a magnetorheological robot gripper for handling of delicate food products with varying shapes. *J. Food Eng.* **2010**, *98*, 332–338. [CrossRef]
14. Nguyen, Q.H.; Choi, S.B.; Lee, Y.S.; Han, M.S. Optimal design of a new 3D haptic gripper for telemanipulation, featuring magnetorheological fluid brakes. *Smart Mater. Struct.* **2013**, *22*, 015009. [CrossRef]
15. Scilingo, E.P.; Sgambelluri, N.; Rossi, D.; Bicchi, A. A Towards a haptic black box for free-hand softness and shape exploration. In Proceedings of the 2003 IEEE International Conference on Robotics and Automation, Taipei, Taiwan, 14–19 September 2003.
16. Blake, J.; Gurocak, H.B. Haptic glove with MT brakes for virtual reality. *IEEE/ASME Tans. Mechatron.* **2009**, *14*, 606–615. [CrossRef]
17. Ahmadkhanlou, F.; Washington, G.N.; Bechtel, S.E. The development of a five DOF magnetorheological fluid-based telerobotic haptic system. In Proceedings of the Modeling, Signal Processing, and Control for Smart Structures, San Diego, CA, USA, 9–12 March 2008; Volume 692604. [CrossRef]
18. Walker, D.S.; Thoma, D.J.; Niemeyer, G. Variable impedance magnetorheological clutch actuator and telerobotic implementation. In Proceedings of the 2009 IEEE/RSJ International Conference on Intelligent Robots and Systems, St. Louis, IL, USA, 11–15 October 2009; pp. 2885–2891.
19. Diaz, I.; Gill, J.J.; Sanchez, E. Lower-limb robotic rehabilitation:literature review and challenges. *J. Robot.* **2011**, *2011*, 759764. [CrossRef]
20. Kikuchi, T.; Oda, K.; Yamaguchi, S.; Furusho, J. Leg-robot with MR clutch to realize virtual spastic movements. *J. Intell. Mater. Syst. Struct.* **2010**, *21*, 1523–1529. [CrossRef]

21. Kikuchi, T.; Oda, K.; Furusho, J. Leg-robot for demonstration of spastic movements of brain-injured patients with compact megnetorheological fluid clutch. *Adv. Robot.* **2010**, *24*, 671–686. [CrossRef]

22. Jiang, N.; Sun, S.; Ouyang, Y.; Xu, M.; Li, W.; Zhang, S. A highly adaptive magnetorheological fluid robotic leg for efficient terrestrial locomotion. *Smart Mater. Struct.* **2016**, *25*, 095019. [CrossRef]

23. Saito, T.; Ikeda, H. Development of normally closed type of magnetorheological clutch and its application to safe torque control system of human-collaborative robot. *J. Intell. Mater. Syst. Struct.* **2007**, *18*, 1181–1185. [CrossRef]

24. Ahmed, R.M.; Kalaykov, I.G.; Ananiev, A.V. Modeling of magneto rheological fluid actuator enabling safe human-robot interaction. In Proceedings of the IEEE International Conference on Emerging Technologies and Factory Automation, Hamburg, Germany, 15–18 September 2008. [CrossRef]

25. Ahmed, M.R.; Kalaykov, I. Semi-active compliant robot enabling collision safety for human robot interaction. In Proceedings of the 2010 IEEE International Conference on Mechatronics and Automation, Xi'an, China, 4–7 August 2010.

26. Fauteux, P.; Lauria, M.; Heintz, B.; Michaud, F. Dual-differential rheological actuator for high-performance physical robotic interaction. *IEEE Trans. Robot.* **2010**, *26*, 607–618. [CrossRef]

27. Shafer, A.; Kermani, M.R. Design and validation of a magneto-rheological clutch for practical control application in human-friendly manipulation. In Proceedings of the 2011 IEEE International Conference on Robotics and Automation, Shanghai, China, 9–13 May 2011.

28. Shafer, A.; Kermani, M.R. On the feasibility and suitability of MR fluid clutches in human-friendly manipulators. *IEEE/ASME Trans. Mechatron.* **2011**, *16*, 1073–1082. [CrossRef]

29. Hoyle, A.; Arzanpour, S.; Shen, Y. A novel magnetorheological damper based parallel planar manipulator design. *Smart Mater. Struct.* **2010**, *19*, 055028. [CrossRef]

30. Wiltsie, N. A Controllably Adhesive Climbing Robot Using Magnetorheological Fluid. Masters' Thesis, Department of Mechanical Engineering, Massachusetts Institute of Technology, Cambridge, MA, USA, September 2012.

31. Yue, M.; Liu, B.Y. Design of adaptive sliding mode control for spherical robot based on MR fluid actuator. *J. Vibroeng.* **2012**, *14*, 196–204.

32. Schmitz, G.W. Hydraulically Energized Magnetorheological Replicant Muscle Tissue and System and a Method for Using and Controlling Same. U.S. Patent No. 6168634, 2 January 2001.

33. Kashima, S.; Miyasaka, F.; Hirata, K. Novel soft actuator using magnetorheological elastomer. *IEEE Trans. Magn.* **2012**, *48*, 1649–1652. [CrossRef]

34. Zimmermann, K.; Bohm, V.; Zeidis, I. Vibration-driven mobile robot based on magneto-sensitive elastomers. In Proceedings of the 2011 IEEE/ASME International Conference on Advanced Intelligent Mechatronics, Budapest, Hungary, 3–7 July 2011.

35. Flexible Solutions. Available online: http://www.dynalloy.com (accessed on 10 July 2018).

36. Sun, L.; Huang, W.M.; Ding, Z.; Zhao, Y.; Wang, C.C.; Purnawali, H.; Tang, C. Stimulus-responsive shape memory materials: A review. *Mater. Des.* **2012**, *33*, 577–640. [CrossRef]

37. Jani, J.M.; Leary, M.; Subic, A.; Gibson, M.A. A review of shape memory alloy research, applications and opportunities. *Mater. Des.* **2014**, *56*, 1078–1113. [CrossRef]

38. Follador, M.; Cianchetti, M.; Arienti, A.; Laschi, C. A general method for the design and fabrication of shape memory alloy active spring actuators. *Smart Mater. Struct.* **2012**, *21*, 115029. [CrossRef]

39. An, S.; Ryu, J.; Cho, M.; Cho, K. Engineering design framework for a shape memory alloy coil spring actuator using a static two-state model. *Smart Mater. Struct.* **2012**, *21*, 055009. [CrossRef]

40. Flemming, L.; Mascaro, S. Analysis of hybrid electric/thermofluidic inputs for wet shape memory alloy actuators. *Smart Mater. Struct.* **2013**, *22*, 014015. [CrossRef]

41. Zhang, X.; Hu, J.; Mao, S.; Dong, E.; Yang, J. Design and property analysis of a hybrid linear actuator based on shape memory alloy. *Smart Mater. Struct.* **2014**, *23*, 125004. [CrossRef]

42. Leng, J.; Yan, X.; Zhang, X.; Huang, D.; Gao, Z. Design of a novel flexible shape memory alloy actuator with multilayer tubular structure for easy integration into a confined space. *Smart Mater. Struct.* **2016**, *25*, 025007. [CrossRef]

43. Shim, J.; Quan, Y.; Wang, W.; Rodrigue, H.; Song, S.; Ahn, S.H. A smart soft actuator using a single shape memory alloy for twisting actuation. *Smart Mater. Struct.* **2015**, *24*, 125033. [CrossRef]

44. Sheng, J.; Desai, J.P. Design, modeling and characterization of a novel meso-scale SMA-actuated torsion actuator. *Smart Mater. Struct.* **2015**, *24*, 105005. [CrossRef]

45. Rodrigue, H.; Wei, W.; Bhandari, B.; Ahn, S.H. Fabrication of wrist-like SMA-based actuator by double smart soft composite casting. *Smart Mater. Struct.* **2015**, *24*, 125003. [CrossRef]

46. Paik, J.K.; Hawkes, E.; Wood, R.J. A novel low-profile shape memory alloy torsional actuator. *Smart Mater. Struct.* **2010**, *19*, 125014. [CrossRef]

47. Paik, J.K.; Wood, R.J. A bidirectional shape memory alloy folding actuator. *Smart Mater. Struct.* **2012**, *21*, 065013. [CrossRef]

48. Shin, B.H.; Jang, T.; Ryu, B.J.; Kim, Y. A modular torsional actuator using shape memory alloy wires. *J. Intell. Mater. Syst. Struct.* **2016**, *12*, 1658–1665. [CrossRef]

49. Song, S.; Lee, J.; Rodrigue, H.; Choi, I.; Kang, Y.J.; Ahn, S.H. 35 Hz shape memory alloy actuator with bending-twisting mode. *Sci. Rep.* **2016**, *6*, 21118. [CrossRef] [PubMed]

50. Thayer, N.; Priya, S. Design and implementation of a dexterous anthropomorphic robotic typing (DART) hand. *Smart Mater. Struct.* **2011**, *20*, 035010. [CrossRef]

51. Tadesse, Y.; Hong, D.; Priya, S. Twelve Degree of Freedom Baby Humanoid Head Using Shape Memory Alloy Actuators. *J. Mech. Robot.* **2011**, *3*, 011008. [CrossRef]

52. Chu, W.; Lee, K.; Song, S.; Han, M.; Lee, J.; Kim, H.; Kim, M.; Park, Y.; Cho, K.; Ahn, S.H. Review of Biomimetic Underwater Robots Using Smart Actuators. *Int. J. Precis. Eng. Manuf.* **2012**, *13*, 1281–1292. [CrossRef]

53. Wang, Z.; Hang, G.; Li, J.; Wang, Y.; Xiao, K. A micro-robot fish with embedded SMA wire actuated flexible biomimetic fin. *Sens. Actuators A* **2008**, *144*, 354–360. [CrossRef]

54. Wang, Z.; Hang, G.; Wang, Y.; Li, J.; Du, W. Embedded SMA wire actuated biomimetic fin: A module for biomimetic underwater propulsion. *Smart Mater. Struct.* **2008**, *17*, 025039. [CrossRef]

55. Rossi, C.; Colorado, J.; Coral, W.; Barrientos, A. Bending continuous structures with SMAs: A novel robotic fish design. *Bioinspir. Biomim.* **2011**, *6*, 045005. [CrossRef] [PubMed]

56. Yuk, H.; Kim, D.; Lee, H.; Jo, S.; Shin, J.H. Ketner Shape memory alloy-based small crawling robots inspired by *C. elegans. Bioinspir. Biomim.* **2011**, *6*, 046002. [CrossRef] [PubMed]

57. Jin, H.; Dong, E.; Xu, M.; Liu, C.; Alici, G.; Jie, Y. Soft and smart modular structures actuated by shape memory alloy (SMA) wires as tentacles of soft robots. *Smart Mater. Struct.* **2016**, *25*, 085026. [CrossRef]

58. Jin, H.; Dong, E.; Alici, G.; Mao, S.; Min, X.; Liu, C.; Low, K.H.; Yang, J. A starfish robot based on soft and smart modular structure (SMS) actuated by SMA wires. *Bioinspir. Biomim.* **2016**, *11*, 056012. [CrossRef] [PubMed]

59. Villanueva, A.; Smith, C.; Priya, S.A. A biomimetic robotic jellyfish (Robojelly) actuated by shape memory alloy composite actuators. *Bioinspir. Biomim.* **2011**, *6*, 036004. [CrossRef] [PubMed]

60. Marut, K.; Stewart, C.; Michael, T.; Villanueva, A.; Priya, S. A jellyfish-inspired jet propulsion robot actuated by an iris mechanism. *Smart Mater. Struct.* **2013**, *22*, 094021. [CrossRef]

61. Kim, H.; Song, S.; Ahn, S.H. A turtle-like swimming robot using a smart soft composite (SSC) structure. *Smart Mater. Struct.* **2013**, *22*, 014007. [CrossRef]

62. Song, S.; Kim, M.; Rodrigue, H.; Lee, J.; Shim, J.; Kim, M.; Chu, W.; Ahn, S.H. Turtle mimetic soft robot with two swimming gaits. *Bioinspir. Biomim.* **2016**, *11*, 036010. [CrossRef] [PubMed]

63. Wang, W.; Lee, J.; Rodrigue, H.; Song, S.; Chu, W.; Ahn, S.H. Locomotion of inchworm-inspired robot made of smart soft composite (SSC). *Bioinspir. Biomim.* **2014**, *9*, 046006. [CrossRef] [PubMed]

64. Zhang, K.; Qiu, C.; Dai, J.S. Helical kirigami-enabled centimeter-scale worm robot with shape-memory-alloy linear actuators. *J. Mech. Robot.* **2015**, *7*, 021014. [CrossRef]

65. Kim, S.; Koh, J.; Lee, J.; Ryu, J.; Cho, M.; Cho, K. Flytrap-inspired robot using structurally integrated actuation based on bistability and a developable surface. *Bioinspir. Biomim.* **2014**, *9*, 036004. [CrossRef] [PubMed]

66. Jung, G.; Cho, K. Froghopper-inspired direction-changing concept for miniature jumping robots. *Bioinspir. Biomim.* **2016**, *11*, 056015. [CrossRef] [PubMed]

67. Colorado, J.; Barrientos, A.; Rossi, C.; Bahlman, J.W.; Breuer, K.S. Biomechanics of smart wings in a bat robot: Morphing wings using SMA actuators. *Bioinspir. Biomim.* **2013**, *7*, 036006. [CrossRef] [PubMed]

68. Festo. *BionicOpter—Inspired by Dragonfly Flight*; Festo: Esslingen am Neckar, Germany, 2013.

69. Dimitris, C.L. *Shape Memory Alloys: Modeling and Engineering Applications*; Springer: Berlin, Germany, 2008.

70. Mohammad, H.E.; Hashem, A. Nonlinear control of a shape memory alloy actuated manipulator. *J. Vib. Acoust.* **2002**, *124*, 566–575.

71. Price, A.D.; Jenifene, A.; Naguib, H.E. Design and control of a shape memory alloy based dexterous robot hand. *Smart Mater. Struct.* **2007**, *16*, 1401. [CrossRef]

72. Choi, S.B. Position control of a single-link mechanism activated by shape memory alloy springs: Experimental results. *Smart Mater. Struct.* **2006**, *15*, 51. [CrossRef]

73. Taril, N.T.; Ahn, K.K. Adaptive proportional–integral–derivative tuning sliding mode control for a shape memory alloy actuator. *Smart Mater. Struct.* **2011**, *20*, 055010.

74. Hannen, J.C.; Crews, J.H.; Buckner, G.D. Indirect intelligent sliding mode control of a shape memory alloy actuated flexible beam using hysteretic recurrent neural networks. *Smart Mater. Struct.* **2012**, *21*, 085015. [CrossRef] [PubMed]

75. Lee, H.J.; Lee, J.J. Time delay control of a shape memory alloy actuator. *Smart Mater. Struct.* **2004**, *13*, 227. [CrossRef]

76. Ma, N.; Song, G. Control of shape memory alloy actuator using pulse width modulation. *Smart Mater. Struct.* **2003**, *12*, 712. [CrossRef]

77. A Revolution In Motion! Available online: http://migamotors.com/ (accessed on 20 July 2018).

78. Selden, B.; Cho, K.; Asada, H.H. Segmented shape memory alloy actuators using hysteresis loop control. *Smart Mater. Struct.* **2006**, *15*, 642. [CrossRef]

79. Ma, N.; Song, G.; Lee, H.-J. Position control of shape memory alloy actuators with internal electrical resistance feedback using neural networks. *Smart Mater. Struct.* **2004**, *13*, 777. [CrossRef]

80. Lynch, B.; Jiang, X.X.; Ellery, A.; Nitzsche, F. Characterization, modeling, and control of Ni-Ti shape memory alloy based on electrical resistance feedback. *J. Intell. Mater. Syst. Struct.* **2016**, *27*, 2489–2507. [CrossRef]

81. Nagai, H.; Oishi, R. Shape memory alloys as strain sensors in composites. *Smart Mater. Struct.* **2006**, *15*, 493. [CrossRef]

82. Weinberg, B.; Nikitczuk, J.; Patel, S.; Patritti, B.; Mavroidis, C.; Bonato, P.; Canavan, P. Design, Control and Human Testing of an Active Knee Rehabilitation Orthotic Device. In Proceedings of the 2007 IEEE International Conference on Robotics and Automation, Roma, Italy, 10–14 April 2007; pp. 4126–4133.

83. Nikitczuk, J.; Weinberg, B.; Canavan, P.K.; Mavroidis, C. Active knee rehabilitation orthotic device with variable damping characteristics implemented via an electrorheological fluid. *IEEE/ASME Trans. Mechatron.* **2010**, *15*, 952–960. [CrossRef]

84. Kim, J.H.; Oh, J.H. Development of an above knee prosthesis using MR damper and leg simulator. In Proceedings of the IEEE International Conference on Robotics and Automation (ICRA), Seoul, Korea, 21–26 May 2001; pp. 43686–43691.

85. Park, J.; Yoon, G.H.; Kang, J.W.; Choi, S.B. Design and control of a prosthetic leg for above-knee amputees operated in semi-active and active modes. *Smart Mater. Struct.* **2016**, *25*, 085009. [CrossRef]

86. Furusho, J.; Kikuchi, T.; Tokuda, M.; Kakehashi, T.; Ikeda, K.; Morimoto, S.; Hashimoto, Y.; Tomiyama, H.; Nakagawa, A.; Akazawa, Y. Development of shear type compact MR brake for the intelligent ankle-foot orthosis and its control; research and development in NEDO for practical application of human support robot. In Proceedings of the IEEE 10th International Conference on Rehabilitation Robotics (ICORR), Noordwijk, The Netherlands, 12–15 June 2007; pp. 89–94.

87. Kikuchi, T.; Tanida, S.; Otsuki, K.; Yasuda, T.; Furusho, J. Development of third-generation intelligently controllable ankle-foot orthosis with compact MR fluid brake. In Proceedings of the IEEE International Conference on Robotics and Automation (ICRA), Anchorage, Alaska, 3–8 May 2010; pp. 2209–2214.

88. Avraam, M.; Horodinca, P.; Leiter, P.; Preumont, A. Portable Smart Wrist Rehabilitation Device driven by rotational MR-fluid brake actuator for telemedicine application. In Proceedings of the International Conference on Intelligent Robots and Systems, Nice, France, 22–26 September 2008; pp. 1441–1446.

89. Egawa, M.; Watanabe, T.; Nakamura, T. Development of a wearable haptic device with pneumatic artificial muscles and MR brake. In Proceedings of the IEEE Virtual Reality Conference 2015, Minneapolis, MN, USA, 29 March–2 April 2015; pp. 173–174.

90. Matsubara, S.; Okamoto, S.; Lee, J. Prosthetic hand using shape memory alloy type artificial muscle. In Proceedings of the International MultiConference of Engineers and Computer Scientists, Hong Kong, China, 14–16 March 2012; Volume 2, pp. 873–876.

91. Kaplanoglu, E. Design of shape memory alloy-based and tendon-driven actuated fingers towards a hybrid anthropomorphic prosthetic hand. *Int. J. Adv. Robot. Syst.* **2012**, *9*, 77. [CrossRef]

92. Kode, V.R.C.; Cavusoblu, M.C.; Azar, M.T. Design and characterization of a novel hybrid actuator using shape memory alloy and D.C motor for minimally invasive surgery applications. In Proceedings of the IEEE International Conference on Mechatronics and Automation 2005, Niagara Falls, ONT, Canada, 29 July–1 August 2005; pp. 416–420.

93. Giataganas, P.; Evangeliou, N.; Koveos, Y.; Kelasidi, E.; Tzes, A. Design and experimental evaluation of an innovative SMA-based tendon-driven redundant endoscopic robotic surgical tool. In Proceedings of the 19th Mediterranean Conference on Control and Automation 2011, Corfu, Greece, 20–23 June 2011; pp. 1071–1075.

MDPI
St. Alban-Anlage 66
4052 Basel
Switzerland
Tel. +41 61 683 77 34
Fax +41 61 302 89 18
www.mdpi.com

Applied Sciences Editorial Office
E-mail: applsci@mdpi.com
www.mdpi.com/journal/applsci